Overkill!

- the euphoric rush to industrialise Ireland's sea fisheries and its unravelling sequel

Edward Fahy

Edward Fahy

First published in Ireland by Edward Fahy, 2013
edwardfahy@eircom.net

ISBN-13: 978-0-9575218-0-3
ISBN-10: 0957521804

For Barbara
with all my love

Edward Fahy

The fishing of Ireland might prove a mine under water as rich as any under ground...

Sir William Temple (letter to Lord Essex, Lord Lieutenant of Ireland, 22 July 1673)

Whether however the mine of wealth *which has so often been referred to by witnesses before us as existing in an underdeveloped state, off the west and south coasts is a reality or a myth, can be decided only by experience...*

Dalhousie et al, 1884-85

A man and his wife owned a very special goose. Every day the goose would lay a golden egg....

Just think, *said the man's wife,* **If we could have all the golden eggs that are inside the goose, we could be richer much faster.**

So, the couple killed the goose and cut her open, only to find that she was just like every other goose... and they had no more golden eggs.

Aesop's Fables

Contents

Acknowledgements vii

Chapter 1   From inauspicious beginnings… 1

Chapter 2   Harvesting the fauna of Ireland's shelf waters 21

Chapter 3   A matter of size 56

Chapter 4   For want of capital… 67

Chapter 5   Widening sea, narrowing ocean... 87

Chapter 6   A brief flirtation with reality 97

Chapter 7   The elaborate regime that failed 125

Chapter 8   Good news and subsidies 139

Chapter 9   Guidance, guesstimates and the first casualty of many 161

Chapter 10   Meanwhile, back in the territorial sea 176

Chapter 11   Fish producers' organisations take possession 206

Chapter 12   A fleet for all seasons 229

Chapter 13   The scramble for fish.. 251

Chapter 14   The official fate of some fish stocks 270

Chapter 15   Two last casts of the dice 288

Chapter 16   A rot on the coast 307

Chapter 17   The treacherous reassurance of fisheries science................ 334

Chapter 18   Anatomy of a take-over 352

Chapter 19   When did you meet a happy fisherman? 385

References   409

Appendix 1   Abbreviations and acronyms... 421

Appendix 2   List of commercial "varieties" landed by the Irish fishing fleet: compiled from recent fishing years. 422

Appendix 3   Direct subsidies to the fishing industry between 1952 and 2009.. 424

Appendix 4   Development of the Irish fishing fleet, 1929-2009 427

Appendix 5   Composition and value of landings by the Irish fleet.......... 431

Appendix 6   The official fate of four whitefish species......................... 434

Index   436

Chapter notes   457

Acknowledgements

Overkill! is the culmination of several years' endeavour. Many people contributed to the end result, among them my colleagues and collaborators in Ireland's Marine Institute, Inland Fisheries Ireland, the government Department responsible for the industry and other agencies. I would particularly like to thank David Meredith who generously shared his knowledge of fisheries economics and put me in contact with others who have similar interests. The library staffs of the Marine Institute, an Bord Iascaigh Mhara and the National Library in Dublin were unfailingly helpful and courteous. The *Irish Skipper* which reported developments since the mid-1960s was an invaluable archive and that journal gave me the opportunity to write about many significant aspects of the industry over a twenty year period when I reported under the pseudonym Clemente de Callao about marine fishing in Spain. The *Marine Times* allowed me to air some of my ideas which related more directly to the industry in Ireland and the cover photograph of the book was originally used in one of their issues. Although I no longer work in the laboratory or at sea, the industry is so inescapably powerful that its impact ripples through every aspect of the marine environment. I am glad to be a member of the Irish Specimen Fish Committee which, indirectly, monitors the consequences of over-fishing on a range of species and I am grateful to its secretary, Willie Roche and chairman, Trevor Champ for their support. I must record my thanks to John McArdle who encouraged my writing the book and Lucille Redmond and Jean Harrington who gave practical assistance and advice on the business of getting it published. Declan MacGabhann, whose knowledge of the Irish industry is encyclopaedic, devised Appendix 2. Barbara Cashin was the proof-reader whose common sense and facility with the English language did much to knock the manuscript into shape.

Finally I wish to express sympathy with commercial fishermen, particularly those who go to sea in small craft, with many of whom I worked during my career, and from whom I learned a great deal.

Edward Fahy

Chapter 1

FROM INAUSPICIOUS BEGINNINGS...

Those who believe that fish should be free to all will be surprised to learn that under Brehon law, the indigenous system which lasted from Celtic times to the early seventeenth century, when English law supplanted it, there was no public right of access to freshwater fisheries. From the eighth century in early Christian Ireland, the monasteries and private groups controlled the most valuable inland waterways, erecting weirs to capture salmon, trout and eel. Some relationship between parent and juvenile fish stock was recognized and rudimentary management was practised: a free gap of one third the width of a river was stipulated to allow a breeding escapement of migratory species.

Restricted access to inland waters obliged the poor to seek their fish elsewhere, along the coastal margins and in the shallow marine inshore. Fish was, however, well established in the national diet. By the fourteenth century, overlords had put restrictions on what could be sold; in parts of Europe species of higher protein value, such as small whales and seals, were claimed as the rightful property of the nobility.

By the end of the medieval period there were two established traditions in Europe, one land-based and rural, the other rooted in trade and sea fishing. A century later, sea fishermen had become specialised, targeting a succession of species and depths in different seasons.

The colonisation of Ireland was essentially the confiscation of resources or their purchase from those who exchanged power for money. It was also a terrestrial rather than a marine experiment. Fishermen were lumped together with smugglers and pirates (possibly with some reason) and branded a threat to the settlers. But the infiltration of Ireland's marine territory had begun before the terrestrial plantations when the waters around Ireland had been a *mare liberum*, literally, an open sea.

The previously shared nature of our fisheries left a contentious legacy to the modern industry. Fleets from Spain, Holland, France, Portugal and North Africa began to fish close to the south west coast from the twelfth century. They paid tribute to Gaelic overlords to secure protection from pirates and for shore access to process their catches. The overlords included the O'Donnells, (Donegal), O'Malleys (Mayo), O'Flahertys (Galway), O'Driscolls and the O'Sullivan Beare (west Munster). In the mid-fifteenth century there were 500 Spanish vessels working in the vicinity of Baltimore, their former presence now acknowledged in place names along the west coast. This presence was then so pervasive that houses in Dingle were constructed in the Spanish style. A century later 600 Spanish boats were reported fishing off the south west. Hake, the species most sought after by Spaniards, and one for which they continue to trawl the globe, was a staple of the Irish populace in the south west in the sixteenth century. At the same time the Gaelic overlords themselves managed substantial fleets, exporting large quantities of fish.[1]

The Anglo Normans in the twelfth century brought a number of innovations and new species (including freshwater fish) with them to the country and they are credited with altering the diet of the native people with whom they merged.[2] The next incursion from Britain, the plantation of Munster, four centuries later, transferred new technologies to Ireland, notably the seine net fishery for pilchard on the south coast, which was operated by

English settlers.[3] Packing and preserving fish for export became a priority and in due course the use of ice assumed an important role in fish preservation. In the eighteenth century ice was being imported from as far away as Norway when weather conditions in Ireland did not favour its creation here. The technique was subsequently lost and had to be rediscovered in the 1950s. So unaccustomed to using ice was the industry at that stage that there was consumer resistance to purchasing fish with ice crystals sprinkled on it.

The *mare liberum* of the pre-colonial period had made Irish waters safely accessible to various adventurers who established business and gradually rose to prominence in the coastal towns before the plantations came into existence. The plantations themselves were about the creation of a terrestrial, Arcadian landscape whose occupants regarded themselves as civilised, and morally and intellectually superior to the usurped vagrant "heathenish rebels, lawless fishermen and fearless smugglers" who were displaced to the mountainous and coastal wilderness.

At the end of the seventeenth century, Europe had a thriving marine economy: some 50,000 gross tonnes (GT) of fishing vessels exploited the Baltic and North Seas and twice as many people were employed at sea as in agriculture. Irish waters were, however, fished mainly by Scottish and English boats.

The avaricious English, anxious to weaken the power of those they sought to replace, attempted to deprive Gaelic overlords of continental revenue and secure fishing grounds for their own vessels. Their objective, the creation of a *mare clausum*, a closed and regulated sea, was the antithesis of pre-occupation times. The continental connections' ability to introduce weaponry without constraint was an additional security incentive to do so. At one stage the invaders considered redeploying Dutch pirates to harass foreign fleets and undermine their hosts' claim to protection money. And then, in the sixteenth century, the Newfoundland cod fisheries commenced, drawing much European fishing effort to the other side of the Atlantic. Enterprising Roman Catholics, constrained by penal laws, invested in the Newfoundland fisheries rather than closer to home.[4] The Irish have had a long and honourable place in maritime history.[5] Very often however, it was in the service of other lands and navies.

A topic of great relevance is the position of the fisherman in society, recently elucidated by Jim Mac Laughlin.[6] The sixteenth century witnessed the birth of civic organisation in Europe, the cleansing of streets, the hygienic regulation of markets, the gentrification of the built environment and the development of bourgeois sensibilities. The disagreeable odours of fish and slaughtered fowl and mammals were relegated to certain areas of towns, apart from more prosperous society; fishermen became akin to an untouchable caste. That century also saw the Munster plantation and the seventeenth its successor, the Ulster one, which dealt an even more grievous blow to fishermen in Ireland.

If the native Catholic Irish were unable to make the most of their resources, the attraction of good fishing off south west Ireland for others remained strong and, in the eighteenth century, 300 French vessels were reported working close to the south coast.

Just how elusive the facts of fishing history are is illustrated by the loss of precise information on the frequency of days of abstinence from meat eating in Ireland during medieval and colonial times. This is not a mere academic point. In Spain during the seventeenth century, meat eating was forbidden on 36% of calendar days, providing a considerable market for fish and a financial incentive to fishermen.[7] Such strictures have been removed by the Roman Catholic church in Ireland but, as recently as the 1960s, the miseries of the forty days of Lent were positively relished by the resurgent fishing industry and Thursday, in anticipation of the meatless day after, was the only opportunity in the week on which buyers for the catch were guaranteed.

Indigenous interest in sea fishing did not disappear although it continued in more primitive and less competitive mode. At the end of the eighteenth century within the "congested districts" of the Atlantic seaboard, 50% of the income of coastal families came from fishing. Poor catches meant "fish famines" which were devastating to communities whose members were obliged to seek employment elsewhere. There were occasional glimpses of what might have been, as when the Southern Fisheries Association of Kinsale in 1818 employed 25,000 people, funded with a levy on imported herring. Such levies were to be encouraged in Scotland, but not in Ireland and the employment scheme was wound up, along with the levies, in 1829. In Scotland 1/7 of the population was employed in fishing, in Ireland 1/200. The unequal treatment of two member states of the Union with England was denounced at a meeting of the Home Government Association of Ireland addressed by the lawyer Isaac Butt in 1871.[8] It is the first instance of a subsidy making a tangible difference to the fishing industry of this nation, a phenomenon which would recur in the following century when financial assistance from government was expected for virtually all fisheries-associated development.

In the nineteenth century, Ireland's fishing fleet consisted of a large number of flimsy craft, the majority constructed of wooden frames covered with canvas or animal hides, whose operators were extremely skilled. Occasionally, they were able to fish very successfully when the fickle herring shoals moved close inshore, but they were very vulnerable to weather. Numerous men and boys were engaged in fishing then although we have no idea of the number of full time job equivalents provided by the activity.

The great potato Famine of the 1840s, the only famine until that time outside Asia to claim two million lives, might have been expected to revive the fortunes of the industry. Instead, fishermen sold their gear to purchase meal. Many succumbed to hunger and disease, others emigrated. The years following saw something of a revival but also ominous signs of marine over-fishing. Fishing companies, a forerunner of later marketing developments, put canvas-covered curraghs out of business, oysters were dredged to extinction in the Irish Sea and there was much talk of what could be achieved with greater fishing power, an indication that former abundance was beginning to wear thin.

The marginalisation of fishermen continued in the years following the Famine. As a group they had more in common with fishermen elsewhere and tenuous terrestrial connections. "The deep sea fisherman's religion was a mixture of superstition, doubt, disbelief and fantasy". He did not fit in, was socially inferior to the rural landowner and, in the early years of the twentieth century, an unhappy reminder of the poverty of the great Famine that everyone was anxious to forget. Fish was dropped from the diet in the wealthier parts of Irish cities. Instead, fishing and those who practised it would be romantic curiosities depicted by artists recollecting an Ireland that had passed away. As a group, fishermen played no part in the struggle for independence. Fishermen were also seen to be a primitive part of a social hierarchy as described by influential German political geographers of the time in accordance with Darwinian principles: the survival of the fittest.[9]

When, in the 1930s, the development agency, the Irish Sea Fisheries Association and, in 1952, its successor, Bord Iascaigh Mhara (BIM) attempted to rescue the fishing industry, the absence of a home market proved an enormous obstacle. Fish was eaten without enthusiasm as a penitential requirement. As late as the early 1950s, the annual *per capita* consumption of fish in Ireland was less than 4 kg, much of it imported tinned salmon, compared with 15 kg in Britain and 27 kg in Norway.

The 1930s are a convenient starting point for the story of Ireland's modern fishing industry although its roots go much further back than that. The most widely used fishing method, trawling, was developed in the fourteenth century. From the beginning it was controversial, causing enormous ecological damage, but also highly productive because surplus landings were the raw materials for processing and exports rather than just food for the fishing communities harvesting them.

The second significant development was the introduction of fisheries subsidies in the eighteenth century. Although they had a brief existence at that time, they yielded demonstrable benefits for the quality of product and they incentivised heavier landings. Raising capital to fund the acquisition of gear and boats became a charitable pursuit in the nineteenth century but the state resumed its assistance to the industry in the following one. Subsidies would become an indispensable element in the industrialisation of fishing worldwide during the 1930s and 1940s when they provided "seed capital" for the purchase of more powerful boats and gears. Ireland joined that race. In due course, the provision of financial assistance became an unbreakable habit which would, eventually, lead to the creation of enormous fishing power which was destructive and counter-productive and utterly dependent on the perpetuation of government assistance for its survival.

We can trace the first signs of a management regime for marine fisheries to British legislation in the mid-nineteenth century. An Act, intended for the regulation of salmon, trout and eel fisheries, contained some foundation principles for a modern sea fishing industry. It set up a register of fishing vessels, a basic measurement of fishing effort. It also established the rule that any fishing method would be permitted in the sea unless and until provision was made to curtail its use. A number of regulations to locally control the use of trawl nets were introduced on foot of this Act but they did not long survive the handover of power from Britain to the Republic after independence.

The significant feature of contemporary fisheries development, which made its appearance in the mid-1900s in Ireland, was the impetus to industrialise marine fisheries, the topic with which *Overkill!* is concerned. Until that time fishing was undertaken by local communities in small boats for subsistence. To generate profits, industrial methods were intended to catch larger quantities, and those who practised them had little loyalty to local areas or communities, operating instead where fish were most abundant and then moving onto other grounds once they had exhausted those. After the mid-twentieth century, industrial fishing enterprises, employing heavy trawl gears, would come to dominate the harvest of marine fish throughout the world.

The twentieth century opened with a smaller than hitherto Irish industry, populated by fewer boats, which achieved some financial success at times, notably during the First and Second World Wars when competition from neighbouring fleets was suppressed, although periodic difficulties occurred. Availability of a marine resource, which appeared to be of unlimited, or at least un-measurable, extent, demanded that investment be directed at exploiting it. Initially, efforts were modest: a co-operative state-sponsored approach was attempted in the 1930s and it appears to have met with some success which, in retrospect, would be dismissed as inadequate.

External circumstances conditioned attitudes in Ireland. One was the appearance of foreign fleets up to and often inside the three nautical mile (nm) exclusive fishing limit at the end of the Second World War. There was little new in that except that these vessels, small by today's standards, were the largest and most powerful seen to that time. Attempts to defend the industry's interests within national territorial waters with complex boundaries were inadequate and, in the 1950s, Ireland joined the international movement to devise wider and more easily defined territorial limits which were agreed at the Geneva Convention of 1958. The exercise was a prolonged horse trade and it was essentially a new approach for a nation which had achieved independence within the previous half century. In future, marine fishery entitlements would be debated and agreed internationally, rather than only with our sister island.

The national development agency, BIM, which would dominate marine affairs and funnel finance into industrialising the fleet, came into existence in 1952. It was based on a similar agency in Great Britain which had been set up the year before.

There was no shortage of ideas on how the fishing industry should develop. The question was where to begin. The catches were too small to supply exports and too large to be consumed at home. Exchequer investment in developing the industry was immense and this fact caused resentment within the industry, such as it was. The agency was accused of having too much control. But, while there were numerous squabbles about details of policy - how things were done - there was considerable unanimity about the direction that progress should take and it could be described as "unlimited expansion".

A consequence of these early attempts to find an effective way forward embroiled the agency in an immense array of activities to which it clung tenaciously over sixty years. No industry-related innovation, fish capture method or processing technique was outside its ambit of interests.

During its first decade BIM had managed a number of money-making projects at the heart of the fishing business, all of which lost money. For the remainder of its time, with one brief exception, BIM would function as a non-profit making development agency. Throughout its history it sourced finance, initially from the exchequer and then the European Economic Community, and channelled it into fishing enterprises, and many of those, hardly surprisingly, were doomed to financial failure too. But desired progress was also made. Money was invested in processing to confer additional value on landed fish. Grants were lavished on increasing the consumption of fish in Ireland, the demand for which would be partly satisfied with imports.

When it came into existence, BIM's main purpose was the provision of a fishing fleet for the nation. Initially the fleet was populated by a very characteristic pea-pod shaped design of wooden vessel of small dimensions but as time went by, individual boat size began to expand and, more significantly, their engine sizes grew disproportionately. That may have been a sign that fish stocks in waters close to shore were beginning to thin out. The fleet was intended to exploit the resources of the continental shelf but, in the final years of the twentieth century, the largest steel vessels ventured deeper, onto the slope.

Another warning sign which was not given sufficient attention was the growing problem with repayments on vessels. Rather than heed these signs, the development agency devised a series of schemes to make boat repayments easier and to raise extra finance to write off bad debts. BIM proved adept at spending money and the demands on its supply were insatiable.

A rule of fishery development which has been witnessed throughout its recorded history tells us that bringing a particular method of harvesting into existence creates an overspill of fishing effort into other methods. Successive technical innovations in trawling were accompanied by explosive expansion in the use of other fishing gears. In the 1950s and 1960s, innovation was accompanied by the development of a more sophisticated fishing technology; tackle built of lighter and stronger synthetic materials. Bigger, more powerful, vessels put their more elderly antecedents out of business and some of these re-entered fisheries equipped to harvest fauna which, hitherto, had been the prey of fish exploited for human consumption. Thus, fisheries for prawn and whelk were established.

As a gatherer of funds, predominantly from the exchequer and the European Community, to be dispensed among a variety of projects, BIM became enormously influential within the industry. The more money it distributed in grants, the more dependent on it the industry became. There was no shortage of criticism of its activities but discontent rarely rose to more than a faintly audible grumble. Even in 2010, when the fortunes of fishing were clearly on the wane, its grant bank substantially exhausted and its total income halved from the year before, the funding agency was able to boast of having assisted over 700 vessel owners, approximately one third of the fleet.[10] At that stage the industry was irrevocably hooked on subsidy so that criticism might have been considered an expensive self-indulgence by anyone who risked it.

Just as agreement on the Geneva Convention was reaching fruition, another negotiation was beginning to take shape. Ireland along with the UK, Norway (which later withdrew its application) and Denmark applied for membership of the European Economic Community (EEC) in 1969. These, and the six nations which had already acceded, had all been part of the Geneva Convention negotiations which enlarged Ireland's territorial sea and demands by the industry for a fifty nm exclusive fishing zone, expressed soon after the general acceptance of new territorial limits, were dismissed. The next extension of territorial boundaries took place on foot of another international agreement. The limits of the Exclusive Economic Zone (EEZ), was finalised after Ireland had become a member and they circumscribed waters belonging to the EEC, rather than Ireland.

The Common Agriculture Policy would precede its fisheries equivalent by 25 years and the problems confronting the two were quite different. Agriculture had to cope with surplus, Fisheries with reconciling catching power and expectations with already declining resources. The Common Fisheries Policy (CFP) required that all member states of the Community would have a fixed proportionate share of resources which, although renewable, would not get any larger. Entitlement to fish within the new Community would, logically, be determined on what each nation had harvested up to the time it became a member. All fishing nations who joined the Community had ambitions to catch more; market demand for fish would increase while technological improvements, subsidised by Community funds, would harvest greater quantities. One way of securing agreement among acceding member states was to provide grants for fleet enlargement, and that jeopardised the resource on which the industry depended. The CFP came into existence on 1 January 1983.

There is a popular belief, nurtured within the industry, that the CFP was a plot which an innocent nation was gulled into accepting. In fact, the Irish civil service and senior Irish politicians played a major role in devising it. The Fisheries brief was regarded as the most awkward to resolve at the formation of the EEC, and Ireland was not the only disappointed member state. But Ireland's proximity to rich fishing grounds and her niggardly allocations of fish have been a bone of contention ever since. The ensuing frustration fuelled a disregard for fisheries management, and co-operation with the Commission (the civil service of the EEC) by Irish agencies and the Department, has been less than wholehearted.

While the industry remained sullen about having its activities regulated by the Commission, there were some perceived advantages of membership for Ireland's growing fishing industry. Suddenly, a market for hitherto unexploited species was created and Spanish fishermen displayed, although not very openly, the advantages of fishing at greater depths and thus increasing exploitable marine areas. The Commission also assisted with finance for the administration and protection of fisheries off Ireland's coasts, in addition to subsidies for vessels and industrial infrastructure.

In the early 1960s proposals were accepted for five large new purpose-built fishery harbours. The fisheries development agency took on a more dynamic management style and became a very effective sales, propaganda and publicity machine. Its annual reports emphasised positive achievement and shelved mention of inconvenient truths. And, because the fleet was expanding, landings, processing and exports did so also and the consumption of fish increased in Ireland. For a time fleet building became so successful that BIM relaxed its earlier embargo on commercial activity and had one more go at making a financial success of a business enterprise.

As the 1960s and the following decade progressed, more boats were launched and a succession of vocal fishermen's organisations was formed, supplanting one another as their focus of interest shifted. At first they represented the interests of salmon fishermen and others engaged in near-shore waters but their repertoire widened as the average size of new vessels grew and, in keeping with practices established long before, the interests of the owners of larger vessels, towing trawls, gradually gained precedence. Self interest replaced earlier more altruistic motivation. It was not long after Ireland's accession to the EEC that small boat fishermen were neglected. Eventually, the fisheries on which they depended would be unregulated and those whose craft were confined, by their small size, to near-shore waters were left to do as they pleased; many of their boats were not even licensed.

Producers' Organisations (POs) were set up as agents of the Commission in the 1970s with a view to regulating markets and administering fishing entitlements (quota allocations) within the industry. POs were the only form of representation which was acceptable under the terms of the Treaty of Rome. They were the wealthiest representative organisations to date; their membership in Ireland was largely confined to bigger vessels and their owners rapidly forged political links with the dominant political parties. Their creation sealed the fate of the inshore fleet which still employs about 50% of those who work in capture fisheries.

The Commission subsequently tried to make amends by encouraging the Irish authorities to make provision for its fleet of small craft within the exclusively national territorial sea. Those overtures have been rejected; the Department takes the view that smaller craft should be available, were it considered desirable, to be "cannibalised" into larger vessels. Meanwhile, the largest boats in the fleet are free to fish anywhere they wish.

One of the first actions of the POs was the introduction of intervention market support, a mechanism for guaranteeing fish prices by withdrawing, and often destroying, fish which did not reach a threshold price at auction. Its captors were then compensated. Seen from another perspective, intervention was destruction of human food and damage to fish stocks. Soon after the formation of POs, many fishermen were fishing specifically for intervention payments to the concern of both the public and the Commission. Harvesting fish merely to dump them in exchange for compensation coincided with demands by the industry for larger quota allocations. But, intervention was not the only wasteful practice of a rapidly expanding industry. Equally short-sighted was a fishery which captured juvenile food fishes, like herrings, to keep fishmeal factories in business, thereby depleting stocks of one of Ireland's most valuable food fishes.

In the mid-1970s, before the CFP was launched, things began to go wrong. Signs of over-fishing became more obvious, although there was reluctance to acknowledge them, and the annual headlines of record volumes and values of landings became scarcer. Financial crises in vessel repayments followed. Its last commercial enterprise having failed, BIM sold its boatyards and cultivated an interest in aquaculture which swelled to equal and surpass its investments in capture fisheries and deflect attention from them. Its staff shrank by half but its influence in marine matters survived.

Capture fisheries would have to be worked intensively in the future to feed a fleet which was, at this stage, enormous: seven times the displacement of the fleet in 1930 (it would reach a maximum declared tonnage of ten times the 1930 value) and possibly 100 times greater in engine power. Supplying it with fish would be fraught and manic. Every discovery of a hitherto unexploited species or stock was welcomed with enthusiasm undimmed by previous disappointments. In exploratory fishing, the emphasis was on locating protein which was trawled to exhaustion before even a perfunctory assessment of it had taken place. A variety of methods of sieving water to extract the last fauna was devised and no species was too insignificant to be ignored. Comparatively tiny resources, like shellfish beds, were fished down in no time. In 1991 it was openly admitted that a desperate search for fish resources was under way.

As the end of the century approached, a new departure, fishing the waters of the continental slope, made accessible by the purchase of the largest steel vessels in the fleet, was greeted as the latest salvation. The continental slope proved far less productive than the shallow shelf seas and its possibilities too were quickly exhausted.

In 1995 a scheme to re-deploy some of those engaged in fishing was funded by the Commission, a frank admission that marine resources were over-worked. The Commission, anticipating what lay ahead, capped the size of the fleet. It phased out boat purchase and construction. But for the development agency in Ireland, the business of arranging finance for the industry had become an established way of life and it sought inventive ways of maintaining the service. Alternative routes of channelling finance to the industry were devised: vessel modernisation, grants for safety equipment, fleet renewal, decommissioning and the purchase of vessels to pursue non-quota species in which the Commission had no stated interest were some of them. The trend towards bigger boats continued and when grants were distributed the latest schemes would disproportionately favour the owners of the largest in the fleet.

The trend towards larger vessel size was accompanied by a power shift in the administration of the industry. The POs, now the largest fishermen's organisations, became closely associated with the state's development agency and the government Department which oversaw the industry. A group-think consensus favouring industrialisation germinated and infected other state agencies. Significant conditioners were self- interest, self-delusion and greed.

Today, few of the executives of the POs come from a fishing background. Most are university educated, performing as lobbyists who served their apprenticeship in the professions or in one of the agencies belonging to or associated with what has become the politicised fisheries establishment. Senior personnel move among its component organisations and sit on the boards of state bodies which provide services to the industry. The fisheries establishment demonstrated strong affiliations with Fianna Fail but the latest Fine Gael minister has effortlessly picked up the baton. Fisheries scientists monitor fishing activity and issue warnings about the consequences of uncontrolled exploitation but their various recovery plans, designed to restore depleted fish stocks to good health, were ignored by the industry which is openly resentful of anything that stands in the way of its efforts to make money. In an attempt to continue working with the industry, state scientific services have become apologists for it.

There was a rationale to the way the industry behaved at the end of the twentieth century. As more and more fish were removed from the fishing grounds only the largest and best technically equipped vessels could continue to make a living by fishing. Within the fisheries establishment the emphasis transferred to serving their owners. They had greatest influence and, equipped with ever more modern craft, subsidised by the taxpayer, were enabled to capture the best of whatever remained. Despite the low levels of remuneration which always characterised fishing, a minority of the fortunate ones became immensely rich.

In 2004, a whistle blower revealed the industry was surviving by fraud. Landings of up to eight times the permitted amounts were being put ashore and, apparently, the regulating authorities knew nothing about it; how it was conducted under the supposedly strict controls of the Department was never explained. There was no alternative but to put an emergency decommissioning scheme in place to take boats out of service. The move did indeed slightly reduce the size of the fleet but the scheme which facilitated it may well have made a bad situation worse because an unquantifiable proportion of compensation monies awarded to owners were used to purchase more modern vessels, thus restoring the killing power which had supposedly been taken out.

Although it is smaller in size than in 2004, Ireland's fishing fleet is still too large to survive on the allocation of fish available to it. A new fisheries enforcement agency, set up in 2006, is composed of personnel transferred from its predecessor. It received a hostile welcome from the industry and has taken a very subdued approach to enforcing regulations since.

Overkill! is an account of the expensive and disastrous attempt to create an industry which, in 2010, had a well equipped fleet with few remaining fish to catch. It is not a unique history but it contains familiar elements which we may come to regard as characteristic of Ireland over the past fifty years.

In retrospect, the industrialisation process consisted of a variety of independent initiatives, many at cross purposes with one another. One example which demonstrates this is that at the same time as monies were supplied by the EU Commission to divert employment from capture fisheries, and thus remove fishing pressure from over-exploited stocks, grants were provided to make the fishing fleet more powerful, cancelling any benefit from the redeployment initiative. The process of creating a modern fleet was un-coordinated and even reckless. The result was a lemming-like rush to disaster which swept up many, but Ireland was not the only nation to be seduced by its fishing industry's desperate race for success which deteriorated into a fumble for financial survival and culminated in unregulated and illegal activity. Nor was the fishing industry the only national pursuit to collapse under the single-minded enthusiasm of a population hell-bent on making progress, regardless; there are similarities between this story and other debacles, such as the collapse of the post "Celtic tiger" property boom.

Occasionally, attempts were made to forecast how aspects of the industry would develop in the short term when the programmes for economic expansion were introduced in the 1960s and, later, as a prelude to fleet expansion. Their predictions were wide of the target and the exercises were neither continued, nor were their conclusions dwelt upon. And this has been the pattern throughout the past sixty years: subsidies were won to create jobs which never materialised. Funding is still being sought on this pretext. Over the medium and long terms, jobs have been shed by this industry and that continues to be the case, accelerated by a declining resource, capitalisation and improved technology.

The amount of money contributed to the industry through infrastructure, administration and grants has been enormous. Recent analysis suggests that in the recent past, for every €4 worth of fish landed, the taxpayer (Irish and European) may have contributed as much as €3.

As the industry grew, a progressively greater variety of fish species was commercialised. In 1903 some 23 were listed; in 2000 the number was 150. Some of those are the invertebrates on which our whitefish species, like cod, once fed. The finfish have gone and their food organisms are now harvested for human consumption. Those whitefish that remain are of small average size and low average age. In the meantime, a number of once-listed species are no longer included in the statistics while the capture of others is prohibited by the EU for conservation reasons. Species which grew slowly, matured at a later age and produced few young were most obviously at risk from intensive fishing. More stark however, is the quantity of once abundant fishes, like cod, whiting and plaice, whose biomass has been reduced over a century by, it is estimated, at least 90%. The value of fish landings peaked in 2001 (taking inflation into account) and has trended downwards since. The heyday of fishing is over. And some of the organisms which are becoming abundant in marine waters, like jellyfish, may owe their expanding numbers to the absence of finfish which prey on their larval stages.

The industry has wreaked and is still inflicting enormous damage, dragging heavy nets across the seabed, sieving more and more water to extract whatever fish remain. Small boat operators are now lucky to have one species to concentrate their efforts on rather than the succession of invertebrates and finfish that kept them occupied throughout the year half a century ago. The amount of fuel expended per tonne of fish landed is likely to be in excess of one tonne of diesel. The latest report on the future of the industry, published in 2006,[11] recommended spending almost as much again as had been invested by Irish and European sources in the course of its development. It also acknowledged that the whitefish fleet would require a 45% increase in landings to remain in business.

So, how has an industry which generates less than one percent of Gross National Product, claimed so much exchequer finance and favourable political consideration? One explanation suggests it is disproportionately loquacious and troubled and given to occasionally organising blockades. The fisheries brief is a difficult one to deal with and politicians favour quiet lives. The easiest course is acquiescence. On the other hand, politicians fear public opinion more than an industry which generates only bad news and it is surprising that the public is not more effective in obtaining a change in policy. The industry also fears public opinion and is sensitive to the fact that there is widespread unease about the destructive way marine fisheries have developed and the undeniable damage they have inflicted on fish stocks and the marine environment.

Commercial fishing has a warm spot in the people's hearts and the way it operates is profoundly misunderstood. It is a traditional way of life, and a hard one, and there is sympathy for those who eke a living from the sea. But the industry is careful to cultivate images of itself that are at variance with reality: men in small boats dispossessed of their fish by the European Union whose other member nations have selfishly victimised them. It is incorrect to claim that Irish fishermen respect fisheries regulations and that illegal fishing is something engaged in only by other nations. The history of industrialisation shows otherwise. Irish fishermen have had scant regard for conservation since mid-last century. The industry in Ireland takes no responsibility for the damage done, despite the undeniable facts of its own misbehaviour. Within the exclusively national territorial sea, where the Commission's writ is less potent, national regulations are rarely enforced. Some were drafted in such a way as to make them unworkable.

Another of the many myths cultivated by the industry is its "family-owned, small boat" image. On the contrary, although the fleet embraces relatively small numbers of large fishing vessels they account for the vast majority of the displacement (gross registered tonnage) and fishing engine power. The industry is represented as a plucky David scrapping with a Goliath Commission from whom fish must be wrested. But Brussels does not have any fish; rather the Commission's regulations are intended to return fisheries to sustainable productivity, unsuccessfully as it turns out. Negotiating with Brussels is not a practice engaged in by the owners of small vessels who have no influence or place in the process. The owners of large ones have a lot. The largest fishing boat in the world of its time, *Atlantic Dawn*, incongruously a member of Ireland's "inshore fleet" (the Department's official definition), had an enormous amount which translated into considerable wealth for its owner. Why such lucrative links had been forged with the political establishment is still a matter of speculation.

In the next chapter I want to review the produce of the shallow shelf waters surrounding us, the area at which most fishing effort has been directed, provide a brief account of the main categories of marine fauna and describe the technology used to harvest them. The method of fishing of greatest relevance to this story is trawling and I will provide a brief history of its controversial development in Irish waters.

Chapter 2

HARVESTING THE FAUNA OF IRELAND'S SHELF WATERS

Waters of the shelf deepen gradually from the shoreline to the continental margin, the break from which the slope descends more steeply to the abyssal plain. The shelf edge is arbitrarily demarcated at a depth of 200 m. Ireland's shelf area covers approximately 160,000 km², some 18% of its marine territory. The shelf to the south is labelled the Celtic Sea; north of Galway the waters are referred to as "west of Scotland". The third division, the Irish Sea, accounts for less than a third of the total. It is a sheltered marine area, not exceeding 100 m in depth. A map showing the marine features in Ireland's vicinity is provided in Fig 1.

The dominant current influence is the Gulf Stream, part of the North Atlantic gyre, which gathers in the Caribbean and the Gulf of Mexico whence it is forced towards the Arctic by wind and the earth's rotation, the Coriolis effect. The stream reaches depths of 600 m and dissipates its heat in an easterly direction under the influence of prevailing westerly winds.[1]

The edible fauna of these marine divisions, distributed from the high tide mark seawards, is referred to in conversational terms as

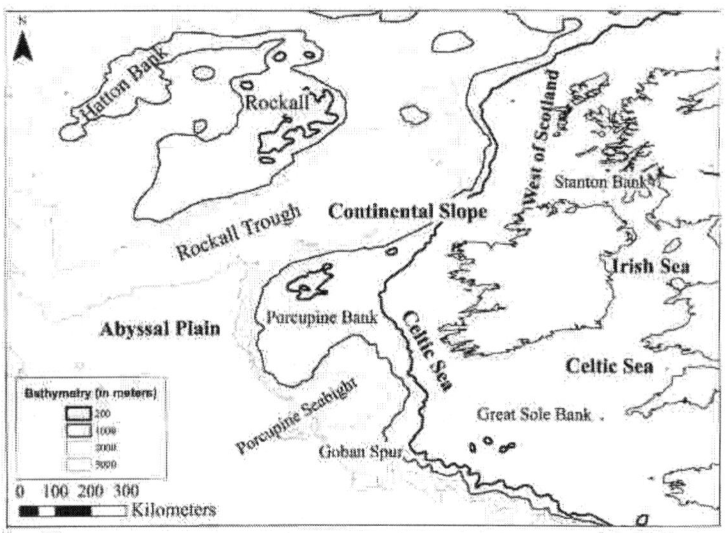

Fig. 1 Map showing bathymetric features and fishing grounds in the vicinity of Ireland.

"fish", a term which embraces a variety of biological forms and species; "wet fish" are freshly landed, without further processing. More precisely, *shellfish* are invertebrates with a hard exoskeleton: crustaceans (shrimps, crabs, lobsters, crawfish, all of which will be encountered in the book), echinoderms (purple sea urchin), molluscs (mussels, cockles, razor and surf clams) are all embraced within the term but not necessarily within its legal definition.

Finfish, more accurately referring to vertebrates with gills, include two large groups: the bony teleosts – like cod and mackerel - of which many species are relatively short-lived and produce large numbers of eggs, and the cartilaginous elasmobranchs, the sharks and rays which can live considerably longer but whose rates of reproduction are slow.

Fish species which swim, often in large shoals, high in the water column, are labelled *pelagic*. These species are referred to elsewhere as "blue" fish. Their colour scheme camouflages against background - a dark dorsal surface so that they blend with the water below when viewed from above and silver underneath so that they do not contrast sharply with the sky when seen from below. They have buoyant, oily flesh, and include herring, tuna, mackerel and salmon among their species. The other division comprises *demersal* or *white* fish - whose flesh is flaky - such as cod, haddock and plaice. They live close to or on the sea bed with whose background their dorsal colouration blends. The term *whitefish* usually refers to teleosts but it can include elasmobranchs such as spurdog which have been grouped with them on occasion. A list of species landed by the Irish fleet today, together with their scientific names and the categories to which they belong, is set out in Appendix 2. The remaining category of food fishes, which we will come to later, comprises species from deep water.

Other terms which are encountered in the course of the book are *prime* fish, meaning white fish species which have high money value. The identity of prime species has changed from time to time in response to market demand; the contrary term is *coarse* or rough species but that is not generally applied to marine food fishes nowadays.

When T.H. Huxley, the first professional marine biologist, delivered the inaugural address to the Fisheries Exhibition in London in 1883, he observed that the oldest monuments of antiquity show us the fisherman in full possession of the implements of his calling.

...even those tribes of savages who have reached neither the pastoral nor the agricultural stages of civilization are skilled in the fabrication of the hook, the fish spear and the net.[2]

Today his generalisation is in one sense true and in another could not be wider of the mark. The basic principles of fish capture are much the same as they ever have been but a better understanding of fish behaviour accommodated their refinement, in some cases to a high degree of sophistication which has been further facilitated by better designed gear constructed with stronger, lighter synthetic materials. Much innovation has been directed towards the capture of animals which are more widely dispersed and at lower density because intensive fishing has thinned out their numbers. A significant aspect of the modern industry is its increasing use of larger fishing nets and more hooks and traps.

The purpose of this chapter is not to provide a universal or comprehensive list of fishing gears and techniques, rather to set out a simple scheme for classifying methods in use in Ireland, referred to later in the text, and briefly describe their development.

How fish are caught

The simplest method of harvesting marine produce is to gather it along the shoreline at low tide. Equipment is rudimentary. A garden fork or rake is a useful means of exposing buried bivalves; salt poured over the burrows of razor clam prompts them to forsake the substratum. Possibly the most common accessory is a bucket in which the animals are collected.

Moving down the shore, into the tide, fishing gear becomes indispensable. Fishing techniques belong in one of two families: should the animal come to the gear, the method is described as *passive*. Thus, a *hook* is a passive gear because the fish pursues and takes it. A fish hook belongs to that gear grouping even when it is part of a lure drawn behind a moving vessel. Hooks are also set as elements of a long-line which may have many thousands. *Long-lines* are set on the bottom for certain species, such as hake, or at the surface for others like tuna or sharks.

Traps, pots and creels, set mainly for shellfish in our waters, also operate on the basis that target species are enticed by bait to enter them. Typical species sought by this method are the crustaceans, edible crab, shrimp and lobster and the mollusc, whelk.

A number of nets belong within the passive gears grouping. The most common are *gill nets*, curtains of meshing stretched vertically just under the surface or on the sea bed, in which fish become jammed or *gilled* (a single mesh grips the body behind the gills) as they attempt to swim through the obstacle. Gill nets are fairly selective, the smallest fish get through, the largest individuals bounce off them. The capture mechanism is "clean", straightforward, usually involving a single mesh. Species sought by gill nets include spurdog, mullet and gadoids (members of the cod family).

Several types of net operate by gripping animals in a more complex way involving several meshes. *Tangle nets* are literally bundles of net, loosely rigged (not set taut) and deposited on the bottom. They are unselective and a variety of crustacean and finfish species and sizes become enmeshed as they attempt to cross them. Target species for this method include crawfish, rays, and turbot.

The most elaborate passive net design in our waters is the *trammel net*. It is constructed of two sheets of meshes, one mesh size smaller than the other, stretched side by side between *head* (top) and *foot* (bottom) ropes; trammel nets are set vertically. In common with tangle nets, trammels have the ability to enmesh a wide variety of species and sizes. Like tangle nets, their mechanism is complex and it often takes time to remove animals captured in this gear. Species harvested in trammels range from flat fish, like plaice, to edible crab.

All of the above are also referred to as *fixed* or *static* gears, literally anchored to the sea bed. Gill nets can perform anchored or they can be left to move with the tide in which case they are referred to as *drift nets*. At times they may be legally required to operate in that way. Species captured by drift nets have included Atlantic salmon, herring and albacore tuna.

When, in 1958, the DuPont Corporation invented monofilament fishing line, a new era of possibilities was created for passive gears. Plastics of various kinds were subsequently used to manufacture nets and ropes. Plastic is lighter than natural materials like hemp, so that larger quantities could be carried to sea. It could be manufactured in a form less visible to fish, hence more effective. It was easier to handle, it was non-biodegradable so it required less care and maintenance - hemp had to be dried after use, a labour intensive exercise. When plastic nets were lost or abandoned they went on "ghost fishing".

Other inventions were introduced at about the same time. Before the introduction of propulsive power, generated by steam and mineral oils, most work aboard was carried out by hand. This included the back-breaking toil of rowing smaller vessels, manipulating sails, hauling static gears and nets aboard and reeling in warps attached to nets by, on the larger vessels, a hand-operated winch or capstan, which spooled rope in a horizontal plane. Obviously, there were limits to the amount of labour a small crew could expend. On its introduction, a proportion of steam or motor power was diverted to winches which took much of the physical strain out of recovering gears and bringing them aboard, but the smallest vessels had to await the development of small hydraulic equivalents to lift gear off the sea bed. In the 1960s these became widely adopted. The early models hauled horizontally, like their prototype capstans. Later, a vertically mounted hauler was developed to retrieve pot lines. It could be affixed to the superstructure of a boat at head height allowing crustacean traps to be winched from the sea bed and then swung inboard. In the case of the Norfolk crab fishery, geographically close to Ireland, all pots were hauled by hand until the 1960s. In one particular instance, a single boat fitted an automatic pot hauler in 1967 and three years later all 40 boats frequenting the port in question were equipped with them.[3] During and following the 1960s, technical innovations which were adopted elsewhere in Europe quickly became part of the repertoire of methods and practices in Ireland.

Other technical advances which assisted all fishermen, including the operators of small boats, were the development of sonar which could be used to image depth and substratum, enabling more accurate positioning of fishing gear and, of course, modern navigational aids which include the ultimate global positioning system (GPS), developed by the United States Department of Defence. Prior to its introduction, a navigational system for the larger vessels in the fleet was installed in the early 1970s; provided by Decca Navigator Co Ltd, it consisted of a master station in Galway and three slave stations in Cos Donegal, Kerry and Waterford.[4] Before it came on the scene, operators had to rely on sextant and chronometer. Small boats were confined to near shore waters in which they spent a large amount of their time wandering in search of static gears left soaking over night. Dependent on landmarks at best, poor visibility might stymie fishing altogether.

The combined effect of these improvements was an increase in the amount of gear that could be handled. The numbers of pots and lengths of static nets were not usually recorded but occasional studies indicated an explosive increase in the amount deployed. Along one section of the coast of Co Wexford, the number of pots fished increased by 241% in the thirty years following 1968.[5]

The synergistic interactions of various scientific and technical advances in fishing methodology is known as *technology creep* or improving technological efficiency. The various elements which contribute to it are difficult to isolate and quantify but its overall effects have been estimated to raise catching capacity on a global scale by 4% annually.[6]

Trawling

The second large family of fishing gears is labelled *active* methods because the gear pursues the fish. Familiar examples would include a child's minnow net or a trout angler's landing net which encloses a fish. Commercial examples range from shrimp push nets operated by a single individual in shallow waters to the trawl nets of the largest super trawlers, reputed to have a mouth area of eight football fields. For obvious reasons, these methods are also described as *mobile* or *towed* gears.

At this point it is appropriate to introduce the terminology for estimating fishing power which is obviously relevant when nets are dragged over the sea bed or through the water column. The earliest accounts of fleet size in Ireland, dating from the nineteenth century, grouped boats by size categories, usually length, for the smaller craft[7] and weight[8] for the larger ones, and recorded their distribution along stretches of coastline. Increasingly, after the 1940s indications of engine power were provided[9]. Three statistics - vessel number, size and power - are used as crude indicators of *fishing ability* or *capacity* of the fleet.

Mobile gears designed to target bottom assemblages of fish are referred to by their descriptive label, demersal; gears which pursue pelagic species are known by that term.

One of the simplest enclosing net designs is the beach *seine* net, a sheet of meshing which can be set by boat from the land to surround a shoal, after which the net containing the fish is hauled ashore. Such nets were set for herring as early as the tenth century.[10] Similar nets constructed of tarred hemp were used for the capture of pilchard in Baltimore by the English settlers who where kidnapped to Algiers by Barbary pirates in the seventeenth century.[11]

An adaptation of the beach seine allowed it to be operated from an anchored fishing vessel from which the net was paid out in a circle surrounding a prospective fishing area. The circle was closed when the far wing of net was returned to the boat which might tow the gear behind it for a time before winching it in using the anchor to hold the vessel in position, hence the description of the method as "anchor fishing". Seining was popular in Ireland during the early years of the last century. The method allowed an enclosing net to sieve the water for its fish content without consuming large amounts of fuel. However, because it was fairly laborious to use and covered a comparatively small area of seabed– relative to the potential of a mobile towing vessel – its success was dependent on the availability of high densities of fish.

Both beach and vessel seines may have reinforced centres designed as "bags" or "cod ends" whose stronger meshes are less likely to burst in the course of making a haul. Vessel seines have been used to capture demersal species, beach seines small pelagics. *Purse seines* are sheets of uniform mesh size which can be up to several km in length and of considerable depth. The head (float) and foot (lead or "purse") ropes are of equal length and the purse line is threaded through rings which, when drawn tight, close the bottom of the net which assumes the shape of a purse from which the fish are transferred to the catching vessel. Purse seines are regarded as the most lucrative method of capturing pelagic shoaling species, mackerel, herring and tuna, among others.[12]

The physically smallest and relatively heaviest commercial towed gears in use today are dredges used to harvest molluscs. Their designs are various. One consists of a metal frame to which a net bag is attached to scrape mussels off the seabed; another, a metal frame to which a bag of chain mail is attached, is designed to collect scallops. The heavy weight of the apparatus is intended to dig into the substratum and scallop dredges are equipped with tines (teeth) to agitate the fishing ground and disturb any swimming bivalves in its path. Individually small, as many as 34 scallop dredges have been fished simultaneously, attached to two metal beams drawn on either side of a single vessel in Kilmore Quay, Co Wexford.

Other designs are heavy metal cages which operate by ploughing softer sediments to a depth of 30 cm and exposing molluscs like razor clams. Their progress can be eased by pumping water through the frame of the apparatus and fluidising the ground it cuts through. Some cockle dredges operate on the same principles as vacuum cleaners, sucking up the sandy substratum and washing it through a grader to separate it from the shellfish.

The largest towed gears are trawl nets, a method of fishing first developed in the fourteenth century. It subsequently became universally adopted, undertook a number of variations in design and underwent a number of innovations. Initially trawls may have been operated from row boats with the net hauled aboard by hand. Later the vessels were driven by wind power, and steam became the method of propulsion in the late nineteenth century.

In the second half of the twentieth century diesel-propelled trawlers penetrated waters greater than 1 km in depth for the first time. The introduction of each phase of the development of this technique was followed by an expansion of fishing effort.

The technical development and spread of trawling has compellingly shaped the modern fishing industry, profoundly altering the species composition of landings and the marine environment, its ability to harvest large quantities of fish altering even the way marine produce is marketed. Acknowledging these realities, trawling claims high priority in this account of fishing in the last century.

Although the technique was invented some six hundred years ago, the practice of trawling spread slowly at first. By the late seventeenth century trawlers were common only around the port of Brixham in the English Channel and the entrance to the Thames estuary. In these primitive trawls the net mouth was held open by a transverse wooden beam that was 3 - 3.5 m wide. At the time the market was adequately supplied with hook-caught fish. The difficulty of disposing of large catches was an obstacle to the expansion of trawling and this may have slowed its more rapid adoption.[13]

In the early nineteenth century beam trawlers from the south of England ventured into the North Sea to harvest brill and turbot, which were known in London as "west end fish", and sold to the wealthy. By the mid-nineteenth century beam trawls had grown in size. The wooden beam was $11 - 12$ m long (three to four times as long as the earliest example cited here). The "foot" line was a weighted rope. The trawl was shot over the side of the towing vessel and drawn along for some five hours before being winched aboard using a hand-operated capstan.[14]

The English trawling fleet, propelled initially by sail, later by steam, expanded from 130 vessels in the early 1840s to 800 twenty years later. The techniques it replaced, employed by all other fishers, included baited hook lines, fixed or drifting nets and traps or creels.

Beam trawls grew in size along with the vessels that towed them. Early in the nineteenth century, sailing trawlers were 20 - 30 GT. Fifty years later they were 70 - 80 GT propelled by large spreads of canvas. By the 1870s there were 1,600 - 1,700 British trawlers, 70% of them working in the North Sea.

In the 1850s steam-driven tugs appeared on the scene. Their first role in the fishing industry was to haul sail-propelled trawlers in and out of harbour. Within a decade they were towing them over the fishing grounds, the trawl gear down. Sail-driven trawlers were dependent on wind to operate and, when it was available, it determined the direction of fishing. The use of steam to propel the towing vessel may have begun in France as early as 1865[15] although the method was not widely utilised at the time. In 1877 a steam tug owner in North Shields decided to operate a trawl directly from his vessel, to the considerable amusement of on-looking sail boat personnel whose mirth probably turned to consternation soon afterwards. The first purpose built steam trawler, *Zodiac*, purchased by the Great Grimsby Steam Trawling Company, was constructed in 1881. Capable of reaching nine knots and burning four t of coal a day, it proved considerably more productive than sailing trawlers. An obvious advantage was its ability to tow gear in any direction, independent of tide and wind.[16]

By 1887 a thousand wooden steam-driven trawlers had been constructed in Britain. Gradually, as the new century advanced, iron, and later steel, hulls replaced the wooden ones. By the turn of the century the standard steam trawler was 35.5 m long with a beam of 6 m. By today's standards the engine power was still miniscule: 50 – 60 HP (37-45 kW).

A census of the numbers of steam trawlers in the ports of western Europe carried out by the British Board of Agriculture and Fisheries in 1907 placed almost 60% of them in England and Wales where the coal industry gave the method an impetus by supplying it with cheap fuel. Scotland was next with 13% while France and Germany each had 10%.[17]

Displacement of sail by steam propulsion allowed the size of towed gear to enlarge further. The beam extended to 15 m in length and further innovations were introduced. Earlier beam trawls were equipped with foot lines of old rope which fragmented if they became snagged on the sea bed. Later, chains were wrapped around the foot-lines, steel cables replaced rope warps and steam drove the winches, relieving the crew of much back-breaking work recovering the net.[18]

These technical achievements provoked a more general overspill of fishing effort, a recurring phenomenon in the course of Ireland's fleet expansion. Herring drift netters switched from sail to steam power. Long-liners increased the number of hooks they set. It is likely that the older, smaller, wooden trawlers were made obsolete by competition with larger, more modern vessels and that their owners resorted to using pots and fixed nets. Coincidentally, potters set greater numbers of creels. Fishing pressures increased everywhere.

Greater quantities of gear resulted in larger captures. It is likely that many of the species taken on board were utilised for human consumption but the list of those officially recorded in the last century in Ireland was short. Presumably, these were the species of greatest commercial value. As time elapsed, and sizeable quantities of a wider variety of organisms entering towed gears became available, markets were sought for them. The list of commercial species has extended substantially over the past century.

Beam trawls were particularly successful - they still are - at capturing flat fish species. Their technical enhancement has involved the manufacture of heavier beams and the addition of tickler chains and, most recently, the attachment of electrodes to the foot rope, to disturb fish buried in the substratum.

A new departure in trawl design was introduced in the 1880s as a result of experiments undertaken by an Irishman.[19] Instead of using a wooden cross beam to keep the mouth of the net open, two hydroplane *doors* were attached, one each side of the net to serve this purpose. The doors were rigged at an angle so that the forward motion of the towing vessel drove them apart. Its height not being constrained by the rigid beam across the mouth of the net, the *otter trawl* proved more adept at the capture of round (as opposed to flat) fish species like cod and haddock. Otter trawls were, in time, adapted to rough ground by the addition of rollers, or bobbins, to their foot rope, permitting them to bounce over rock obstacles rather than snagging on them. *Rock hopper* foot ropes exposed larger areas of hitherto inaccessible sea floor to mobile gear. The immediate result was another increase in landings. In time the hydroplaning doors would be attached at greater distance from the net where they would operate as "herding" devices, funnelling fish into the net mouth.

A further advance was the installation of paraffin-fuelled engines to propel vessels towing gears. One such, the *Ovoca* was fitted with a paraffin engine as an experiment. Paraffin became the fuel of preference, except for small craft which were fuelled with petrol. The report of the Department of Agriculture and Technical Instruction for 1910 noted:

The development of motor power in fishing craft has continued. The demand for large fishing vessels has been so great that the building yards have been unable to keep up with it and more first class boats have been built for east coast ports than were built in the previous fifteen years.

Between 1910 and 1920 the GT of the motorised fleet is estimated to have expanded fifteen fold.

Following the Second World War, British ships were constructed for the northern North Sea. Refrigeration enabled them to keep firm fleshed round fish, like cod and haddock, of interest to the fish and chip trade, a major market, in good condition. The Soviet Union took developments a step further, pioneering construction of large factory ships which were like small towns, complete with medical facilities and cinemas. They only docked rarely in order to re-provision and discharge their landings.[20]

Other innovations from approximately the 1960s included deck design features like stern ramps which allowed a net to be hauled directly onto the deck at the centre of a vessel without, as in earlier times, taking it on board over the side often causing the vessel to swing, with added risk, across the wind.[21] Manipulation of gear was further assisted by the use of power block and derrick to lift the "bag" or cod end, burdened with fish, out of the water, an operation which, to that time, had been accomplished by hand.

After 1960, the Irish trawling fleet undertook a number of developments at much the same time as other European nations. Pair trawling, the combination of two vessels to drag a single net, made its appearance in the 1950s. The first mention of it here was in the trade press which reported the adoption of this method in Spain. Reflecting the ethos of the time, it was reassuringly reported to be a social co-operative movement which

...is something that meets with the warmest approval of Christian social teaching...

In support, the papal encyclicals *Rerum Novarum* and *Quadragesimo Anno* were cited and the pope was said to be in favour.[22]

As vessels increased in size and engine power, they were able to tow, not just one, but two trawls simultaneously. Twin rigs are a significant source of landings in the capture of *Nephrops* at the present time.

The intrusion of trawl nets into deep water (more than 1 km deep) is credited to the fleets of the Soviet Union in the mid-1960s when research vessels from Russia, Poland and East Germany explored the "flanks of continents" in the Atlantic and Pacific Oceans in search of Greenland halibut and roundnose grenadier. Initially trawls captured 15 - 30 t per hour and there was a ready market for these species in the Soviet Union. Most of the ground they fished was a muddy slope substratum but occasional patches of rock or coral which occurred between 800 and 850 m depth shredded the gear. The problem was overcome by fitting heavy steel and rubber bobbins on the ground rope which tore through or rolled over the obstacles.

Modern trawling is highly automated and very precise. Sonar systems in the most advanced vessels display the shape of the seabed on a screen on the bridge and convey information on its texture. The net can be manipulated and shaped so as to avoid obstacles. Sensors relay information on the proportion that has been filled. Some trawls also carry power units which enable their operators to adjust the gape of the entrance. Nets designed to blaze new trails in deep water have been described as:

...(the foot rope) fronted by steel balls 60 to 80 cm across. These bloated descendents of early bobbins allow nets to penetrate rocky, coral-filled canyons and thunder across the reefs and ridges of the deep. Trawls are towed by boats whose 10,000 horse power (7,350 kW) engines can drag rocks (or reefs) 3 m in diameter. The nets are held open by trawl doors that weigh 5 t and are aptly named Canyonbusters *by the manufacturers...*[23]

Ireland's largest trawlers began to join the fleet in the late 1970s when five skippers committed themselves to purchasing "tank" also known as "refrigerated sea water" (RSW) vessels. These, the largest boats at the present time, then up to 530 GT, were to engage in pair trawling, two in combination drawing an enormous pelagic net between them. Each was equipped with refrigerated tank capacity to accommodate a mixture of 80% fish and 20% water. This section of the fleet would expand to more than twenty vessels and policies accommodating their interests would be accorded high priority in the future development of the industry.

In the late 1920s Ireland's fleet consisted of some 4,000 – 5,000 vessels with an estimated combined tonnage of 20,000 GT approximately.[24] Towards the end of the first decade of the new millennium, numbers of boats had more than halved but their GT had risen more than four-fold. At its maximum, the power of the Irish fleet was 200,000 kW whereas in the 1920s it is unlikely to have exceeded 3,000 kW. Much of the motive power at that time was provided by wind and oar. For the moment, we can accept that over the best part of a century the Irish fleet came to consist of fewer, larger, more powerful boats.

Allowing that fleet size is a fundamental factor in the performance of any fishery, there is considerable debate about what exactly the fleet statistics signify. Management policy has operated by grouping fishing vessels according to the methods they use and the fish they target. It may be convenient but it is very simplistic.

I take the view that all fishing effort is inter-related. An increase in, for example, a particular gear targeting a certain species, may divert effort by some boats into other fishing methods, and we will encounter examples of that having occurred. Elements of marine fauna are also inter-related through "food web", environmental and other associations so that removing a large tonnage of a particular species is unlikely to be without consequences for others.

The irresistible rise of trawling despite attendant controversies

Any method is capable of over-fishing and can be problematical. Hook and line, gill and tangle nets and crab traps have all been implicated in reducing fish and other faunal biomass or marine diversity but towed gears are far more environmentally destructive than static ones. Dredging and beam trawling are probably the most intensively damaging, incidentally disturbing and killing associated non-target fauna, but they are not so widespread as otter trawling which generates enormous quantities and varieties of discarded fish for which there is no market because they are too small, or their capture is prohibited, and they also wreak considerable environmental alteration, reducing vast tracts of sea bed to sand, mud or gravel.

Ever since its invention in the fourteenth century trawling has proved controversial. A commoners' petition in 1376 to ban the invention, submitted to the English King, Edward III, survives. The petition referred to an instrument known as the *wondyrechaun*, constructed in the manner of a modern oyster dredge but wider, to which was attached a net so close-meshed that any fish that entered it would not be able to escape.

And that the great and long iron of the wondyrechaun runs to heavily and hardly over the ground when fishing that it destroys the flowers of the land below water there, and also the spat of oysters, mussels and other fish upon which the great fish are accustomed to be fed and nourished. By which instrument in many places, the fishermen take such quantity of small fish that they do not know what to do with them; and that they feed and fat their pigs with them, to the great damage of the commons of the realm and the destruction of the fisheries, and they pray for a remedy...[25]

The response to the petition was a parliamentary commission which described the gear. It turned out to be a beam trawl, tiny by today's standards, a mere 3 m wide and 5.5 m long. At the ends of the wooden beam were two iron frames which functioned, along with the beam, to hold the net mouth open. The mesh was nailed to the beam but the foot rope dragged along the bottom and was weighted with stones to scare fish into the net. The commission decided the gear should be used only in "deep" water and it has been observed that this decision was not followed by legislation so that there was probably agreement by both sides to live with the compromise.

The case illustrated the elements of the argument against trawling: a wasteful fishing method involving the sacrifice of huge by-catch, destruction of the substratum and damage to the food-web.

Much of what we know about early trawling arises from documents associated with complaints against it. Callum Roberts cites two from the fifteenth century: one from southeast England in 1491 for the use of "unlawful engines" and small meshed nets to catch juvenile fish; another from Flanders in 1499 where trawls "which rooted up and swept away the seaweeds which served to shelter the fish" were banned. In 1583 the Dutch prohibited trawling for shrimp in their estuaries. France made trawling a capital offence the following year. In England two fishermen were executed for using chains as the foot line on their trawls (something that is common practice today). But trawling was too lucrative to be abandoned.[26]

Exactly when trawling began in Irish waters is not known with certainty. According to one source, it was first practised in Dublin Bay by Brixham fishermen in 1815 when French privateers interfered with their activities and displaced them from the Devonshire coast.[27] A community of trawling folk subsequently settled in Ringsend, close to Ireland's main fish market in Dublin. However an act of thirty years earlier was intended, among other purposes, to prevent the destructive practices of trawling for fish in Dublin Bay.[28]

An act to regulate the Irish fisheries whose influence in formulating policy in Ireland is still discernible was brought into existence in 1842.[29] Representing the interests of the landed classes, the riparian owners, the legislation was preoccupied with the management of inland fisheries, especially for species of greatest economic value, salmon, trout and eel, all of which may spend a significant portion of their lives in the sea. For administrative purposes all were regarded as freshwater species, a custom which endures. Some ownership rights were extended into estuarine and marine waters but some significant regulations for marine fisheries were part of legislation, although they were few.

In the sea itself, property rights were minimal. The crown and all other persons had the right to use the sea shore. Fishermen might use "beaches, strands and wastes" on or adjoining the sea in pursuit of their interests, which included beaching boats and spreading nets to dry, provided they did not erect any fixed structures there.

The legislation displayed an early sensitivity to the problem of discarding by commercial fishing methods. Any gear consisting of materials (hide or canvas was alluded to) which resulted in the capture of juvenile fish would not be permitted unless it was exclusively for the capture of shellfish.

The coasts would be divided into fisheries districts whose demarcation persists to this day although the need for economic rationalisation has reduced their number by amalgamation. The jurisdiction of maritime districts extended seawards. The coastline would thus be divided into small areas administered by the local coast guard or custom house.

One of its most far-seeing provisions was for the registration of fishing vessels which had to be marked with identification. The name of the owner, place of anchorage and the number of men employed on the craft would be recorded. There was a penalty for using an unregistered boat. This was an essential exercise in the management of marine fisheries. As recently as a decade ago, regulations of this kind had fallen into neglect and were not being enforced.[30]

Administratively, the commissioners of public works in Ireland were allocated responsibility for implementing the legislation, for whose purposes they were referred to as the commissioners for fisheries who would achieve their objective through the appointment of inspectors. The new commissioners of fisheries made their first report on 1st May 1843. Progress was reported on the demarcation of fishery districts and the establishment of a marine fishing register of vessels, and one significant policy departure was signalled. Resolving conflicts among fishermen using methods which impeded others had been the motive for many regulations over the years but the 1842 Act announced a new departure:

The introduction of trawls and trammel nets has for years been regarded with great jealousy by the fishermen along the coast of

Ireland, who felt inclined to continue their long established habits, or were unable to pursue the new system.

Preceding Acts of Parliament, so far conceded to their views as to admit of trawls and trammels to be used as an exception; that is in such places as might be formally authorised.

The present Act is founded on a different principle, and makes the restriction of any mode of fishing in the sea the exception and consequently admits to these implements being used everywhere, excepting where they shall be formally prohibited by bye-law.[31]

Thenceforth any method of fishing might be allowed until such time as an objection required it to be restricted was upheld. It is a principle that applies to this day in the waters of the Republic of Ireland, to the detriment of fish stocks and marine ecology generally. Elsewhere in these islands there has been a more cautious, precautionary approach which examines the implication of a new method for target species and the environment before its operation commences; that however, might be attributed to custom and practice. Similar legislative provisions introduced in Britain some 25 years after the 1842 act in Ireland are regarded as having had catastrophic consequences.[32]

Although the 1842 Act preceded two royal and one parliamentary commission on the consequences of trawling, there was a perception twenty years before the earliest that this method of fishing was having deleterious consequences. When, in 1848, Wallop Brabazon published his prospectus for the establishment of a commercial fishing company, he advised it be located on the west coast, because of the perceived damage done to "spawn and small fish" by trawling which impinged particularly on the activities of line fishermen in the Irish Sea.[33]

The primitive beam trawls towed by sailing ships could be highly productive when first exploiting undisturbed fish stocks. In such circumstances trawling was highly profitable and the mobile gear cut through set-lines, nets and traps and carried them away. In order to fulfil one of its purposes and maintain order among fishermen, bye-laws were introduced to exclude trawling from sensitive areas. The first, introduced in 1842, removed this gear from parts of Dublin Bay, the same year the authorising act passed into law. Similar bye-laws were drafted in 1851, 1879 and 1900 for other parts of the country.

As trawling became a more established practice, tension between the operators of mobile and static gears intensified and in the 1850s there were violent protests against the rapidly expanding methods in Britain and Ireland. Fishermen also complained that trawling ruined the sea bed and destroyed the patches of mussels and whelks on which they depended to provide bait for their lines and that towed nets broke up shoals of pilchard and herring and crushed moulting crab. Eventually, in 1863, the British government set up a royal commission to inquire into these claims.[34]

The commission consisted of three persons, two of them parliamentarians, hence unpaid. The third, the only one with a professional interest in biology, received a fee. T.H. Huxley, then notable as a palaeontologist already celebrated for his energetic defence of Charles Darwin's theory of evolution, would be particularly associated with the commission and its outcome. The commissioners toured the nation, visiting 86 fishing communities in England, Wales, Scotland and the Isle of Man, including 21 locations in Ireland, and taking more than a thousand pages of evidence from hundreds of witnesses.

Two case histories from Ireland were carefully examined. One, from Galway, recorded that trawling had been practised prior to 1820 when "the violence by which the native fishermen against that mode of fishing manifested itself."

Initially the practitioners were amateurs whose success consistently excited the jealousy and opposition of the fishermen by whom the commissioners of fisheries in Ireland were at length induced to pass a byelaw prohibiting trawling within a line drawn from Barna Pier to Glenninagh Castle.

About 1852 trawling on a large scale was introduced. It was opposed, including with violence but persisted. Galway men themselves purchased (trawling) gear with funds subscribed to them but as their gear wore out it was not replaced and fishing companies (capitalists) took over this method of fishing.

At first, the trawler men were paid by wage but that changed to payment by share of the landings in order to increase output. The Claddagh (Galway) men then complained to the authorities that trawling had reduced their catches. They also cited damage to the spawn of fish by trawl netting.

The commission examined the evidence carefully and concluded that the people of the Claddagh had for thirty years

...remained substantially the same poor and improvident community, pawning their goods and household furniture for food, becoming indebted for their gear and ready at all times to have recourse to violence and intimidation against their neighbours...in defence of their real or imagined interests.

Evidence from Galway Bay was not given much credence by the commissioners.

The second instance, the case of trawling in Dingle Bay, was examined on the basis of evidence collected initially in 1835 when

The introduction of trawling excited the fears and jealousies of many of the fishermen though the fisheries appear to have been in a wretched condition before trawling began...between the trammels and seines the bay has utterly failed...Trawling was once attempted in Brandon Bay but the trawl was soon forced by the local fishermen to cut and run.

Trawling was next introduced in 1848 by a company strongly opposed by fishermen, especially those in Annascaul where representations to the board of fisheries persuaded it, four years later, to pass a bye-law which cut the trawling ground in half "and is said to have had the effect of greatly diminishing the supply of soles and turbots". The 1852 regulation that did so was subsequently rescinded and the commission approved of that development, stating:

...indeed it is difficult to imagine a stronger case against the opinion that trawling in May is injurious to the supply of fish, that that of Dingle Bay...

The commission worked conscientiously. Information was sought on the quantity of fish carried by the railway companies (one of the few reliable sources of data available to it) and on the quantity of captures by individual vessels.

In the 1860s knowledge of marine biology and the dynamics of fish populations was scant. Fishermen gave witness to their observations, as in the cases of Galway and Dingle, but there was little background information against which the information could be interpreted. It is not unknown today for one group or fishers to blame another for their situation – the history of the industry is riddled with such occurrences - and allegations of this kind might be interpreted as biased and self-serving "assertions that can be neither proved or disproved". The commissioners decided that

...fishermen, as a class, are exceedingly unobservant of anything about fish that is not absolutely forced upon them by their daily avocations...consequently, not only prone to adopt every belief, however ill founded, which seems to tell in their own favour, but they are disposed to depreciate the present in comparison with the past...

Armed with these views, the commissioners prepared their report which they completed in 1865 and published the following year. The commission pronounced on the three propositions posed for it; first, whether the supply of fish was increasing, stationary or diminishing. In the absence of statistics it was simply impossible to reach any decision on this question. There were no clear trends in abundance and, whereas one species might decline in a sea area, others increased. Instead, the commission was impressed by the quantity of fish trawlers were delivering to the Billingsgate Market in London where 90% came from this method of fishing. Line caught fish being larger and probably of better quality, fetched higher prices. The commission was concerned lest restricting trawling might inflate prices and thus reduce the supply of fish to the poor.

Next, the commission had been asked to investigate whether any of the methods of catching fish involved a wasteful destruction of fish or spawn and whether legislative restriction would increase the supply of fish. The commission decided these allegations were unproven. There was no shortage of evidence of juvenile fish in trawl nets but it was dismissed as a very minor source of mortality compared with what happened in nature. As to the question of spawn — fish eggs — being destroyed by trawling, the commissioners decided that much of what fishers believed was spawn was a misidentification of various invertebrates ("various soft organisms") which were incidentally scooped up in the net.[35]One witness had referred to trawling performing in the same way as does a plough on land. Agitating the substratum would increase rather than diminish the supply of fish.

Finally, the commissioners were asked whether any existing legislative restriction operated injuriously. Their reply is worth quoting at length:

(Were) any trawling ground (to) be over-fished, the trawlers themselves will be the first persons to feel the evil effects of their own acts. Fish will become scarcer, and the produce of a day's work will diminish until it is no longer remunerative. When this takes place (and it will take place long before the extinction of the fish) trawling in this locality will cease, and the fish will be left undisturbed, until their great powers of multiplication have made good their losses, and the ground again becomes profitable to the trawler. In such circumstances as these, any act

of legislative interference is simply a superfluous intervention between man and nature.

.....We advise that all Acts of Parliament which profess to regulate, or restrict, the modes of fishing in the open sea be repealed; and that unrestricted freedom of fishing be permitted hereafter.

The supposition that a reduction in fish populations would activate corrective behaviour was misplaced. Fishing, as future years demonstrated, depletes fish biomass but the reaction of the industry has always been to intensify fishing effort and to enhance its methods of accessing the surviving fauna.

Concerns about trawling were not quelled by the 1866 commission and in 1878, a parliamentary commission examined the use of the "trawl or beam trawl in the inland seas and territorial waters of England and Wales".[36] The use of seine nets on the coasts of Cornwall and elsewhere was also examined. Its terms of reference were not very different from those of the 1866 enquiry: whether the use of trawls resulted in the destruction of fry and spawn of sea fish in the estuaries of rivers and whether, in consequence, the supply of fish on the coasts of England and Wales was decreasing. The inquiry also looked into whether fishing methods conflicted with one another and which legislative improvements would promote the welfare of fishermen engaged in other fisheries and increase the supply of fish to the benefit of the public. The outcome of the 1878 enquiry failed to implicate trawling as a source of destruction.

Ten years later and almost twenty years after the first royal commission, another was established to examine similar issues. There was little change in its procedure although there had been small but significant additions to knowledge in the meantime. The Scandinavian, Professor G.O. Sars, had established that the eggs of haddock and cod were pelagic and so could not be damaged by demersal trawl. Professor W.C. McIntosh was asked to undertaken observations on board trawlers but they proved inconclusive.

The 1883 royal commission was

...to inquire into the complaints of line and drift-net fishermen owing to the use of trawl-net and beam trawl in the territorial waters of the United Kingdom, whether such complaints were well founded...

Its terms of reference indicated how it should report:

...and what legislative remedy could be adopted without interfering with the cheap and plentiful supply of fish...

There were four members of this commission, chaired by the Earl of Dalhousie.[37] Huxley, in poor health for much of it, did not sign its report.

The 1883-84 commission met at various locations in England and Scotland. Trawler-men defended their methods; users of passive gears complained about them. Two trawling companies, Granton Steam and the Smack owners Association of Hull, acknowledged their vessels did deplete areas of fish after which their boats moved elsewhere.

The most plaintive evidence reported that in some places old men and small boys could no longer obtain a living from small boats. Trawler men disagreed and stated there had been no decline in catches, which was very probably true. A rapidly evolving technology captures increasing quantities of fish for a period even if their populations are declining.

Opinion canvassed in Norway, Sweden, Denmark, Germany, France, Belgium and the Netherlands yielded some noteworthy facts. Certain species were reported to be disappearing from sea areas; small meshed nets used for shrimp were said to be doing damage to fish species while Germany reported a decline in the average size of plaice landings which was attributed to trawling.

We now know from a wealth of examples that declining fish size is a by-product of heavy exploitation.

There were also some contributions from Ireland. An experiment in Galway Bay was reported to have yielded a "considerable number of immature fish in the trawl" but they were not the most valuable commercial species. Therefore, it was concluded, there was no evidence of wasteful destruction of immature commercial fishes. There was also a reference to the prohibition of trawling in Dublin Bay in 1842. Haddock stocks had not recovered in the mean time. Haddock is a species which periodically has large recruitments or "year classes". Its landings, indicative of abundance, plotted against time, show up as "spikes" on a graph. The observation on haddock may not have had any significance.

The commission sought opinions from fisheries districts within the British Isles: had there been a falling off in fish landings attributable to trawling and were its effects different in inshore and offshore waters? The most consistent account of decline in Irish inshore waters came from the east coast. In the majority of fishery districts there was uncertainty, and change was attributed to weather. In Dingle and Ballinskelligs Bays trawling was reported to have driven the fish into deeper water from which drift nets prevented their return. In a south west fishery district the decline was explained by French vessels eviscerating their landings and discarding the offal. Dogfish moved inshore to eat it and their presence scared other species into the deep.

The commission distinguished between evidence collected for inshore and offshore fisheries and concluded that coastal waters had been more heavily impacted. Steam trawling was more powerful than trawling by sailing vessel and, because steam propulsion was replacing sail power, was likely to have been influential.

Ireland had a mechanism for excluding trawl fisheries through the legislation of 1842, which had not been introduced to England where the commission now recommended a similar system be set up. It also agreed that mobile gears inflicted considerable damage on static ones, specifically on drift nets, haddock (hook) lines, "great-lines" set for cod, halibut and skate and crab creels or nets.

The commission also decided that there had been a decline in flat fish and haddock resulting from trawling at locations between the Moray Firth and Grimsby. There had not been a decline, other than of sole, in the North Sea. Beam trawling might diminish fish in "narrow waters" but trawling inflicted insignificant damage on the food of fish while there was no evidence it caused injury to the spawn of herring.

Roberts summarised its findings –"something of a whitewash" – as follows:

The spread of trawling caused the greatest human transformation of marine habitats ever seen, before or since. The descriptions of witnesses....chart the shift from biologically rich, complex and productive habitats to the immense expanses of gravel, sand and mud that predominate today. The change came first to Britain and parts of Europe but by the 1920s would spread to reach the Americas, Africa, Australia, New Zealand and seas beyond. Today, restless shifting sands drift where once an oyster empire spread across the southern North Sea.....

The power of the trawl to dig up the seabed was much increased by steam. With chains on the ground rope, steel cables and engine power, trawls could drag and roll rocks across the bottom, crushing, pulverising and stripping the living matrix and liberating the mud and sediment beneath...[38]

The conclusions of the 1884-1885 commission were at variance with the volume of criticism of trawling. Conclusions included one that there was evidence for a decline in only a few species of fish close to the coast, that the eggs of cod and haddock (which are pelagic) were not affected, that there was no proof of damage to herring spawn, that the injury to the food of fish was insignificant and that there was "no wasteful or unnecessary destruction of immature food fishes". There was wide resentment at these conclusions. Roberts quotes a pamphlet published in Aberdeen in 1899:

Dozens of witnesses swore before the commission that the trawl destroyed the food of fishes, the spawn of fishes, and the immature fishes; many declared from their personal experience that it was enough to make an angel weep to see the awful signs of destruction brought up on deck by the trawl.[39]

Reconciling the despair among those fishermen who did not use mobile gears to capture fish with the equanimity of the commissions appointed to examine the consequences of trawling is not straightforward but the phenomenon may be more common where the exploitation of natural resources is concerned than an outsider would initially suppose. On the one hand, there was evidence against the practice which was largely anecdotal and certainly unquantifiable, and on the other, the obvious benefits of cheap and abundant food harvested by this fishing method. Scientists were unwilling to stand in the way of progress for a case that was less than watertight while the ever increasing landings of an expanding technology, whose destructive consequences would become undeniable only some time later, overcame any scruples they may have had.

Some of the observations of the 1884-1885 commission can be described as timeless and of the kind which foment scepticism about observations by those involved in fishing:

Fishermen are apt to rely on tradition rather than observation, and are naturally opposed to any method of fishing that may interfere with their own livelihood...

Once again, as in 1866, there were no statistics to support or dispute opinion. The 1866 royal and the 1878 parliamentary commissions had recommended their collection. Some insightful comments which acknowledged this deficiency and recognized the obstacles to interpreting a changing technology were proffered:

The increase in number and size of fishing boats and in the efficiency of their apparatus make it difficult to say how far the supply of fish brought to market is an accurate test of the increased or diminished productiveness of the fishing grounds. Nor do the grounds themselves always remain the same...

The same bland findings had been reached as in 1866 but the commission concluded that trawlers inflicted considerable injury on drift net and hook and line fishermen and that the introduction of steam trawling had made a bad situation worse.

Recommendations were concerned with assuaging the impact of mobile gears on the activities of those who did not use them. A "competent person" might examine a specific instance of competing gears and issue a provisional order prohibiting the use of any trawl net or beam trawl in any of the territorial seas and this order would come into effect after its approval by parliament. Competent persons would be appointed by extending the powers of inspectors under the salmon fisheries acts to all fisheries.

Although he did not sign the 1884-1885 report of the royal commission, due to poor health, and was acknowledged not to have participated in it as intended, Huxley remained an important influence throughout. He had become inspector of fisheries for the English government and he held steadfastly to the views which he reiterated when he delivered the inaugural address at the opening of the fisheries exhibition in London in 1883.[40] Once more he referred to the sea as supplying "food of almost unlimited extent", in support of which he provided some astonishing estimates of its productivity. His speech anticipated the development of aquaculture and a globalised fish market facilitated by steam and refrigerating apparatus. But for fishermen he retained his prejudices saying

...the practical fishermen, as a rule, knew nothing whatever about fish, except the way to catch them...

He continued, revisiting the conclusions of the 1866 Commission and then he posed the question

Are fisheries exhaustible? That is to say, can all the fish which naturally inhabit a given area be extirpated by the agency of man?

The answer could not be delivered categorically. There were fisheries and fisheries. A salmon fishery, for example, would be managed on much the same principles as a sheep farm. The same reasoning applied to all river fisheries. But what of the sea? An oyster bed could be dredged clean and the way of managing it would be to estimate the numbers of shellfish on it and to remove only those justified by the estimate. This approach is common to fish harvest strategies today but Huxley could not make the leap of imagination from the limited size of a shellfish bed to the vast, then unfathomable, shoals of finfish frequenting the seas surrounding the British Isles, exploited, in his time by a still small, although growing, trawling effort.

I believe...that the cod fishery, the herring fishery, the pilchard fishery, the mackerel fishery, and probably all the great sea fisheries, are inexhaustible; that is to say that nothing we do seriously affects the number of the fish. And any attempt to regulate these fisheries seems consequently, from the nature of the case, to be useless.

Huxley is greatly criticised because, as a man of science, he appeared to ignore much of the evidence. In his defence, it has been observed that Huxley's remarks applied only to pelagic fisheries which co-exist with trawling more successfully than demersal species and that the comments were made in the context of the technology of his day.[41] But Huxley did include cod in his brief list of the great fisheries and that is a species which has been devastated by demersal trawling.

At variance with the conclusions of the various commissions on trawling was the co-existing concern for the capture of small fish – what would subsequently be described as *discards* – expressed within the industry. As early as 1819 an act for the Encouragement and Improvement of the Irish Fisheries had introduced a number of administrative regulations for the management of fisheries for "herrings and fish…more especially the white herring fishery…including whales" which stipulated, among many other details, that a mesh size of less than 89 mm (3½ inches) measured from one knot to the next should not be used in "any drag or other sea net" other than for the capture of "herrings, pilchards, sprats and shrimps". Another example of the recognition of the damage that could be done by mobile gears was the enactment in England in 1881 of the Sea Fisheries (clam and bait beds) Act which empowered the Board of Trade to prohibit the use of beam trawls within any area of territorial waters if, after enquiry by an inspector, it was satisfied that an injury was being inflicted on those fishermen who depended on the bed to bait their hooks.

One outcome of the deepening concern and agitation arising from trawling and the obvious dearth of data with which to quantitatively assess it, was the establishment of marine laboratories in various parts of Europe. One of the first investigators to become involved was W.C. McIntosh, zoology professor at St Andrew's in Scotland who conducted studies on the effects of trawling on spawn and immature fish and had contributed to the second royal commission.[42]

In the end, it is difficult to escape the conclusion that lucrative and highly productive – albeit only in the short term – methods of fishing have a persuasiveness all their own. The phenomenon is apparent at various junctures in the unfolding story, including its most recent episodes.

At the end of the nineteenth century, the Irish fishing fleet consisted of a large number of wooden and canvas vessels. The 1842 Act[43] had made registration of fishing vessels compulsory and there were two repositories of detail: the register of general shipping included fishing craft and that list was supplemented by the coastguard's annual detailed census of the fleet's disposition: location at anchorage, number of crew per boat and use to which the vessels were put.[44] The fishing fleet was made up of a variety of sailing craft, the legacy of invasion (Scandinavian designs of *yawls* and *cots*), fishing expeditions from Spain or the Netherlands (*hookers*), importations from England (*nickys* and *nobbys*) and vessels displaced by superior technology from Scotland (*fifies* and *zulus*).[45] Their design features suited them for various purposes but the annual census of the Department of Agriculture and Technical Instruction gathered them into size categories irrespective. In 1900, the entire fleet of Ireland, north and south consisted of 6,341 boats crewed by 26,013 men and boys.

Two issues vied for precedence in policy formulation. Officially, productivity was respected, even favoured. The fishing method which yielded most, trawling, was reported in greater detail than the use of passive methods, indicating it was held in higher esteem. Fishing was not merely a method of providing subsistence and employment, it might also generate a surplus to feed industry. The second issue, which proved less persuasive although was by many more people supported it, enunciated the democratic will of the industry. Fishing was a traditional occupation, hence a conservative way of life. Innovation conflicted with established practice – the obvious example being the damage inflicted on static nets, pots and lines by trawling. Despite objections, mobile gears gradually gained ground everywhere, as had been anticipated by the 1842 legislation.

In 1900, fewer than 500 Irish vessels employing 1,700 personnel fished by trawling: small proportions of total vessel numbers and personnel. The majority of the fleet would have been sail-propelled, a handful of steamers making their debut.

The impetus towards greater use of trawls was acknowledged in the early years of the twentieth century. The report of the Department of Agriculture and Technical Instruction for 1908 observed:

Notwithstanding the efforts which have been made through the agency of bye-law to restrict the operations of large steam fishing vessels to the outer areas, which they are especially well qualified to work, there can be little doubt that the fishermen who are compelled to trust to sails or oars work more and more at a disadvantage...change in their methods must be brought about gradually if success in fishing is to be secured. The cost and upkeep of a steamer are, however, so great that unless many favourable circumstances are present the maintenance of such craft is an impossibility...

Initially, attempts were made to segregate trawls and static gears in Irish waters. Between 23 July 1896 and 25 October 1910 24 bye-laws were enacted forbidding access by steam trawlers to certain coastal waters in all maritime counties of what is now the Republic. The regulations languished on the books for many years after the last of these vessels ceased operations and, in 1976, their revocation was recommended. At the same time, advice was proffered that small craft which were unable to venture far from shore should be protected against the activities of larger vessels. The straightforward way of doing so would have been to transfer the bye-laws from steam to motor-propelled vessels but there appears to have been a reluctance to stand in the way of progress.[46]

There was only one bye-law to curtail use of beam trawling and that, dated 27 November 1917, referred to the coasts of Cos Wicklow and Wexford. During the 1970s fishermen made representations to the Department[47] to have beam trawling prohibited altogether. They were unsuccessful but scientific opinion at the time favoured inshore waters being closed to this fishing method when operated by larger vessels.[48]

Between 1857 and 1944 a series of bye-laws was introduced with the intended effect of prohibiting trawling of any kind in certain bays and along the coasts of southern Irish counties. The purpose for most of these was to protect herring drift nets from the destructive movements of trawlers. One might view this is as a struggle between small inshore operators and large vessels. After the Second World War, drift netting for herrings diminished in importance as trawling became the principal method of fishing them. On that basis and because of difficulties enforcing the regulations, the revocation of these bye-laws was recommended.[49]

There was however, one exception: Bye-law no 12 of 23 August 1860, prohibited trawling in Brandon Bay, Co Kerry. It was one of the oldest regulations on the statute book and exemplified its1842 parentage. A review of its intention in 1976 described it in these terms:

...(It) was made at a time when trawling was thought to be harmful to fish stocks, particularly in small bays, which were then regarded as recruitment or nursery areas...

The reviewer had doubts about the wisdom of retaining the bye-law and recommended it be revoked.[50] Twenty years after that advice, depletion of flat fish species in the region prompted a rehabilitation programme to restore the number of flatfish juveniles through the use of hatcheries. Inshore conditions had to be made more hospitable for the fish. Reintroducing a prohibition on trawling would have been unthinkable; instead, less enforceable and totally unreliable voluntary agreements were obtained from fishermen that they would not trawl Brandon Bay.[51]

At this point we should have sufficient information to understand how sea fisheries create value and how their harvesting technology evolved. One further matter which merits examination is the way fishermen organise themselves. It is crucial to understanding how the industry in Ireland developed and we will review it in the next chapter.

Chapter 3

A MATTER OF SIZE

Two models of fishermen's organisation recur throughout this story. To illustrate their defining characteristics I refer to prototypes in other places and other times because they more formally illustrate the principal distinctions in the way sea fisheries are managed.

The key defining trait dividing them is the size of fishing vessel. Small boats must work close to land. They are too frail to venture far from port and must remain within rapid recall distance should weather conditions deteriorate. Little boats do not have the capacity to accommodate large quantities of gear or alternative fishing methods or substantial numbers of crew. Small crew size imposes the need for everyone on board (numbers may be as low as one to three, often members of the same family) to share duties, hence there is little specialisation of labour nor, in a small boat is there need for it: the crews are unspecialised, artisanal. Nowadays these operations are known as "small scale coastal" or "inshore" fisheries.

The principles by which larger boats organise are the antithesis of small craft. Being more able to withstand heavy weather, they are less dependent on home waters outside which they can operate with confidence. As a result, they are less tied and loyal to a particular area; they move to where the fish are. Larger boats have capacity for more and alternative gears and their crews are bigger. Within these crews there is division of labour, an emblem of their industrial nature. In the case of deep sea Spanish vessels for example, there are two skippers: one responsible for navigation to the fishing area (*el patron de la costa*), and one who takes over when fishing commences (*el patron de la pesca*).[1] Some of the key characteristics of earlier organisations live on in more recently created institutions.

In France and Spain inshore systems of fisheries management have venerable origins. At the time the system in France was reviewed, there were 34 *prud'homies* in existence on its Mediterranean coast. These had functioned for more than a thousand years, officially recognised in the fourteenth century, legalised by royal ordinance in 1681 and they survived the dissolution of the privileged guild system during the French revolution.[2] Prud'homies have four key attributes. They operate in a specific area which they administer as a communal marine tenure system, a form of ownership. To operate in such an area a boat owner must be a member of the local prud'homie and abide by its rules. Their second characteristic is the nature of their authority which is based on an elective system whereby men of experience and integrity (prud'hommes) are chosen to represent the users of gears which are associated with inshore activities. Prud'homies have the right to exclude new members and they apply principles of equity when assigning access to fisheries. They have tended to operate conservatively and oppose the introduction of new gears which might jeopardise the continuance of existing fishing methods. Finally, they have extended powers of conflict resolution which permit them to make rules which have the force of community law without being sanctioned by other forms of jurisdiction. These regulations are not subject to appeal.

In effect, the regulatory systems developed by the prud'homies are a codification of traditional customs and to some extent such practices are probably less formally in existence in inshore fisheries everywhere. Their operation in the French Mediterranean ensures that benefit from available fish stocks is shared by those who participate in them, much as on the lines of a shellfish co-operative in Ireland at the present time. By their application of rules for the conduct of fisheries – decisions on when to fish and what gears might be used, for example – they have the incidental benefit of conserving and managing the resource. Which is not necessarily to regard them as paragons of conservation; but prud'homies tend to be among the better management systems in existence, at least partly due to the fact that small inshore craft use less and physically lighter gears than larger vessels and so are capable of operating without inflicting so much damage on the environment.

In Spain the *confradías de pescadores* perform a similar function to the prud'homies. Like their French equivalents they are rooted in the medieval guilds; specifically they are successors of the medieval Brotherhood of Mutual Assistance. Although they are seen to be primarily concerned with inshore fisheries management, confradías appear to represent a broader spectrum of industry workers; 70% of those in the small scale coastal fisheries and, indeed, those who operate without vessels, gathering cockles at low tide for example, are said to be members. They also appear to have a wider brief which ranges between making representations for improvements in social welfare and regulating the fishing activities of their members. A review of their activities in 1997[3] numbered confradías at 225 (each having between 600 and 1,500 members), grouped in 21 provincial federations. Everyone from boat owner to deck hand along with workers in associated industries and services was entitled to join. Unlike the prud'homies, the confradías also operate on Spain's Atlantic coast but they are regarded as most influential on the Levante coast of the Mediterranean on which 88 were then located. In theory at least – the article describing their activities was based on an appraisal by the confradías themselves – this inshore sector had adopted a number of conservation measures prior to legislation being enacted. The fleet they administered was of smaller vessels which, nonetheless, included trawlers, purse seiners and long-liners, in addition to boats fishing fixed nets and traps, grouped under the term *artes menores* (literally "minor" fishing methods). They also fixed minimum prices for their products.

Local representative artisanal groups have had a declining role in fisheries administration within the European Union. The fact that they are inseparable from the locations in which they operate militates against them. The Union prefers to manage products rather than locations. In theory at least, nationality is attributed lesser importance in the founding treaties of the EEC and subsequent amendments. Local agencies which regulate fisheries on behalf of the Commission, *Producers' Organisations* (POs), were designed to purchase and distribute fish products in bulk, in the cheapest and most efficient way, to obtain best value for the consumer.

POs could not operate as cost efficiently in circumstances where they were required to buy up small quantities of landings from widely dispersed artisanal fleets. Instead, they more naturally lean towards purchasing from big vessels which harvest large quantities. POs in Ireland display strong similarities with the fishing companies of the nineteenth century and we will look at one of them next.

Wallop Brabazon's murder machine

Referring to a text of the mid-nineteenth century as an exemplar of high seas fisheries management is relevant because it describes the kind of organisation which has since brought success to many an enterprise of its kind. This is the model which has gained dominance in fisheries management in Ireland over the past half century, although it might be regarded as an extreme example where delicate sensibilities are concerned.

When Mr and Mrs Hall analyzed the scenery, history and economy of Co Mayo in the early 1840s they described it as having 65% of its surface cultivated, the remainder either unprofitable mountain and bog or under water. Much of the land holdings consisted of "huge estates, valuable only as preserves of game". This was hardly the concentration of population that would provide a market for a marine fishery, consisting as it did of striking but barren scenery. Yet the little town (it was hardly more than a village) of Newport, at the north east corner of Clew Bay, was, in the words of the Halls,

...intended to be a better town than it now, unquestionably is...Few towns on the coast are more fortunately situated...At the quay a vessel of up to five hundred tons may unload...The town is rapidly rising into an importance that will, in a short time, render it second to none on the western coast...

And it was this little town that Wallop Brabazon, less than a decade later, nominated as the theoretical hub of a commercial fishing company. Newport was favoured for its sheltered location, its quay facilities and because there was a flax mill in the vicinity that could be used for the manufacture of canvas. The vessels used would be sufficiently robust to be independent of weather and so that they were not reliant on, for instance, spawning herring shoals coming close inshore. As Brabazon envisaged it, commercial fishing would make the most of whatever was already available on the coast, rather than creating new opportunities.

In his prospectus, Brabazon warned against trying to better the fishermen of the west coast in hook and line fishing; they were acknowledged masters of their craft and the fishing company would purchase what they caught for curing and thus provide them with employment; the company would also sell gear to them. Brabazon's objective was to make money, buying from and selling to existing operators. In this, his conception of fisheries was at variance with that prevailing up to the nineteenth century, which was to provide employment and sustenance only, a form of social service rather than an industry.

Brabazon did not claim originality for his ideas

...I beg to impress upon the minds of speculators that they are not entering upon a field lately discovered...this is an old

speculation that has often failed heretofore under good managers. I attribute this want of success to the following simple reasons, first, the inadequate vessels with which the fisheries were attempted during the winter season, at which time the fish are in the greatest shoals along the west coast of Ireland…

Rather, the value of his publication resided in its evaluation, albeit optimistic, of the possibilities of commercial success, taking various aspects of the industry into account: the species available and the uses to which they might be put, the methods of catching them and the legal restraints on the use of those methods, manpower and the way it might be deployed to yield best commercial results.

A market was known to exist because in 1844 almost 128,000 barrels of herring and 880 t of cod, ling and hake had been imported to Ireland, amounting to just short of stg£122,000 in value. And there had been subsidies, in the form of bounties, available for the capture of certain species at the time.

The year 1848, at the conclusion of the great Famine, was also an opportune time. Brabazon hailed it thus:

…The fisheries are the greatest and at present almost the only resource in the west of Ireland. From the loss of the potatoes as a certain means of subsistence, the fishermen would give their undivided time to the fisheries, instead of losing the best seasons by wasting their labour on a high rented potato garden. They would thus throw an immense quantity of cured food such as hake, cod, ling, herrings, haddock and coal fish (saithe) into the interior of the country, while they could dispose of such fish as could not be cured such as turbot, plaice, mackerel, gurnard, john dory, ray and soles, at the market along the coast.[4]

Newport would serve as the main hub for the venture but two other locations were also identified, Killybegs in Co Donegal (prophetically, it subsequently became Ireland's most important fishing harbour) and Shannon Mouth (the Shannon estuary). Each would be provided with a store, curing facilities and an ice house. They were selected as fishing centres for good reason too. At a time when transport of perishable commodities to market was problematic – fish caught off the Inishkea Islands in Co Mayo, less than 70 km away from Westport as the crow flew, would be rotten when it reached its place of sale. If fish were rushed to Ballyshannon it might be transported through the Erne lakes and by canal to the industrial centre of Belfast and distributed to towns and villages along the way. Similar thinking underlay the choice of the Shannon estuary from which landings would be carried on by water to Dublin.

The main financial outlay would be on a fleet of three or more *wherries* or schooners of approximately 150 GT each and a tender of 50 GT to transport landings and gear from one to another and to shore. They would participate in all fisheries throughout the year and use all available methods of capture, including trawling; despite Brabazon's harsh critique of the method in the Irish Sea he recognised its potential as a generator of wealth, although he was also apprehensive that the west coast fishing grounds were too foul (rocky) to have nets drawn over them. The large vessels would stay at sea for prolonged periods and Brabazon advised on the way crews should be managed and periodically relieved.

The most effective way of fishing off the west coast, he advised, was to position large vessels provided with hammocks and provisions for 40 to 50 souls on the fishing grounds where they would remain for long periods. From the perspective of today, the prospects for 40 – 50 men effectively imprisoned on a vessel of 150 GT for any length of time are grim.

Brabazon had some cautionary words on personnel management:

The fishermen generally are a superstitious and self-willed class of men, they know it is by their knowledge of their business that it is either a gaining or a losing concern, and that this knowledge is solely gained by long experience. Their goodwill must be enlisted…You may lead fishermen to do anything, but if it goes to driving them, they will go there is no doubt, but they will take right good care that it is at your expense.

On fishing gears he has numerous recommendations. Herring drift nets in use on the west coast were not deep enough and they were hung from a buoy line at a fixed depth; it would be better if they were adjustable so that they might be raised and lowered to coincide with the depth at which the fish were moving. Being selective, gill nets of various mesh sizes should be stored on board to optimise the capture of fish of different lengths. And there were observations on when gear should be fished. If gill nets were set during the day they would scare herring away from the vicinity; they should only be fished during the hours of darkness.

Processing herring required three types of employees: gutters, to clean them, curers to process the fish and coopers to manufacture the barrels which would contain them. If oil were extracted during the curing process – it served to fuel lamps – it would provide sufficient money to pay the wages of curers.

Long-line or *spilliard* fishing was to be conducted over ground that was too deep for trawling. Cod, haddock, ling, conger and dogfish – Brabazon claimed all were suitable for curing and he provided examples of what he called the *baccalow* or salt cod of Spain and Portugal, *Shetland ling* and *Findon haddock*. The "finest haddock ever seen" he claimed, came from the vicinity of the Inishkea Islands and they reached 15 or 20 lbs (6.8 – 9.1 kg).

A spilliard was up to 450 m long with 200 hooks attached. A row boat crewed by five would fish about 1,000 hooks with a soaking time (the period during which the gear was left to fish) of twelve hours. Each spilliard represented one man's share of the endeavour. The order in which the lines were set, across the tide, was decided by drawing lots. If set midwater, spilliards were supported with inflated dog skins. Bait consisted of small fishes purchased from trawlers, or of lugworm, mussels or scallops collected by families who would arm (bait) the hooks. Brabazon regarded this method of fishing to be satisfactory and said he could not improve on it.

Trawling was a suitable method for flat fishes, turbot, sole, plaice, flounder, john dory, skate or ray and some round fish were also taken by this method; his list included haddock, red gurnard, whiting and conger eels, along with lobsters, crawfish, oysters and crabs. The size of the trawl depended on the dimensions of the towing vessel. The method used was beam trawling and it was imperative to keep the net on the sea bed and prevent it from capsizing. Trawling boats required a large draught to keep them to windward while towing. Only a boat of 70 GT or more could fish by this method at depths of 80 - 140 m. Irish vessels, unlike those operating on the other side of the Irish Sea, used a trawl with an arched beam which facilitated greater capture of haddock. A better trawl design was fitted with reversed, inward-opening pockets to prevent fish swimming out of it.

The large fishing boats envisaged by Brabazon were equipped for all eventualities, including the capture of basking shark, also known as sun fish, for which they carried harpoons. Salmon had traditionally been intercepted in the course of moving upstream in fresh water when they returned to spawn. More recently salmon had been harvested at sea where they were captured in greater numbers and in better condition. A variety of gears was described for the purpose. Hand lines were recommended for the capture of mackerel. Pots, creels or gaffs could be used to trap large crustaceans or drag them out of rock crevices. Crawfish might also be captured in trawls, and dredging was the method for scallops.

"Trammels" referred to nets of various kinds (not as defined in the previous chapter): drift (gill) nets within the meaning of the term were similar to those in use for herring, but had meshes of 1.5 - 2.0 inches (3.8-5.1 cm) measured from knot to knot and designed to curtain the depth of water in which they were set. The nets, corked rather than buoyed as this caused less wind drag, were 60 m long and set across the tide where they proved effective in taking salmon, whiting, pollock and mullet, sometimes in large numbers. The term trammel also embraced seine nets, 100 - 160 m long, 2 m deep at the ends and 3 at the centre. Seine net mesh was 2.5 inches (6.4 cm) knot to knot and the head rope was corked, the foot rope leaded.

The methods in which species might be utilised were detailed: dogfish (a number of species are covered by this term but the one Brabazon apparently referred to was the spurdog) required a steel trace between hook and line. Its flesh was a "rich coarse" one much liked by fishermen for their own use rather than for sale. The flesh could be cured or dried and it was possible to extract oil from it which had medicinal properties "superior to Friar's balsam" for application to wounds.

In describing fishing methods and likely quarry, Brabazon was careful to list regulations so as to provide a comprehensive account of the possibilities and constraints of managing a business. Salmon fishing by traps made out of netting (bag or chamber nets) had to be suspended over the weekend.

One of the species referred to as abundant in Clew Bay was halibut which is very rare in Irish waters today. The greater abundance of all species in his time militated for higher capture rates with lower effort. Gear was deployed in much smaller quantities than in recent times. All gears were made from biodegradable rather than synthetic materials and hence, they had to be carefully dried after soaking to prolong their useful life.

Fishing companies, organised along the lines of Brabazon's were a product of the early nineteenth century. They moved into an area and fished it irrespective of the sensibilities of inshore fishermen whose lesser technology tied them to making a living in the vicinity. The inevitable conflict of traditional fishing with innovation and superior power was captured in an article carried in the *Dublin Journal* which, in turn, reprinted extracts from the *Galway Weekly Advertiser*.[5] This, or events very similar to it, was almost certainly the incident which had been examined by the 1866 commission on trawling.[6]

The Dublin Fishing Company was conducting its business in Galway Bay when its vessels were attacked by fishermen from the Cladagh who assaulted the crews and destroyed the nets "as if they had the exclusive privilege of the ocean". The *Weekly Advertiser* report went on to say

...some notable examples were made of the ruffians and, and the consequence is that fish of all kinds are sold in the market of Dublin at half the former price...

The *Dublin Journal* inveighed against the Galway fishermen, referred to as pirates, describing them as miserable, ignorant and bigoted creatures following the business of fishermen in Ireland, in the very depths of superstition whose calendar had sixty saints' days. They got drunk on fair day and, as a result, lost a week's fishing and they were resentful of anyone else taking the fish they believed should be theirs.

It would have been more appropriate to recognise the Galway incident as a clash between artisanal and industrial fishing cultures. Resentments of this kind survive to the present day and we will see more modern examples of them; the passing of time has done nothing to mollify their pain. True, frustration is no longer expressed in such a graphic way as in 1820; rather it simmers as jealousy of larger vessels which are free to fish anywhere. Their activities are not excluded from any part of territorial waters, the innermost of which should, as a matter of principle, be reserved for the operators of smaller boats. Instead the less fortunate fear openly challenging those favoured by power and authority.

Chapter 4

FOR WANT OF CAPITAL...

Legislation passed in 1720 conferred on the Westminster Parliament the power to make laws for Ireland. At the time, Irish agriculture was inefficient and legislation enacted there was used to maintain Protestant economic advantage. In the wake of rebellion in her American colonies and the intervention of France and Spain, England was obliged to reinforce her troops in the New World and withdraw military forces from Ireland. Fearing trouble closer to home, Westminster adopted a more conciliatory line, removing most of the trade restrictions in 1779; three years later the 1720 Act was repealed. "Grattan's Parliament" eased the most oppressive of the penal laws and, in 1793, Catholics won the right to vote which, however, was limited by a property qualification. Several financial supports for herring fisheries were introduced in 1783: the bounties (rewards, the term covers grants or subsidies) consisted of a 20 shillings per t payment to a wherry-type fishing vessel known as a "buss" and two shillings per barrel on herrings exported. In 1785 the bounties were increased.

The value of these subsidies was lower than in England or Scotland but in Ireland they were available over a longer proportion of the year and they could be expended to purchase fish from small, sub-contracting vessels (not permitted there) so their benefits were more widely spread. The principal advantage of the bounty system was the associated quality control exercised through thorough inspection of cured product before export which established a strong reputation for Irish goods on foreign markets such as the Madeiras and the West Indies, to which herrings were sold. For Ireland, the success was short lived.[1]

New found independence was brief and, in 1800, the Irish Parliament voted itself out of existence, adopting the second Bill of Union with England and Scotland that was put to it.

Subsidy and loan

Financial assistance to sea fisheries was made available in England in 1809 and ten years later the system was extended to Ireland. Its entry into force coincided with the appointment of a board of commissioners in 1819. The system of administration, similar to Scotland's, consisted of twenty unpaid commissioners, a paid secretary, three clerks, four inspectors and twenty local inspectors. Public monies were authorised to assist fisheries development between 1819 and 1829 in a number of Acts of Parliament entitled An Act for the (further) Encouragement and Improvement of the Irish Fisheries.[2] A sum of stg£5,000 was made available but it could be used only for the repair of fishing vessels and the construction of piers. A fishery loan fund worth up to stg£10,000 with which the commissioners would assist poor and industrious fishermen to procure fishing tackle and articles for the repair and outfitting of their boats was also made available; the capital might also be utilised to construct a limited number of "model" boats, specifically for use along parts of the coast which were capable of sheltering vessels.

Under the nineteenth century legislation, bounties were provided for herrings landed, cured and packed according to regulations, and for codling, hake, "glasson"[3] and conger which had been dried or pickled. There were also bounties for vessels and for exports of fish from Ireland to foreign ports.

The success of the administrative system in Ireland could, it was claimed in a House of Commons debate in 1880, be measured by comparing the preceding and succeeding states of the industry. In 1819 there were only 27 vessels and 188 men involved in the industry in Ireland whereas four years later those numbers had swelled to the incredible heights of 27,143 boats and 44,448 men. The eleventh report of the commissioners in 1830, after which they were abolished, provided somewhat altered proportions of boats and employment: 12,611 boats supporting, however, 65,571 men and boys. The phenomenal growth in employment and fleet was due, the House of Commons was told in a debate in 1880, to the energy of the commissioners administering the industry and, possibly more significantly, the contribution of the bounty (subsidy) system which channelled capital to them.[4]

The provision of monies by the exchequer ceased in Ireland on 30 April 1830. The commissioners were abolished and responsibilities for fisheries passed to the Board of Works which employed the fisheries inspectors. The legislation which terminated the Irish Board of Commissioners retained its equivalent in Scotland and maintained the branding (guarantee of quality) system. In its twenty years in existence up to 1929, Scotland, with a population of 2.3 million souls, received stg£1.2 million whereas Ireland, in half of that period, with a population of 7.5 million people, received no more than stg£0.3 million. The withdrawal of these grants or subsidies and, more acutely, their continuance in other parts of the Union, of which Ireland was a constituent part, meant the loss of critical investment in sea fisheries.[5] In 1871 the lawyer Isaac Butt invoked it as a justification for home rule. As part of his argument he quoted a speech by William Gladstone, Liberal statesman and three times British prime minister, delivered in Aberdeen:

...our Irish deep sea fisheries have been for many years declining and are now on the verge of extinction; that this decline is to be traced not to the want of energy or industry on the part of the our fishermen, but to the effect partly of the want of encouragement such as the fisheries of other parts of the

United Kingdom are receiving, and partly to the terrible effects of the famine...

While in 1830, fisheries subsidies were withdrawn from both countries a grant of stg£3,000 annually was made available to Scotland for harbour development and stg£500 for the repair of boats belonging to poor fishermen. The amount paid annually in bounties to Scotland was stg£15,000 which was automatically "charged by law upon the consolidated fund – not subject to annual vote, discussion and cavil".

Six years after subsidies had been withdrawn, a commission formed to inquire into fisheries, recommended financial assistance for the industry in Ireland. In 1838 Lord Morpeth introduced a bill to implement its recommendations but the Duke of Sutherland, heading, appropriately, a Scottish deputation, objected and the legislation was deferred.

Despite the withdrawal of subsidies and assistance, the Irish fishing fleet (the majority of which would have been constructed of canvas or hide covered frames) numbered 19,883 vessels crewed by 113,073 men and boys in the year before the great Famine. In 1868 their numbers had reduced to 9,000 boats operated by 40,000 personnel.

The Famine was undoubtedly to blame and the devastation of its aftermath was obvious:

...Already in many places the coast may be traversed for miles even where good shelter exists and fish abound, without a boat being seen. This deplorable state of things is certain to increase if a helping hand is not extended to save this important industry from perishing. Ten or twenty thousand pounds judiciously expended now – not as a gift but as a loan – would do far more good than a million in half a dozen years hence...

Butt referred to the economist John Stuart Mill (whom he described as "the first living political economist") speaking in the house of commons; Mill attributed the lack of industrialisation in Ireland to England's policies there and argued that Ireland's fisheries were more deserving of financial assistance than those of either England or Scotland. He recommended even risking "...the loss of small sums to advance that industry which we had formerly endeavoured to retard...".

The lack of subsidy in Ireland was exacerbated, Butt maintained, by the fact that France assisted her fisheries and he suggested that Ireland would do a lot better as a province of France.

A House of Commons committee carried out an enquiry into Irish fisheries and recommended assistance by loan in 1849. Butt recalled that a select committee of the House of Commons in 1867 and the inspectors of fisheries in their report in 1869 had similarly recommended support for Irish sea fisheries but all had been refused by the Westminster government. A Mr Barry of the Office of Public Works had observed in his report of 1866 - initially suppressed but presented to the House of Commons a year later: "the markets of Ireland had been restored to Scotland". The balance of trade supported his arguments: in 1830 the export of herrings from Scotland had been 181,654 barrels; four years later, 272,093 barrels were sold abroad, 149,254 of them to Ireland.[6]

The debate was updated to the early 1880s when a Sea Fisheries (Ireland) Bill[7] was considered in the House of Commons.[8] In that year the fleet had fallen to 5,759 vessels occupying 20,726 men and boys. Once more, the purpose of the legislation was to establish a board of unpaid commissioners who would be willing to revive the fishing industry in Ireland. At the time, there were only three inspectors of fisheries in the country but their means were so limited that they were unable to do more than restrain abuses of the law. The intended legislation provided for the establishment of an Irish fishing authority equipped with the means and staff to give effect to its decisions. Again, reference was made to Scotland which, for some time, had had such an arrangement which had rendered valuable service to the industry there.

In addition to a new administration, the bill under consideration would hand over to the prospective commissioners the Reproductive Loan Fund (providing funds for the industry) and increase it by an annual injection from the treasury of stg£30,000. This should fund some 2,000 annual loans to fishermen. Piers and harbours would be transferred to the commissioners who would be granted an annual sum of stg£20,000 for construction and maintenance.

The previous commissioners had observed

...no improvement can be looked for in the sea fisheries until loans are advanced to a portion of fishermen for the repair and purchase of boats and gear...

One of the MPs who participated in the debate was J.A. Blake, representing Waterford. In 1868 he had written an account of the Irish industry whose content informed the debate. Blake spoke with authority and quantified the sums required to restore it to productivity. He pointed out that there were applications from coastal communities for 100 fishing harbours of which 50-60 had been approved. The cost of building them would be stg£150,000.

In one sense the argument for restoring the industry in Ireland to profitability was unanswerable. At the time the country exported stg£0.5 million worth of fish annually. The size of the market in London alone was stg£7 million, as much fish again being sold in Great Britain, outside the capital. The commissioners had estimated that a more competent industry could increase its landings ten fold.

In the course of the debate three justifications for grant-aiding the fishing industry were rehearsed. The first was the need for a greater supply of cheap and wholesome food for the population in general. The second, the service performed by fishing in schooling personnel for the merchant and royal navies. In the course of the debate the following contribution was made:

...(Was it) the decline of the Irish fisheries as one of the causes which increased the difficulty of finding good seamen? Twenty three years ago there were 113,073 seafaring men and boys in these fisheries. Today there was only one sixth of the number...

The British merchant fleet alone employed some 200,000 men and boys, exclusive of captains and superior officers. Apart from providing food and employment, the fishing industry had been encouraged by a number of governments because it served as a nursery for seamen.

The third justification harked back to Butt's speech. It had a menacing edge:

Ireland was not a difficult country to govern when its circumstances and wants were generously considered...

Mr Blake trusted that Irish MPs would receive assurance in the form of action from the government that they did not require home rule to convert what had been a neglected to a flourishing industry.

Once more, Scotland was referred to as having prospered from the availability of government assistance. The fishery board was equipped with a steamer which it used to preserve order among fishermen and to locate shoals of fish, and four gun boats to police its coastal waters. The house was informed that Holland, France and Norway were doing their utmost to encourage their fisheries. Ireland, more than any of them deserved help. Ireland needed larger and better boats.

The MP, T.P. O'Connor, stressed that Ireland did not require subsidies of the kind dispensed to Cornish pilchard fishermen. The pattern in Ireland, hitherto, had been the provision of loans by private individuals which "in no single instance...remained unpaid".

Loans from charitable sources

Another contributor to the 1880 debate, MP Colonel Colturst recalled judicious funding in earlier years. In 1874 the population of Cape Clear in Co Cork was 450. They had only seven small boats (hookers) among them yet much of their income, amounting to stg£400 annually, came from fishing. At night the lights of 500 vessels from elsewhere were visible seaward of them. A local clergyman persuaded a man to borrow money to purchase a boat and in three months he made stg£500, more than the entire population had earned in the previous year. Other loans were obtained and a fleet of new vessels grew to four. Then, in 1878, a storm had wreaked havoc in the area. Some 300 boats were more or less damaged, the Cape Clear fleet among them, because the coastline was so exposed (from Crookhaven to Bantry in Co Cork, a distance of some 90 km, there was no harbour facility). The local economy would have been reduced to penury "but providence came to their aid in the shape of a lady whose name was known over the world for her benevolence".

Angela Burdett was the most notable benefactor in Ireland in Victorian times. Her father had been an inspirational and charismatic figure. In 1802 he became MP for Middlesex but three years later his election was declared void and he was excluded from the House of Commons. In 1807 he was elected radical MP for Westminster and held the seat for thirty years. He is said to have been an eloquent and mesmerising speaker admired by both Byron and Disraeli. He nurtured a romantic passion for Ireland on which he made his maiden speech which consisted of an attack "on the oppression of an enslaved and impoverished people by a profligate government". Of him it was said: "…all the sights and smells of poverty – these were the mainsprings of his politics…"

In 1837 Angela Burdett inherited the Coutts banking fortune along with the family name, which she adopted as part of the package. The Coutts family was at the centre of Victorian society, counting royalty, the aristocracy and practitioners of the arts among its friends and, as were many Victorians, Angela Burdett-Coutts, who later became Baroness, was disposed to establishing and participating in an enormous list of charities. The novelist Charles Dickens was a close friend who inspired her to channel her energies into building model houses, cleaning up slums and promoting sanitary reform. In 1843 she worked with him on the ragged schools scheme to provide education for the poor. Her interests were not confined to fashionable causes; she was solicitous for the employees of Coutts bank and constantly attempted to improve their working conditions.

Burdett-Coutts had absorbed her father's affection for Ireland although the Duke of Wellington, another close ally, discouraged it: on one occasion, in 1846, he told her that a famine relief fund would be injurious - "not an Irishman would work anywhere" - if it were introduced. Burdett-Coutts listened but took little if any notice.[9]

Burdett-Coutts's interest focused on south west Cork, the vicinity of Sherkin, Skibbereen, Baltimore and Cape Clear. At times of food shortage and famine she documented associated hardship and sought government intervention. When she failed to elicit a response, she set up relief centres where corn, flour, meal, sugar and tea could be purchased cheaply. Emigration was an escape for many but Burdett-Coutts was convinced the ultimate solution lay in developing agriculture and fisheries. Money was advanced in loans of stg£250 - 300 to individuals for the purchase of boats and nets and when, in 1887, she sailed in on the yacht *Pandora* to open the industrial fishing school for which she provided finance, she was greeted "Queen of Baltimore" by the ecstatic populace.

Construction of the school began in 1885. It was approved by the government by special Act of Parliament under the Industrial Schools Act which awarded an annual capitation grant to each boy of stg£13. Subscriptions for the school were attracted from a wide range of sponsors who included the Duke of Norfolk, who donated stg£500, and Michael Davitt who contributed stg£50. Before 1888 there were 60 boys in residence from maritime counties all around Ireland. Towards the end of the nineteenth century it became a school of correction which it remained until its closure in 1951. Local fisheries underwent considerable expansion as a result of Burdett-Coutts's interest. The First World War brought unprecedented prosperity to the port and in 1917, fish worth stg£400,000 was landed in Baltimore.[10]

In the aftermath of the famines of the 1840s some landlords attempted to assist their tenants by providing more robust employment opportunities. Brabazon described their actions in these terms:

Several benevolent proprietors of estates, on different parts of the sea shore of Ireland, have announced...their readiness to contribute...towards the expense of procuring good and sensible boats for the poor and industrious fishermen who reside on their properties.

They could, he went on, assist by interesting themselves in suitable individuals, by being vigilant on boat building contracts and by assisting the fisherman in question with repayments.

Charitable donations played a significant part and there had been others preceding and following the famines of the 1840s. During a period of deprivation in the 1820s, a subscription was raised, largely by the Society of Friends in London, and passed to the board of commissioners for disbursement as loans. The balance of monies remaining was saved in "The Irish Reproductive Loan Fund Institution" and placed in the hands of governors with an additional sum of stg£45,000. In 1848 the Fund, administered by the inspectors of fisheries, stood at stg£43,000.[11]

The procedure for obtaining money was as follows: an applicant sought a printed form from the local fisheries inspector, completed it and took it to a magistrate or gentleman or the nearest clergyman of any religious persuasion for signature. This confirmed that the industrious applicant would repay any loan he received. The fisheries inspector would then grant monies to purchase gear, although larger sums would have to be approved by the fisheries board.[12]

Loans were issued at 2.5% interest. In 1874 monies in these funds were vested in the Board of Works and, when the Congested Districts Board was established in 1891 the sum had swollen to stg£93,000. The loss on bad debts was less than 1%. The Congested Districts Board received stg£73,000 and the balance was dispensed outside those areas by the inspectors of fisheries. But the assistance did not go directly to the industry; advances were not put forward for tackle or boats but were confined to the construction and repair of piers and harbours.[13]

The initiative of Baroness Burdett-Coutts in providing stg£10,000 to fund fishermen's loans in 1865 had considerable persuasive impact. The 1921 commission regarded it as the foundation of the modern fishing industry in Ireland. It enabled a new start in the south west of the country. As a result, pickling of mackerel in Ireland for the American market proved a strong competitor with Norway which, until then, had had a monopoly on the trade.[14]

Subsidy as part of a social contract.

By the outbreak of the First World War in 1914, a well-equipped fleet of small craft was able to avail of higher fish prices and pay off its loans quickly. The trawler fleet consisted of some 300 vessels of which 100 had motors, 190 were sail-driven and five were propelled with steam. The gross tonnage of the total is estimated at approximately 4,100 t. But higher prices and readily available finance drew inexperienced fishermen into the industry when times were good; the inevitable slump in 1921 placed fishermen with debts in great difficulty.[15]

The 1921 commission of inquiry into sea fisheries started collecting evidence in 1919, travelling to ports around the country to do so.[16] Around the Irish coasts, it reported, fishing was conducted mainly by English and Scottish boats and Irish fishermen were dependent on shoals of herring and mackerel moving sufficiently close to land to be captured, which was a hit and miss business. Consequently, the fish consumed in Ireland - which was mainly captured by trawl or line - was largely provided by foreign boats "of modern equipment".

The commission suggested:

...In fact, there is no (indigenous fishing) industry at all. The word industry *connotes organised production and an organised attempt to reach markets, neither of which have we observed. It is safe to say that if any other industry were conducted, either in the production or the marketing of its commodity, as the Irish fishing industry is, it would be destined to certain failure.*

The explanation for everything was lack of organisation. Its symptoms were many. Markets had declined. Piers and harbours were (still) inadequate. Larger boats (greater than 15 m oal was the dimension given) were required to venture further from land but safe anchorage was not available for them. The fleet everywhere was changing over from sail to steam and oil was beginning to have a role as a source of propulsion. On the other hand, the commission believed that in many instances boats were too large for their intended purpose and these required a disproportionate expenditure on power. Engines were being fitted to vessels but these also needed to be standardised. Boat building should be undertaken in close consultation with an organised industry. Two of the boat yards in existence then, in the ownership of the Congested Districts Board, at Mevagh and Killybegs (Co Donegal), would provide vessels for the state agencies which were later charged with the development of sea fisheries.

The mechanism for transferring ownership of a vessel to an operator varied slightly from one region to another. In Cos Cork and Kerry, a boat was sold to one person who repaid in fixed annual instalments. In Cos Galway, Mayo and Donegal, a vessel was handed over to a picked crew of joint owners who paid a fixed proportion, generally a third, of net earnings for it. Additional costs of, for example, repairs or gear, were added onto the capital sum so that some prospective owners never had a chance to take full possession of their purchase. Eventually, the method used in Cork and Kerry was applied everywhere. On hearing evidence, the commission decided "fishermen will always punctually acquit themselves of the responsibilities they accept."

There were no concerns about over-fishing, despite earlier protective restrictions in Dublin Bay and the reservations expressed by Wallop Brabazon. "Even in countries where trawling has been continuously conducted it is extremely probable that valuable grounds are neglected" (meaning, yet to be discovered). The Royal Dublin Society had prepared a series of trawling surveys in 1890-1891 but more information was required. The commissioners expressed the view that it would be futile to discuss the future of the industry without accurate information.

A "fresh" market for fish required adequate supplies of ice, observed the Commission. Its provision would become a major issue and something of an industry in future years. Denmark already provided refrigerating plants and cold stores for its industry, a practice which would be adopted in Ireland.

The commission sought information from other nations and reported approvingly of the Canadian experience which encouraged sea fisheries by payment of bounties and seeded the fishing grounds with fish reared in hatcheries. The state also assisted with the cost of transporting fish in order to bring it to market in good condition.

The largest volumes of marine species harvested by Irish fishermen at the time were mackerel and herring. Ireland supplied only three markets with these whereas the Netherlands dispersed its landings among 24. In addition to the home market, an export industry would have to be organised. But the home market would have to be better served with local, motorised transport.

There was a need for processing of the landings to be improved and factories for the manufacture of fishmeal, oil and fish fertilisers to be provided.

A modern industry required production and marketing to be run in tandem. Once the industry was properly organised, fishermen should be encouraged to conduct their business through co-operatives. The need for an educational service for the industry, and minimum prices, were other issues alluded to. Basic research was entirely the responsibility of the state that would bear its costs.

It goes without saying that there was no consideration of a resource having any physical boundaries. Fish were there – and presumably always would be – for the taking. This opinion was shared with many other appraisals of sea fishing prospects.

The commission's report was diligently researched and it made sensible suggestions but it had an additional and unique feature which might be credited to one of its authors, Thomas Ryder Johnson. Born in Liverpool in 1872, Johnson left school at thirteen to work as a messenger boy. An impoverished youth had imbued a passion for socialism and he read avidly. He joined the Independent Labour Party in Liverpool in 1893. Johnson secured employment with Hugo Flinn, a fish merchant, and he became a fish buyer in Kinsale and Dunmore East. He subsequently moved to work in Belfast. In 1912 he established the Irish Labour Party along with William O'Brien and James Connolly. The following year he helped to organise the general strike. He assumed leadership of the party after Connolly's death.[17] His work on the commission was not his only contribution to the organisation of the fishing industry.[18]

Johnson is the likely source of a strong revolutionary flavour in the commission's report but its proposed co-governance by the minister for fisheries and the fishermen's council, was not envisaged as a one-way benefit. A state-supported fishing industry had social responsibilities and

...should be entrusted with the honour of performing certain free public services, such as the provisioning of hospitals and asylums for the aged and infirm...

Suggestions for the provision of finance to the industry were also novel

outside... state grants and loans...the industry be entirely self-supporting...(and) at a date as early as possible, the national assembly (the Dail*) will instruct the fishery council to make provision for its own loans by setting aside, each year, a reserve fund for that purpose...*

It is at that point, in consideration of what actually transpired, that the proposals of the commission left reality trailing far behind.

Any organisation for the fishing industry must take note of its insecurity and uncertainty...

Fish is an important article of food: and in providing that food the fisherman is rendering a social service. The fisherman, therefore, should not be compelled by the community to bear the whole of this risk. The risk to life none can share with him...But society should share his other risks with the fisherman, and give him security in order that he, in his turn, may be equipped and encouraged to provide food.

The state should immediately establish a central fisheries authority or council. This would bulk-purchase supplies of boats and gears and retail them at favourable rates to participants. It would have an educational role and keep the industry informed of technical matters concerning the conduct of its business. The objective

... at which organisation should aim is the creation of a national fishing fleet, self-supporting, self-controlling and self-contained...

This fisheries council would be headed up by the responsible government minister who would prepare legislation in consultation with its members rather than "in consultation with bureaucratic officials":

...(The Minister) would neither be able to introduce legislation, nor to recommend grants, except such as had first received the sanction of the council over which he presided...

Staff working in the department of fisheries would be placed under the direct administrative control of the industry.

Many looking at the organisation of the fishing industry in Ireland today – I would be one of them - would remark that while the above structure was never formally adopted, the effective use of lobbying power by fishermen's representative groups, notably producers' organisations, has secured a very similar end result.

The Irish Sea Fisheries Association Ltd.

A collapse in markets for mackerel and herring and a perceived decline in their stocks in Irish waters prompted a new approach to funding the fishing industry in 1930. Funding by loan having been seen to fail, a different approach was attempted: a co-operative of fishermen, the Irish Sea Fisheries Association Ltd (ISFA), was set up by government in 1930. The purpose of the ISFA was to make gear and boats available to fishermen on hire-purchase terms and to market their catches. Members were required to enter into a co-operative marketing contract and to participate in the general scheme. The ISFA guaranteed fixed prices for whitefish.

The ISFA was administered by a committee of management consisting of eight directors, four elected every three years by the members and four appointed by the minister for agriculture. There was a manager employed by the directors on the nomination of the minister. The directors did not receive any fees. Funding was from the exchequer financial allocation for fisheries. There was a grant for administration and general development work. Capital purchases by fishermen were funded by repayable advances from the exchequer. Fishermen who availed of monies from the ISFA made a capital pre-payment of 20% on the goods and an agreed proportion (usually 25%) of their earnings. Smaller vessels, like curraghs, were made available for small, usually quarterly, fixed payments. [19]

Looking back on the activities of the ISFA as reported to its annual ordinary meeting, it had support from the fishing community. At the end of its first year 650 had enrolled and, when it was finally wound up twenty years later, membership stood at 4,208. That said, pleas and implied threats to those who did not make the once yearly share purchase (cost one shilling sterling) required of members, became a feature of the ISFA's annual report.

The ISFA was involved in the provision of a wide range of support services. The bulk of its financial assistance was allocated for the purchase of vessels, new and re-conditioned, the installation and servicing of engines and the supply of gear.

For some twenty years, the modest annual reports of the ISFA recorded progress on a number of fronts. In due course it would be criticised because progress was seen to be too slow but its achievements, if modest, were significant and it appeared to avoid the pitfalls of over-investment which would characterise future years when the fleet found itself without fish to catch or a market to which it could sell its landings.

The ISFA appeared to be mindful of the need to place larger vessels in ports within reach of markets. Fishermen in more remote areas were supplied with curraghs and oar and sail boats. By 1934, on the eve of the Second World War, 190 vessels had been issued to its members and the ISFA had "under its control" – presumably not fully paid off – 103 motor-driven vessels of the "inshore" type. Almost one quarter of its members (500) had been supplied with boats or gear.

Two years later, unease was expressed that some hire-purchase repayments were not being fully discharged although others continued to be until the end of the war. In 1938 the ISFA's fleet consisted of 131 motor vessels and 394 other craft. Wartime strictures were beginning to bite and what could be dispensed to members would, henceforth, be dependent on materials available to the Department of Supplies which, at the end of the war, reverted to being the Department for Industry and Commerce.

Boats were purchased and re-issued, reconditioned, to members of the ISFA. Boats were also constructed at boatyards in Mevagh and Killybegs, Co Donegal, and Dingle, Co Kerry. There were probably trial and error adjustments to design of successive completions but for most of the time the boats were small (probably under 10 m oal). By 1937 only one person had been able to make a down-payment on a "larger" vessel. At the end of the war however, bigger boats were more seriously considered and in 1949 it became policy to provide a fleet of boats ranging between 38 (11.6 m) and 50 feet (15.2 m) oal. Eleven 50 footers were launched in 1950.

The second principal task of the ISFA was to market its members' catches. Initially, landings were to twelve points on the coast but this number expanded to 45. Some of the ports are noteworthy fishery harbours today but a number will be known only to those who take an interest in history: Loughshinny (Co Dublin), Moville (Co Donegal), Murrisk and Achill (Co Mayo), to name a few. Members were not obliged to sell their produce through the ISFA but, to facilitate those who wanted to do so, a transport ferry from the Aran Islands was inaugurated in 1934. A fast motor service linking Dingle, Cork and Dublin brought landings overland to market. A fleet of three "propaganda" vans distributed inshore catches to towns in the hinterland. To keep fish in good condition, cool storage facilities were provided in the ISFA's premises in Hanover Quay, Dublin, then in Galway, Killybegs, Murrisk and Helvic (Co Waterford). In 1947 a quick-freeze facility was planned.

Shellfish was a high value product which merited additional marketing effort. The ISFA constructed lobster ponds at Gortnasate in Co Donegal and arranged contracts for the sale of this species. Lobster and crawfish were actively marketed on the Continent in 1934 and two years later an employee of the ISFA was sent to the Netherlands to seek outlets for lobster. Periwinkles were sold through the ISFA from 1933 and by 1947 had become a rapidly expanding product line. In 1948 the sales of periwinkles doubled over those of the previous year. In 1944 Connemara scallops were sold in frozen form, a marketing innovation. It was not the only attempt to improve processing either. During the 1940s, the ISFA carried out experiments on filleting, smoking and curing pelagic fish.

Improvements in fishery enhancement and capture methods were also attempted. In 1936 nets for the capture of crawfish were trialled. Mussels were transplanted from seed beds for on-growing. A noteworthy development was the expansion of the mussel fishery at Cromane (Co Kerry) and the establishment of a purification plant there in 1940. Spat were distributed in the oyster beds of Killary and Clew Bays (Co Mayo). The ISFA even undertook exploratory fishing in its vessel, the *Naomh Simon*, in the inshore waters of Cos Galway, Mayo and Donegal. Inshore trawling was attempted and found disappointing.

The ISFA did not claim an educational brief but it did issue instructions on how fish should be handled to secure the highest prices. Young fishermen were provided with practical instruction in long-lining which, together with trawling, was believed to be the mainstay of the industry.

Many of these activities would become the focus of expansion in later development initiatives, notably those of Bord Iascaigh Mhara, the ISFA's successor, but the approach was different in the 1940s. The frugality associated with wartime shortages imposed a discipline in the way gear was cared for. The ISFA frequently reminded its members they should conserve ropes, buoys and other materials which would not be readily replaced. Warnings were issued against the landing of immature fish and the ISFA recorded its satisfaction at the passage into law of the Sea Fisheries (Protection of Immature Fish) Act of 1937. Fishermen were advised to return undersized scallops when there was evidence of the Connemara (Co Galway) beds becoming over-fished in the late 1940s and similar notifications were issued to fishermen harvesting mussels at Cromane.

The success of the ISFA could be evaluated in financial terms. Between 1936 and 1948, when comparable figures for the value of fish marketed on behalf of members was published, the amount handled in a twelve month period rose from stg£12,540 to stg£178,680, an increase of 1,300%, over twelve years. The amount of money distributed to members of the ISFA who chose to have their fish marketed by it rose from 8% of the total landings value in 1936 to 31% in 1950. The amount made available for capital investment between 1940 and 1950 increased from stg£12,240 to stg£72,279, a rise of 491% in ten years. Putting it another way, investment by the ISFA between 1940 and 1950 had averaged 22% of the value of landings annually.

The figures were apparently moving in the right direction but the tide of fortune changed. The Second World War provided fishing opportunities to Ireland's neutral fleet which the British were unable to exploit and that obstacle to monopoly fell at the end of the conflict. The restoration of peace brought competing fleets back into the market once more. Luxury foods, like lobster and scallop, proved difficult to sell in the post-war slump. But there had been successes and the prices obtained for landings in 1948 would still be referred to with envy and as an unsurpassed achievement six years later. The ISFA was wound up in 1952 and its activities were absorbed into a new agency, An Bord Iascaigh Mhara, the Irish Sea Fisheries Development Board. Before looking at that I want to examine one of two international developments in the last century which conditioned the industry to adopt a more impatient attitude to the rate at which development was progressing.

Chapter 5

WIDENING SEA, NARROWING OCEAN

From the eighteenth century, the marine territory claimed by a maritime nation within which it exercised control of its fisheries extended three nautical miles (nm) from its shores, the range of a cannon shot. As the twentieth century dawned, the British Admiralty was policing the seas around Ireland, though not to everyone's satisfaction; the increase in steam trawling in Britain presented obvious challenges to the authority of the smaller island. The formation of the Department of Agriculture and Technical Instruction in 1899 assumed the roles of fisheries research and protection. The SS *Helga*, a 46 m twin screw, steam-driven yacht served both purposes, equipped with a powerful searchlight for night patrols and two booms over the afterdeck for research work. In 1907 the decision was taken to replace her with a substantially similar vessel whose keel was laid in the Dublin Dockyard from which the boat was launched the following year. *Helga* 2 was better scientifically equipped and, to assist with her protection duties, had a "three pounder Hotchkiss quick firing gun". In 1915, during the First World War, the Admiralty commissioned the vessel as an Armed Patrol Yacht. The following year she bombarded the Custom House during the Easter Rising and in 1918 claimed to have sunk a U boat. She was returned to the Department in 1919 but was commandeered in the civil war in 1922 when she assisted with the capture of Youghal by the government forces.[1]

The Anglo-Irish Treaty of 1922 transferred control of fisheries and customs and excise to the Republic. The British coastguard was wound up and the new administration established a "coastal and marine service" of twelve armed trawlers supplemented by other vessels whose duties included fishery protection. They were not to make arrests but to observe and report, and they provided satisfactory service. In 1923 *Helga 2* returned to her pre-conflict duties and shortly afterwards was renamed *Muirchu* – sea-hound, thus starting a tradition of hound names, being succeeded by the research motor vessels *Cú Feasa* and *Cú na Mara*.

The coastal and marine service was wound up in 1924. After independence it had been intended that UK boats infringing the Republic's regulations should be returned to Britain for prosecution but that proved unworkable. UK skippers disdainfully dismissed *Muirchu* as "the Bogeyman" which they ignored. The apprehending gun had been removed and was not restored until 1935 when new legislation conferred powers of arrest and detention. In 1937 the *Muirchu* was joined by the SS *Fort Rannoch*, the two forming the nucleus of a naval service in the republic.

The outbreak of the Second World War in 1939 confronted the neutral republic with the need for naval cover. This was provided within a year by six motor torpedo boats and four other assorted craft for the regulation of merchant shipping and the protection of fisheries. They functioned as the Irish Marine and Coastwatching Service. At the conclusion of the war, this was renamed the Irish Marine Service which, in 1946, was incorporated into the Irish Defence Forces. In that year and the following one, three corvettes were purchased from Britain, the LEs *Cliona*, *Maeve* and *Macha* which would, controversially, provide fishery protection until they were withdrawn from service between 1968 and 1971.[2]

During the early twentieth century, problems caused by poaching were containable. The SS *Helga*s were faster than any fishing vessel so they were effective, and the pressure from competing nations for fish was low. All of the nations currently allowed access to Ireland's section of the later established Exclusive Economic Zone (EEZ), and the six to twelve nm band under EU rules, had traditionally harvested fish from what some persistently refer to, erroneously, as exclusively Irish waters, but their numbers were relatively small. But the Irish fishing industry was also struggling to assert itself: unable to catch sufficient fish to support an export trade but landing surplus to home demand; in 1953 the annual per capita consumption of fish was 5 kg which included herring and mackerel and imported tinned salmon.[3] There appeared to be no limit to the quantities that might be drawn from our waters (stock biomass had not been quantified). A note in the trade press in the early 1950s observed that only 140 men were engaged in trawling and seining between Dingle (Co Kerry) and Bundoran (Co Donegal). Another observed the presence of mackerel shoals which Irish boats could not reach and the Spaniards did not want 12 nm from shore.

But things were changing and change rapidly intensified. One observer recorded the cautious entry of two exotic wooden Spanish vessels to Bantry (Co Cork) in 1946; two years later he counted 127 in the bay simultaneously.[4] Post-civil war Spain was seeking food further afield on the Great Sole Bank (*Gran Sol*), its narrow continental shelf having been quickly exhausted. Heavy fishing in the North Sea also displaced British and continental vessels which turned their focus to waters closer to Ireland. In January 1955, fleets of British and French trawlers appeared off the south west coast. The trade press described it as "the greatest post-war invasion of the fishing grounds". The five Irish 50 footers there found it difficult to operate because of congestion.

Competition for fish throughout Europe, the urge to make money anyhow and a growing fleet at home prompted a number of development strategies which crystallised later in the century. Those who take the view that entry into the European Economic Community in 1973 foisted an alien regime upon the hapless denizens of the Republic might ponder: at that stage there was little new under the sun.

Throughout the 1950s the trade press frequently expressed sentiments like

..these god-given fish-abounding seas that surround our coasts are a constant attraction for the French, Dutch, British and Spanish fishing fleets but they failed to prove an attraction for us...

That was not entirely true. People wanted to make money but appropriate boats and gear were not to hand. However, there were ways of getting them. The concept of joint ventures became controversial when, in 1981, Eiranova set up a joint Irish-Spanish business in Castletownbere. Twenty six years earlier a French-Irish joint venture was launched by the Brandon and Cloghane Farmers and Fishermen's Improvement Association to fish for lobsters out of Co Kerry ports using a mixed French and Irish crew. The vessel, 8 m long, from Cornwall, with a 20 HP engine, seems ordinary enough now but it was apparently novel at the time. And there were other examples of vessels and personnel from outside the jurisdiction coming over to fish in Irish national waters, particularly for lobster for export.

But if such joint ventures were a way of introducing effort to the benefit of the indigenous but ill-equipped industry, the greater concern increasingly focused on the need to keep foreign boats out. The protective armoury of the territorial sea was narrow and, in the days before GPS, its definition was cumbersome. Territorial waters and fisheries limits, both had a single boundary, three nm from the low tide watermark. The boundary tortuously paralleled the shore and many a district court verdict turned on where a defendant might or might not have been relative to it. In 1957 a French trawler, *Ma Bretonne*, was apprehended by LE *Maeve* close to the Bull Rock (Co Cork). District Justice, J. F. Crotty held that limits should be based on the land, not on rocks, and dismissed the case. In fact, as set out in 1933 legislation, the limit extended seawards from any island that was part of the national territory. Crotty's verdict prompted an angry public meeting of the fishing community in Schull demanding greater clarity.[5] Crotty's judgment was a lapse but there were many instances of foreign boats fishing close to the coast and the prosecuting authorities unable to pinpoint their locations beyond doubt.

The most urgent management need was to simplify the coastal outline by substituting a simple map shape for an indented shoreline. *Baselines* normally trace the low water mark but, were the coast was complex a straight line might be drawn across bays and inlets. By the mid-1950s baselines were an established fact in many jurisdictions – if not internationally agreed - but they were not formally adopted in the Republic until 1 January 1960. They are a feature of the irregular west rather than the smooth eastern Irish coast.

The width of the territorial sea was also in need of revision. In 1955, several nations (including neighbouring Chile, Peru and Ecuador) claimed a 200 nm limit which would not become reality for the majority of states under international law until the mid-1970s. Elsewhere, states favoured widths of three, four, six and twelve nm from baselines; there were said to be nine combinations of territorial sea and fishery limits in existence. The resolution of this issue was a textbook example of the way international fisheries legislation would evolve in the future, the compromise it involved and the necessary co-operation with other nations to secure robust agreement. These realities are often forgotten. The justification for an extension of the three nm limit was threefold: there was a need to secure sufficient fish for a maritime state, for fishery protection and for a limited extension of sovereignty. This was necessary because of post-Second World War expansion in fleet capability, a fact which was prompting the industry in Ireland to assert itself.

When the International Law of the Sea Conference opened in Geneva in 1958 it was attended by Irish delegates from the Departments of External Affairs, Defence, Lands (Fisheries Division) and the Irish Legation in Bonn. The conference was dominated by the British Attorney General, Sir Reginald Manningham Buller, who was emphatic that the legal starting point must be three nm from low water mark: because it was already widely recognised and generally agreed. He insisted that this limit should remain in place unless there was two thirds agreement on a new arrangement. When the conference opened, the UK, France and the USA, among others, had three nm limits. But it did not apply to everyone. Some Nordic countries had four; India, Ceylon and Greece had six.[6]

The British negotiating position was informed by the trawlermen of Hull and Grimsby who feared exclusion from traditional grounds by wider national limits. The Netherlands, which recommended that the three nm one should remain, was open to a wider band, if that were adopted by the conference. India favoured six but proposed each nation select its own out to twelve nm. Ireland favoured twelve nm for fishery limits but three for the territorial sea. The 12 nm limit was also championed by Canada, Norway, Iceland and Australia.

Britain suggested a six nm compromise which allowed the passage of shipping, including warships but it was not accepted; the required majority was not reached. But a proposal to adopt baselines was. Concluding, Saudi Arabia proposed there was nothing in international law to prevent a nation claiming a limit of up to twelve nm. Chile, Ecuador and Peru maintained their claims to 200 nm.[7]

The industry in Ireland took to the idea of a twelve nm limit and urged the government to adopt it but events elsewhere demonstrated that the procedure would not have been as straightforward as suggested by Saudi Arabia. No nation had a stronger claim on twelve nm than Iceland because the fishing industry was so important to her economy. Iceland duly announced her intention of claiming it. Between September and November of 1958 British trawlers fished Icelandic waters under naval protection: the first cod war had begun and it involved shooting! But Iceland was not abandoned by her neighbours when the Faeroes and then Norway supported the proposal. And that probably settled the matter for Ireland.

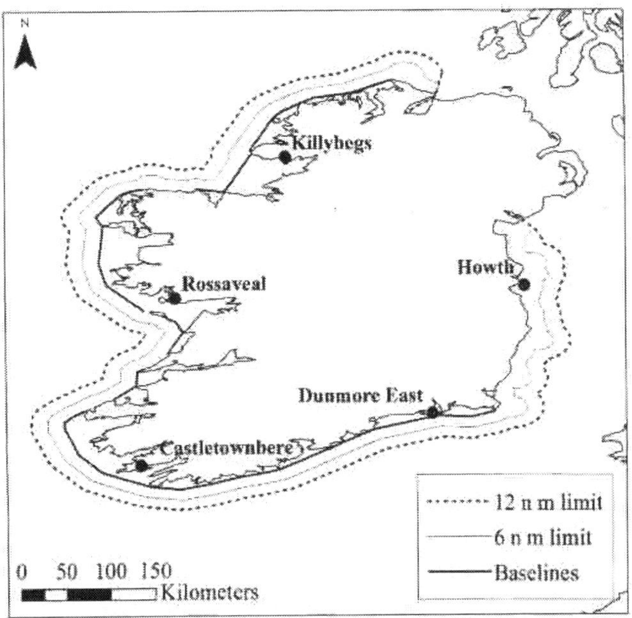

Fig 2a. The Republic of Ireland's territorial sea, sub-divided at 6 n miles and the baselines smoothing coastal indentations. The principal fishery harbours are labelled.

When the Conference on the Law of the Sea again convened in Geneva in April 1960, the heavy lifting had been done as far as Ireland was concerned. Although the majority failed by one vote to gain the necessary acceptance from 82 nations, a number (one of them the Holy See), including Britain and the USA, backed exclusive national limits of six nm with a further concessionary band of six nm for nations which traditionally fished there. Norway agreed a six nm limit with Britain in October 1960; after ten years it would expand to twelve nm. And in February 1961 Britain recognized Iceland's twelve nm limit territorial sea and was allowed access to the concessionary outer six nm band.

The USA suggested informally that all nations that had supported this arrangement might adopt it. All nations fishing close to Ireland agreed, except Russia and Poland. In due course Ireland's new limits, in conformity with those of other nations, came into operation, but controversy did not end there.[8]

Ireland's contemporary territorial waters were agreed in the Geneva Convention of 1958. According to it, our marine and transitional (estuarine waters) are "internal" to the baselines (Fig 2a). "Coastal" waters extended one nm seawards and parallel to the baselines. "Exclusive" national waters band the coast for six nm outside the baselines and a further band six nm in width accepts, in parts, access by the UK, Germany, France, Belgium and the Netherlands to fish certain species. A line, twelve nm outside the baselines marks the limits of the territorial sea.[9]

The areas enclosed by these boundaries are: internal waters, 13.650 km²; baselines to six nm, 13,662 km²; six to twelve nm, 13,824 km²; entire territorial sea, 27,487 km²; and territorial sea plus internal waters, 41,137 km².[10]

Three factors combine to make fishery limits a success or failure: the size of the area to be protected, the strength of the enforcing naval patrol and the hunger of other nations to gain access. On 9 March 1966, Steve Coughlan (Labour) asked Michael Hilliard (Fianna Fail, Minister for Defence) to provide details of the naval establishment. One corvette was operational at the time. Next, Coughlan inquired the size of the extended fishing area. On being told, he responded sardonically "15,000 square miles to be patrolled by one corvette! You are to be congratulated!"

Inadequate protection would be a fractious issue in the years ahead until twelve years later when a solution was proposed: if Ireland dropped her current demand for a 50 nm exclusive fishing limit and accepted a fisheries policy common to member states of the EEC, to which she wished to accede, Brussels would make money available for a stronger naval establishment.[11] It was persuasive.

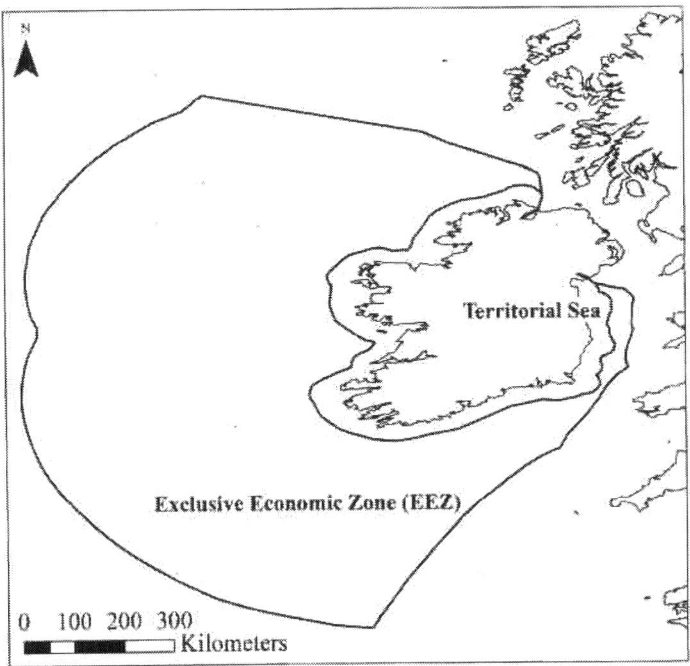

Fig. 2b   Map showing Ireland's territorial sea and its Exclusive Economic Zone (EEZ).

But the Geneva Convention did not conclude the debate on access to near shore marine waters in Ireland. Coinciding with the introduction of new dimensions and rules governing the territorial sea on 1 January 1960, the overall length at which a fishing vessel must be licensed was increased from 10.7 to 22.9 m.[12] Simultaneously, boats of less than this size from Northern Ireland (but not other parts of the UK) were allowed to fish inside the republic's national exclusive limit. The arrangement was not welcomed by the industry which questioned its legality. The decision to make the regulation appears to have been the exercise of personal preference by Sean Lemass who became taoiseach in 1959. Lemass is regarded as one of the greatest leaders of the Republic. He made history when he travelled to Belfast to meet the Northern Ireland prime minister, Terence O'Neill, in 1965. Lemass favoured reciprocal fishing access agreements between the two jurisdictions. His arrangement in 1960 was formalised in 1964, a year before his historic meeting, in an exchange of correspondence between the government departments charged with administering fisheries. It is known as the *voisinage* (translating as "neighbourhood") agreement and has never proved satisfactory.[13]

During the 1970s international anxiety to extend maritime boundaries to 200 nm seawards from baselines rose to a crescendo. The enclosed area, into the territorial seas, would be known throughout the world as the Exclusive Economic Zone (EEZ). Ireland joined the EEC in 1973 so that boundaries which had been extended to 200 nm in 1976 were those of the European Community, not Ireland. After 1 January 1977, when the new EEZ maritime limits came into force (Fig 2b), the authority for the area west of Ireland, from twelve out to 200 nm, resided in Brussels, not Dublin. Thereafter, the EEC Commission would act as broker in all international agreements on fishing matters, including arrangements to exploit capture fisheries made with third countries. Individual states of the community could act to make fisheries regulations within their own jurisdictions provided those did not discriminate against any other member nation. Ireland was primarily responsible for the management of that part of the EEZ contiguous with its territorial sea, and the community would financially assist its effort to police it.

Moving contemporaneously, we will next consider efforts to build a national fishing industry to avail of the opportunities presented by the anticipated expansion in territory which the 1958 Geneva Convention confirmed, and to discourage and compete with other nations whose fishing activities close to Ireland would have deepening material and psychological impacts in the future.

Chapter 6

A BRIEF FLIRTATION WITH REALITY

The surviving records of the ISFA convey an impression of comparative serenity in the way it conducted its affairs. It facilitated the industry wherever it could and was apparently careful not to provide finance for vessels in areas which could not support their activities. However, what we know of the organisation is contained in the lean reports of its annual ordinary meetings. There was no contemporary trade press to critique its activities. De Courcy Ireland who knew its principals, was not entirely supportive of the way it operated. He agreed with those who described ISFA as a "benevolent autocracy" which

...offered the ordinary fisherman even less say in the direction of his industry than its constitution seemed to promise...[1]

De Courcy Ireland was, throughout his long life an enthusiast for the industry and its participants. His life ended as the depredations and consequences of an unregulated catching sector became undeniably obvious. Control of the rate of exploitation is essential where natural resources are concerned and the prudent approach of the ISFA may have had a lot to recommend it.

On the brink of the Second World War in 1938 some 3,400 full time job equivalents were provided by Irish sea fisheries landing

11,000 t, valued at stg£167,000. From that point, wartime shortages made it impossible to get hold of engines and effect technical improvements and when, twelve years later a successor development agency was established, catches had expanded by a mere 17%.

Arguing in favour of a coherent fisheries policy, the Maritime Institute of Ireland, founded in 1941, challenged the widely accepted view that inshore fishing in small boats was the way to proceed; a fleet of middle-distance and deep water vessels should be part of any future development plan. The ideas gained ground in the post-Second World War years and a Commission on Population and Emigration in 1949 proposed that this, among other developments, would be the means of employing up to 100,000 people, afloat and ashore, in a modern fishing industry. Whatever success had been achieved by ISFA, there was an appetite for a more aggressive approach to building a modern industry.

In 1951 Britain passed its Sea Fish Industry Act setting up the White Fish Authority. The Authority would consist of five members appointed by minister and its role would be the regulation of the whitefish industry in the interests of consumers and those engaged in harvesting and selling fish. The following year the Dail passed the Sea Fisheries Act, 1952 which closely resembled the UK legislation. It would however have wider scope, embracing pelagic, whitefish and shell fisheries, and a more extensive mandate. The Irish equivalent of the five-man Authority was a Board, specifically *An Bord Iascaigh Mhara* (BIM), made up of six rather than five members to be appointed by the minister for agriculture.

BIM was set up on 24 April 1952, assuming the responsibilities and liabilities of the ISFA. The various activities of its predecessor were continued by BIM whose brief was to exercise closer supervision and control of sea fishing operations, giving greater attention to the production and marketing needs of the industry.

Structural differences between the white fish Authority and BIM may have been minor but their experiences were quite different. Following consultations with the interests involved, a White Fish Advisory Council, consisting of representatives of the industry appointed by the minister, had been set up for the purpose of providing intelligence to the Authority. The Irish legislation created, in addition to BIM, an advisory group *Comlachas Iascaigh Mhara*, to be known as *the Association* which would be managed by a committee for the same purpose. Its membership contained four representatives of fishermen, one from the retail fish trade, one from the wholesale wet fish trade and two from the distributive fish trade. They would be elected.

The challenges to establishing a thriving fishing industry were enormous. The dilemma was incessantly rehearsed in the trade press of the time. The sea was a wasted opportunity

...clamour(ing) on our coasts, offering a never ceasing supply of fish that ensures health and wealth...

Our fisheries, like our national language, are nearing extinction, and we are gathering round the expiring remnants making a great clamour which, of course, we refuse to recognise as a caoine *but which, nevertheless, we feel is truly one. Response.* Why doesn't the government do something *and* why doesn't the government take its grip off our fisheries?[2]

Everyone in the industry, and many besides, had a vision of what might be achieved. The question was where to begin. The populace regarded the consumption of fish as a penitential exercise confined to Fridays and Lent, which was the principal marketing opportunity in the year. Without a market there was little point in going fishing. Fish handling was primitive, the use of ice to maintain quality unusual, and quick freezing technology, while not unknown, was not widely in use and consequently wet fish had a brief shelf life. And there were gluts of fish, particularly herring which was captured in large quantities whenever it became available, which could only be absorbed by fishmeal plants, as yet un-built.

On the other hand, there was little point in stimulating the appetite for fish if it were not available. A major challenge for BIM was to satisfy and thus maintain consumer demand when Irish boats – regarded as "inshore" up to 21 m oal – were confined to harbour by adverse weather conditions.

The first BIM board seemed to have the right credentials. It was chaired by an engineer from the fisheries division of the Department of Agriculture. Seamus Ó'Mealláin was the son of Michael Mallin, executed for his part in leading the 1916 rising. But, despite including a highly respected herring fisherman from Killybegs, the board of BIM was rapidly seen to be out of touch with industry concerns. From the outset, unfavourable comparisons were made between the White Fish Authority and the agency. BIM was funded exclusively by the exchequer whereas the Authority drew its finance from a levy on the sale of whitefish and on members' subscriptions. This may have bound the Authority more closely to the industry whereas BIM was seen to be remote, even beleaguered, and, not without reason, subject to political influence.

A major source of contention was the hire-purchase system of financing boat purchase. Both BIM and the Authority used it. The Authority provided a grant of up to 30% (to a maximum of stg£5,000) while BIM offered none. The loan to purchase a vessel amounted to 90% of its cost in Ireland and had to be repaid within ten years. In Britain equivalent figures were 55% and twenty years. Annual interest on repayments was 5.5% in Ireland but it could be as low as 2.5% in the UK.

More unpopular still was the method of ensuring repayments were made. Experience in the nineteenth century recalled that loans had been "more or less punctually repaid"[3] but there had been some slippage in the meantime; the ISFA incurred bad debts and BIM intended to avoid repeating its errors. Unlike the ISFA, BIM was subject to the hire-purchase Act of 1946 and individual agreements contained arrangements for repayment through a proportion of the earnings generated by the owner. All landings by vessels on hire-purchase from BIM automatically became the property of BIM. The fish was sold (incompetently according to the industry) by the agency; monies due for vessel and gear repayments were retained, and the balance was passed on to the fishermen. To be party to a hire-purchase agreement with BIM was denounced as a form of indentured labour.

BIM saw its role as providing vessels to the industry and promoting fishing throughout the range of scenarios: inshore, offshore and for shell, pelagic and white fish. Furthermore, there was a general expectation within the industry that BIM would do whatever was needed, down to standardising fish boxes in the ports. Whether BIM was merely anxious to oblige or cannily grabbed every opportunity presented to it is not relevant. Its actions were described by its critics in the Association in the following terms:

...(BIM) had entered into practically every department of the industry to set up its own enterprises; very often in competition with existing industries and at great expenditure of public money...

Its extensive, and widening, brief meant that BIM would tread on vested interests sooner or later. The stage was set for conflict and strife. On the other hand, there was no going back. The first programme for economic expansion would be published later in the decade.[4] There was a demand for a more aggressive approach to setting up a viable industry, whatever that meant.

There is a popular conviction that fisheries are capable of greater development and that they should be developed...[5]

more or less sums it up. Attitudes had become more critical and less tolerant since the leisurely progress of the ISFA, or so it appeared from the commentary on its activities by the Association. Exasperation overflowed in such outbursts as:

The (ISFA) tinkered with the operation of steam trawlers, tinkered with the marketing of lobsters, of periwinkles, of mussels, tinkered with ideas of co-operative marketing until it ultimately degenerated into a form of surrender marketing...[6]

The budget for BIM was large and that too was a cause of resentment. In the accounting year 1 April 1955 to 31 March 1956, 63% of all monies allocated to fisheries (including the running of the responsible government department and the administration of inland fisheries), was devoted to the running of BIM or to its gift for distribution among its projects.

The Association was the appointed forum for discussion but the 1952 Act stated that the Minister might not choose to consult it before making regulations if "...in his opinion it is impractical to do so..." Mistrust and resentment are never in short supply in the fishing industry and the Association, feeling excluded, suspected the worst. The Association also contained opposing interests, particularly among those who sold or distributed fish and who loudly complained about unfair competition from BIM.

The Association operated on a proportionally tiny budget, dispensed to it through BIM, although it had some additional finance contributed in subscriptions from its members. It was articulate and published a journal entitled *Irish Fishing and Fish Trades Gazette* which trenchantly excoriated the state body. A flavour of the inter-relationship between the two is typified by one of its editorials:

In days gone by

A document has come to light which may prove of interest to the historian of fisheries. It is entitled the second annual report of An Bord Iascaigh Mhara *and contains some – but not much – information on the actions of that dilatory body in the period from March 31*[st] *1953 to March 31*[st] *1954.*

The modern reader may wonder at the extraordinarily diverse ways in which the Board contrived to lose money in those far off days; and wonder more purposefully, if the Board is still exercising its spendthrift versatility today.

Some of the innocent amusement to be had from seeing how barren of fulfilment were the prognostications in last year's Old Moore's Almanac may be had by reading the preface to this report in which cautious references are made to future developments in the industry.[7]

A gulf very rapidly opened between BIM and the Association which was meant to be advising it. In the first three years of its existence, according to one of its most prominent committee members, the Association was not asked to advise BIM on any occasion. In retaliation, the Association railed against the amateurish approach of BIM's political appointees:

It is remarkable that in an industry in which it is so difficult to find skilled operators – boat wrights, engineers, skippers, diligent fishermen – a Board of competent and able directors can be drawn together almost by pulling names out of a hat...[8]

The Sea Fisheries Act, 1952, had been ushered into existence by Fianna Fail whose Galway West Deputy, Gerald Bartley, was parliamentary secretary (junior minister) for Fisheries, 1951-1954. He was replaced by Oliver J. Flanagan, Fine Gael deputy for Leix-Offaly. By the time Flanagan took over in 1954, relations between the Association and BIM were so fraught that he was obliged to arrange periodic joint meetings between them.

Irish Fishing and Fish Trades Gazette was written in an entertaining and stylish manner. Whether its commentary was entirely fair I am not in a position to say but its record is influential and, because history is largely the property of those who record it, the *Gazette* conveys a strong impression of BIM as an unimaginative and less than competent bureaucracy, stolidly pursuing an unproductive course.

Three costly cutters

The nature of the rolling skirmish between the Association and BIM is well illustrated by the case of "offshore" vessels intended to extend operations into deeper water. In 1952 BIM purchased three steel German trawlers, the *Loch Lein* (31 m), *Loch Laoi* and the *Loch Lorgan* (each 27 m oal). Although small in size by today's standards, the vessels were large for their time. They were intended to maintain supplies of fish in order to sustain demand on the home market in Ireland and to cut out the need for imports. The three fishing cutters, as they became known, would perhaps best epitomise the fumbling of BIM as described by the Association, which criticised them incessantly. The first parliamentary question on their activities was aired within ten months of their operations beginning, on 27th October 1953. It was followed by a number of other enquiries which made it clear that the cutters were not living up to expectations.[9]

There were difficulties recruiting senior crew who had to be sourced abroad. The vessels broke down and went out of service for prolonged periods. It became increasingly obvious that the boats were not making money and further, they were accused of preventing private operators from doing so. By 1955 the three were referred to as the *Loss Lein*, the *Loss Laoi* and the *Loss Lorgan*. Being offshore boats they went on seven to ten day cruises at the end of which their landings were in less than pristine condition. These landings were auctioned alongside those of the hire-purchase fleet whose owners claimed that BIM insisted on selling the produce of the three cutters first, resulting in lower prices for fish landed by the boats tied to, and effectively working for, BIM, which was later moved onto the auction floor after the produce of the cutters had been sold.

In the early autumn of 1954 skipper Donal O'Driscoll of Baltimore, fishing fifteen nm east of Howth had his nets fouled by the *Loch Lein*. What, demanded the Association, was an offshore trawler doing so close to land when it should have been fishing deep waters?[10]

Irish Fishing and Fish Trades Gazette on 25 September 1954 published an article titled: "State trawlers - a trifle of great importance" setting out the case against the three cutters in measured terms. Their purchase had not required an exorbitant sum, their running costs could be borne and the quantity of fish they landed was too small to distort the market one way or the other (all damnation with faint excuses) but the principles underlying their operation were not insignificant. The state was competing with private enterprise and this amounted to "doctrinaire socialism". The terminology was inflammatory at a period in Irish history when communism was characterised in some circles as a major threat menacing the state.

The Association contended that the purchase of the vessels had been an error; that they were not designed to draw heavy nets through deep water. Subsequently, the role of the cutters underwent some re-interpretation. A staff member of BIM stated that they were never meant to be excluded from inshore waters. However, Bartley, the parliamentary secretary, closely associated with their purchase, maintained they were. When BIM suggested the cutters were supplying a wider range of species to the consumer than the remainder of the fleet, the statement was greeted with demands that their secrets should be made known to the industry which BIM was meant to be serving. In any case the Association dismissed the values attributed to their landings by BIM, values which, it claimed, could only be realised if the reported weights consisted only of black sole!

By mid-1956 the establishment, in the form of the two parliamentary secretaries from opposing political parties, united to defend the cutters. It was proposed that their purpose had been, in addition to harvesting from deeper waters, to provide training for apprentices, a claim that was ridiculed by the Association.

BIM's annual accounts to 31st March 1957 revealed that the vessels had accumulated losses of stg£47,000, almost as much as it cost to purchase them. In operational terms the cutters had lost stg£2 per 50 kg of fish landed when the average sale price for whitefish was only marginally higher. When, in 1956, Oliver Flanagan, parliamentary secretary, declared he had not said the three cutters would be sold, it was clear the writing was on the wall for them. They were finally disposed of in 1961 when they went on to fish successfully under the direction of new owners in the North Sea.

The void at the heart of policy

In 1956 the eight-man committee of the Association unanimously decided against seeking re-election. They explained that their efforts had been ignored by BIM and by the government department responsible for fisheries. They saw no point in carrying on.[11] In any case, private industry was becoming more assertive and emerging trade organisations would in time represent their interests. In due course two of their number were appointed to the board of BIM. Their co-option by the establishment did not entirely suppress criticism however.

As this is being written, BIM has been in existence for sixty years during which time its edges have been smoothed and it has come to terms with or found a way around its early awkwardness. One way of doing so was by distributing cash. Although the sums are modest by today's standards, they were immense for the greater part of its existence. In the early 1960s exchequer funding to the agency exceeded 60% of the value of landings, the principal source of earnings within the industry at the time (Appendix 3).

The Association's publication survived the resignation of its board and continued its barrage of criticism. It referred dismissively to BIM's annual report for 1957 as "the fourth instalment of the Board's tale of woe" and alluded to its content: "The report is very attractively produced with clever (though meaningless) graphs".[12] Expensive, jazzy presentations, of which its annual reports were good examples, maintained a prestigious authority which throughout its existence attracted – or simply commandeered - a variety of responsibilities. Accruing them enabled BIM to become the dominant force of influence within the industry, a position it long retained. No history of sea fisheries in the Irish Republic would be adequate without an assessment of the activities of BIM.

Critical examination of the annual reports reveals a shy organisation crouching within glossy covers. The reports were voluminous but repetitive. Their later format recycled the same facts at least three times: as *foreword* (or *chairman's statement*), as *highlights* and as accounts of the agency's *divisional/sectional* achievement. The management structure was in perpetual flux, new division or section titles serving, like newspaper headlines, to proclaim the agency's topicality. The accounts were inscrutable; this is not to say that anything was incorrect but that their presentation was not sufficiently consistent to follow the financial progress of an activity over the history of the organisation. Sometimes the facts, such, for example, as divisional salaries, were reported as part of the divisional accounts in question, at other times aggregated as part of the general administrative budget. On occasion, a new layout of the accounts was for the stated purpose of making them more comprehensible. Although the names of board members were known from the birth of the agency, divisional managers were not named in the annual reports until later in its history. When the practice of quantifying salaries and wages was adopted as part of yet another effort at clarity, those of contract staff were not included so it was not always feasible to tell how many people were actually employed by BIM.

Throughout its history, BIM was anxious to claim success. Whenever new economic strategies, such as the national plans for economic expansion or the later national development programmes, were announced, their budgetary targets were stated and then quietly forgotten. Some useful tables of figures were included in the annual reports – the table of landings was one, the table of fleet size and associated employment another. As long as these statistics were growing the tables were updated annually but when the trend reversed, the tables were dropped.

The major question in the 1950s, for those who bothered to ponder the matter, was who devised policy for BIM. J.K. Clear, an opinionated member of the Association and, for a time, editor of its journal, who on one occasion described himself as "something of a visionary", put it like this:

...it appears that (BIM) very far from being an autonomous body is really some mongrel adjunct of the Minister's office, neither civil service nor independent board, but ingeniously combining the defects of both; without the freedom of action which is the raison d'être *of all state boards; without the direct Ministerial responsibility which characterises the (civil) service.*[13]

The Association made it clear on occasion that the technical expertise within BIM was first rate. A fish processing factory costing stg£50,000, constructed in Galway, was "well designed" but it was placed in an area "where there (was) virtually no fish". The problems within the organisation were at the top.[14]

Presumably the Association would have avoided the pitfalls it identified had it been in charge, but was its vision of the future any more enlightened?

While conservation was not as urgent an issue as it is today, the trade press referred to the need for prudent and responsible fishing, and the ISFA, years before, had shared those concerns. Yet,, in 1956, the Association published an evaluation of marine fisheries as "an industry based on raw material that is limitless", an unfortunate opinion in view of what happened later. In 1958 Clear, speaking at a Rotary luncheon was equally ebullient:

Once you enter the international fish trade, there is virtually no upper limit...there is no reason why we should not dream of an Irish-owned factory ship fishing, let us say, the Red Sea and marketing her catch amongst the teeming populations of Africa or India.[15]

Something similar did indeed occur half a century later when the Irish-owned *Atlantic Dawn,* then the largest fishing vessel in the world, trawled the inshore waters of Mauritania and sold its landings to the people of Africa. Perhaps Clear was a visionary. But there was no suggestion of prudent progress in the ambitions he described. Nor was there any shortage of such mirages. T.H. Huxley in 1882 had predicted

...steam and refrigerating apparatus combined have made it possible for us to draw upon the whole world for our supplies of fresh fish...My son, or at any rate my grandson, when he goes to buy fish, may be offered his choice between a fresh salmon from Ontario and another from Tasmania...[16]

And Huxley, as we have seen, could not grasp the vulnerability of marine fish stocks.

Likewise for Clear, as for BIM, progress may well have been about getting the maximum out of the resource as quickly as possible without any thought for the consequences in the medium, not to speak of the longer term.

When Clear observed in 1956

State policy for fisheries development in Ireland is still largely a matter of keeping the fishermen alive...[17]

he was not far wrong. Policy was and has since been, if not to keep fishermen alive, then to pacify the industry and silence its strident demands by throwing money at it. BIM became the conduit for the concentration and disbursement of those funds.

Once it had been set up as an autonomous body, BIM became the state in the minds of fishermen. While the minister retained some remote control, the policy of BIM became the policy of the state.[18]

Thus the industry and BIM embarked on a symbiotic relationship which would drastically change both of them.

The first decade

Providing a modern fleet absorbed much of its budget in the early years. BIM operated its own boat yards at Killybegs and Mevagh (Co Donegal), Dingle (Co Kery) and Baltimore (Co Cork) and continued with the construction of vessels, as the ISFA had done before it. Vessels were also bought in from outside yards.

Exchequer funding to BIM, mainly to finance the boat purchase scheme, increased dramatically over its first decade. The founding act[19] allowed the minister of finance to advance sums of stg£500,000 as repayable loans on the recommendation of the minister responsible for fisheries. Four years later the maximum allowed was raised to stg£1 million[20] and three years after that the limit was increased to stg£3 million.[21]

In the accounting year 1952 - 1953[22], there were 61 vessels on hire-purchase, all less than 50 feet (15.2 m) oal. Their combined horse power (HP) was 4,500 (3,308 kiloWatts (kW)) and their value at launch approximately stg£200,000. In 1962 comparable figures were 150 vessels, 33 of them greater than 16.8 m with a combined engine power of 8,504 kW, whose value at launch was stg£1,040,259.

The surrender of vessels by owners who could not keep up repayments or their repossession by BIM was commonplace; the boats were generally cleaned up, sometimes re-engined and reissued to another applicant. Until the late 1950s this fate happened to one or two boats each year but in 1959 five were repossessed and the year after the figure rose to nine. The value of production in the boat yards fell by 22% between 1960 and 1961 and gross profits from vessel construction declined from stg£17,500 to stg£10,200. Technical advance continued and a new prototype for a vessel of just under 20 m was introduced; but the number of orders declined and BIM's annual report expressed concern about the obvious symptoms of slowing demand. Grants and advances for vessels authorised in the year 1962 were 40% less than five years before.

According to BIM's annual reports, the number of surrendered or repossessed vessels on hire-purchase was a relatively small proportion of the fleet at any time but the repayments problem was more intractable than suggested by this indicator. At the end of March 1963, provision for bad debts had climbed to £430,000, 33% of the value of BIM's assets.

A major concern was the demand to fit vessels with larger engines, something which should have triggered more general debate. In 1955 a vessel of between 14.9 and 16.8 m had an average engine power of 62.5 kW; six years later an engine fitted in a similar vessel was 67.6 kW. Smaller vessels of between 10.7 and 14.9 m, formerly fitted with engines of 36.8 kW, were now supplied with motors averaging 44.1 kW. The increases of less than 10% were small but nonetheless significant. When fish stocks are plentiful they can be harvested with relatively low engine power; a larger effort is required to accumulate fish dispersed over a wider area at lower density.

Bad weather was often cited as a reason for poor landings. When the national landings statistics of 1961 turned out to be lower than the year before, BIM blamed under-performance on poor boat maintenance. In due course, owners would be relieved of the cost of maintenance which would taken on by the exchequer; in 1962 the cost of inspecting the vessels and keeping them in good working condition amounted to stg£7,000.

Within the industry itself there was little doubt about the source of problems. The loan terms were exorbitant and fishermen were unable to bear them. It was therefore appropriate that BIM itself took a hands-on approach to managing a number of vessels in order to maintain the supply of fish to the market.

In the mid-1950s the exchequer provided finance to grant aid vessels in the Gaeltacht (Irish speaking) regions. The Gaeltacht boats were manned by skippers originating there but BIM monitored and reported their operations in greater detail than for the fleet it had supplied to skippers elsewhere in the country.

The first two boats in this BIM-managed fleet came into service in 1957. They were 17.2 m oal, powered with engines of 83.8 kW. The fleet grew to four boats of these dimensions, four vessels of 5.6 m and four of 7.9 m oal by 1958. In every year of its operation from 1957 to 1962 this fleet made an operational loss, beginning at stg£3,273 and climbing to stg£7,474 in 1960. The total lost on the venture was over stg£40,000.

A similar fleet directly managed by BIM was brought on stream in 1960, consisting of two boats of 17.2 m and two of 13.7 m oal. They made losses too although the annual report for 1962 observed that the losses had reduced due to the rising price of the landings; they were losses nonetheless. In 1961 the two smaller vessels were sold. The larger ones, *Ard Colm* and *Ard Mhuire* were diverted from commercial fishing to exploratory fishing, surveying and accommodating instructional courses.

Stabilising the fish market was undertaken by importing white fish to satisfy demand when the home fleet was unable to supply it. In 1952, 228 t of white fish were imported but this quantity declined annually to 28 t in 1957 after which no further imports were reported. The decline might have been a consequence of the growing Irish fishing effort. The imports had provoked concern within the Association about distorting the market to the detriment of its members' interests. The annual reports of 1961 and 1962 respectively noted that imports of white and smoked fish to supply the home market had ceased.

The catches which BIM handled from its tied hire-purchase fleet and from boats directly under its control amounted to 3,953 t in its first year of operation, 1952. All wet fish landings (excluding shellfish) in that year totalled 10,150 t so that BIM was responsible for 39% of those. In 1957 BIM more than doubled the quantity of fish it disposed of on behalf of its fleets to 10,834 t or 83% of the national landings. Clearly BIM's efforts were accounting for a sizeable proportion of the marine harvest. However, agitation by the Association was effective and in 1958 fishermen were given the option of having their fish sold by BIM or disposing of it through other outlets. Their response was not as dramatic as might have been supposed, in view of the intense criticism of BIM's auctioneering efforts. In 1958, 52 fishermen chose to seek alternative outlets. In that year, 9,422 t of wet fish were disposed of by BIM and the decline continued until 1962 when the quantity thus sold was reduced to 5,878 t. Although fish auctioneering had made some profits for BIM in its early years, declining volume ate into these despite rising unit value.

Landings of wet fish generate the first sale income to personnel in the catching sector and repay the mortgages on boats and gear. Added value and employment are created by processing the landings, converting them to frozen, vacuum packaged, smoked or ready-cooked meals. BIM undertook those tasks also, and further progressed the efforts of the ISFA in that regard.

The disposal of surplus catches associated with the harvest of herring was a constant preoccupation for the industry and BIM. Discussions in 1953 examined the possibility of securing 50 t per day to supply a pilot plant in Killybegs. For its first three years the Killybegs plant ran with losses of stg£4,687, stg£4,531 and stg£5,135 respectively; these were justified because the plant was "experimental".

The storage of frozen fish got underway in 1954. Three locations, Schull, Galway and Killybegs, were selected for processing.

The Killybegs factory also manufactured other marine food products and, in common with its sister factories at Schull and Galway, it also failed to make any profits on them. In 1959 losses of stg£8,958 were recorded at Killybegs, stg£5,887 at Galway and stg£2,482 at Schull, an accumulated total of stg£17,327. In 1960 the accumulated total rose to stg£20,632, declining in 1961 to stg£13,751 and reaching the highest losses of stg£26,075 the following year. These totals masked variable performances by the individual factories.

In 1961 the Killybegs factory had it best frozen fish output of 174 t, and the smoked fish output of 113 t was also the highest recorded to that time. Other products included marinated herring, 64 t, salted herring, 21 t and 19 t of fish meal. In the same year the Galway plant had its highest production of frozen fish to that time. Schull produced frozen fish and scallops.

In 1960, faced with unrelenting losses, consideration had been given to closing one of the factories, and the following year Schull was selected for that fate.

A single explanation served to cover the lack of success at the three plants: supplies of fish had not been adequate to sustain them and all operated below capacity. BIM's annual report for 1962 reassessed the venture:

... there is an element of obsolescence due to the fact that (BIM) did not engage in direct competition with private enterprise...

The Association would have fully concurred; it always maintained that BIM operated in isolation from the realities of the industry it was meant to be mentoring.

...The future development of these factories as product development units will however ensure their operation and expansion for the benefit of the industry...

And, however unfortunate its experience, BIM had no intention of entirely abandoning any aspect of this complex industry in which it had become involved. In 1960, BIM purchased the Atlantic Fish Industries plant at Killybegs with the intention of having fish meal manufactured and fish oils extracted there by private tender. That factory had been opened in 1958 by a German-owned business, supposedly inspired by the success of BIM's pilot plant[23]. It had a potential capacity of 100 t per day. The following year the Atlantic Fish Industries factory was leased by BIM to a Danish enterprise. In 1961 BIM undertook a feasibility study on locating a fish meal plant on the east coast.

The ISFA had been closely involved with establishing the Cromane mussel fishery in Castlemaine Harbour (Co Kerry) and BIM continued to support it. The extensive mussel beds were served with a purification plant which cleansed the product before export. In 1959 there was a changeover from fresh mussels to canned and bottled product which resulted in a loss of stg£871 to the plant; the following year that loss grew to stg£1,360 and in 1961 to stg£2,198.

Another approach to production, the transplantation of seed mussels for on-growing was considered. Initial trials to the value of stg£450 were funded by BIM which paid stg£200, the co-operative at Cromane contributing stg£250. Transplantation as a commercial process was undertaken in the spring of 1961. Again BIM made the greater part of the investment, stg£720, while local subscriptions raised stg£280. Some 900 t of seed mussels were expected to generate 2,700 t of finished product. Production methods at Cromane were becoming more organised and intensive as demand grew. Towards the end of its first decade there were reports that the fishery was under-performing. Cromane had been known to produce 747 and 926 t of mussels in the years 1941 and 1943 respectively; when, in 1962, the yield declined to 600 t "premature" (over-) fishing was identified as the cause.

BIM would be associated with the introduction of modern technology but, as in the case of the ISFA, one of its robust lines of production was periwinkles bought in from hand gatherers. In 1961 39 t were purchased for on-sale; the following year 36 t and in 1962, 49 t.

Not all of its initiatives had been unsuccessful then, and in 1957 BIM secured contracts to supply frozen haddock to the armed forces of the United States of America and to sell frozen herring to Belgium. By trial and error BIM was gradually discovering a role for itself. Sales, marketing, promotion and "development" were areas in which success or failure was not so brutally diagnosed by scrutinising a balance sheet.

The scientist S.J. Holt had written of the benefits to industry of exploratory fishing which meant, in the Irish context, locating supplies of fish, rather than assessing their capacity to sustain exploitation. Holt's article had been frank:

...The common incentive to the search for new grounds has been the lack of fish on the old ones and the tremendous spread of the European trawling industry at the end of the nineteenth century and the beginning of the twentieth took place largely in this way.[24]

Depletion of fishing resources as a result of greater fishing effort was a recognized fact in the 1950s. Holt went on to observe that the tradition in Europe was that industry and government sponsored exploration, and he chose as an example the British scientist E.G. Hickling who, in 1928, worked with the Fleetwood Fishing Vessel Owners' Association and the Development Commission of the UK to jointly sponsor a commercial trawler to extend hake fishing into the Atlantic off Northern Ireland. In Australia, Holt reported that prospecting by government was followed up by fishing companies.

However, Holt stated

...one suspects that most of the real advances (on new fishing grounds) are made by fishermen in the everyday course of their work, going out a little further from shore here, fishing a little deeper there. This incentivises a higher price for the catch as well as depletion of the grounds.[25]

For a period BIM had acted as shore manager of the government research vessel *Cú Feasa*. The kind of work it did proved a useful diversion for the two larger vessels in the small loss making BIM-managed fleet which was deployed on survey work and funded by the exchequer rather than by the sale of landings. BIM also chartered commercial vessels for these purposes, thus forging financial links with the industry. In 1961 the *Ard Colm*, one of BIM's directly managed fleet, in co-operation with the state agency Gael Linn, Carna (Co Galway), trawled the 50 fathom (91 m) line west of Inishmore and North of Slyne Head where it located "small stocks" of lobster and crawfish.

Experimental fishing undertaken in a joint venture with the *Cú Feasa* and four 15.2 m vessels examined the feasibility of drift netting for herring between Clogherhead (Co Louth) and the Isle of Man. Mid-water trawling was later evaluated using six 17.1 m vessels at Dingle while the possibilities of dredging for scallops were explored at Galway, Schull, Dunmanus Bay and Kenmare River.

In addition to any investigative function, these exploratory cruises were also credited with an educational role. In the course of its first decade BIM had organised demonstration fishing trials of a new trawl design at Portmagee and Caherciveen using the *Naomh Simon*. A Swedish skipper taught the method of pair pelagic trawling to skippers at Dunmore East and Killybegs and an Icelandic expert was brought to various ports to encourage the practice of trawling.

Education was reckoned an important aspect of BIM's brief. The difficulty finding skilled labour to crew vessels associated with BIM's activities highlighted the shortage of suitable training. In its first yearof existence, a course in marine engine maintenance was organised at Howth in association with the Department of Education and the Dublin Vocational Education Committee. The annual report for 1953 announced that a trainee system had been established for skippers. In 1954 inducements were offered to deck hands to attend nautical college.

Early in its history, BIM began to assemble a library of instructive films. A film projector had been acquired and would be moved about the country, thus introducing a service which would greatly expand in the years ahead. Other promotional work consisted of having a stand at the Spring and Horse Shows on the Royal Dublin Society's premises. Various campaigns, like the "Eat more fish" one encouraged the public to do just that. A "Fish Scholarship" was devised in conjunction with the Irish Country Women's Association. BIM also established itself as a fixture at international trade and food fairs. Fish cookery competitions were organised through secondary schools and BIM supplied materials to the Vocational Education Committees to familiarise the public with fish and to popularise it as a food. Some of these early campaigns would be run annually throughout the history of the agency.

If the first decade of BIM's existence demonstrated anything, it was the industry's insatiable demand for funds and the ease with which they were disbursed. In addition to those provided by its parent government department, it was perpetually on the lookout for other sources. The Gaeltacht vessels were financed from funding which did not exist when BIM was established. In January 1959, a fishing industry development fund was initiated by the Minister of Lands. It would be funded by "voluntary" donations by fish importers; the term voluntary is curious because there were prescribed levels of contribution depending on the nature of the import;[26] funds generated in this way were used to provide services such as transport of fish landings to processing factories, as recommended by the Fishing Industry Development Committee, made up of representatives of the catchers, retailers and BIM. In March 1960 it had a balance of stg£15,849; the following year that increased to stg£30,332 and in March 1962 reached stg£40,952.

Dealings between BIM and the industry were evolving into a mutual survival mechanism. It was in the interests of the agency to grow the industry, and the easiest way to do that was to provide it with grants and loans or to facilitate their transfer from other state agencies. During its early years, funding originated in the exchequer and, additionally, later, in the European Commission which thus redeemed promises to reward Ireland's acceptance of the Common Fisheries Policy.

A view from the outside and limited reform

While the Association ceaselessly nagged at BIM about operational detail, there appears to have been little criticism of its developmental strategy. External views were sought and they too were largely supportive.

J.S. McArthur, a consultant from the Food and Agriculture Organisation of the United Nations (FAO), spent nine weeks in Ireland in 1959 making enquiries about the way sea fisheries were organised. His report described an industry gradually taking shape; he anticipated aspects of its future and identified issues which would subsequently assume much greater significance. The fleet in 1957, the most recent account available to him, consisted of 2,211 boats, 605 with engines, 426 propelled by sail and 1,180 row boats; 135 were greater than 25 GT and most had been constructed in BIM and its predecessor's yards. Full time fishermen numbered 1,613, the remaining 4,500 were engaged part time. The fleet fell neatly into three divisions: curraghs and row boats harvested shellfish, close inshore. A fleet of vessels 7.4 - 12.2 m oal also targeted shellfish; some of these boats worked part time, others were fully occupied. The fleet of large motorised vessels 15.2 - 21.3 m oal were multi-purpose, using seines, trawls, drift nets and crustacean traps.

The multispecies nature of the fisheries was a characteristic of the industry at the time; a vessel moving from one to another seasonally. McArthur believed there was a role for larger vessels specialising in single species. He may have had mackerel in mind. At the time of his visit, Irish boats landed less than 2,000 t annually from drifting gill nets. The French took twenty times as much. However, Ireland's most valuable finfish species was Atlantic salmon which dwarfed all others.

McArthur reviewed the resources available to the industry. He expressed no concerns about the ability of stocks to sustain increasing exploitation although he observed that whiting, whose landings had doubled between 1954 and 1958, was "not so capable of (further) expansion as other fisheries". Landings of whiting were either sold fresh to the home market or exported, frozen, to the UK where they were incorporated into cat food. He also advised that scallop stocks should be assessed prior to exploitation. And he expressed the view, which was then generally accepted, though not acted on in Ireland, that

...Any programme for the long term development of the fishing industry must be based on a sound knowledge of the resource from a biological point of view...

McArthur anticipated the introduction of detailed reporting of landings and effort in the future. Although the landing of some few species was reported at the time, the list was far from exhaustive. During the 1950s, a maximum of 36 "varieties" (mainly species) was recorded; in the 1990s, the number exceeded 100. Fishing effort (the amount of time devoted to the practice) was not recorded until the European Community insisted on it twenty years later.

In some countries fishermen are obliged under licensing arrangements to submit daily, weekly or monthly reports but such a method is hardly suited to Irish conditions. Similarly, it can be said that any highly developed system, such as the collection of sales invoices covering every first sale of fish would prove difficult under present day Irish conditions though this method, if complete co-operation is obtained, does provide the best statistics of landings and values.

The headings under which McArthur evaluated the industry were becoming commonplace in BIM's approach to its organisation: expanding production, providing infrastructure and facilities, processing and distributing landings, expanding markets, creating demand. His own background as chairman of the Fisheries Prices Support Board and chief administrator of the Fishermen's Indemnity Plan of Canada lent authority to his observations on the way the industry was financed.

He was critical of the hire-purchase system of providing boats. As they stood at the time of his visit, a 5% down payment was required and a grant of 15% was contributed by the state. The life of a vessel, hence the period of repayment, was ten years and interest was charged at 4%. Repayments were deducted from the sales and they amounted to at least 25% of landings value. There were special arrangements for vessels in Gaeltacht areas.

Remuneration of crew was on the basis of pre-determined shares in the landings rather than by wage, a system which was believed to incentivise production. The allocation of shares among crew could vary from one vessel to another but a proportion of the total was always reserved for the vessel and used to fund its mortgage and upkeep. Because BIM boats had high mortgages, McArthur argued, there was less to distribute among the crew and this adversely affected morale and productivity.

The current arrangements were supported by his finding that the interdependence of state and industry appeared to be unavoidable, and McArthur believed that government should maintain the incomes of fishermen at levels compatible with occupations requiring similar levels of skill. One way of doing this was by government purchasing landings or providing price supports, and McArthur had observations on the stabilization of the market which were salient:

There will always be heavy pressure from fishermen for fixed minimum price guarantees but every effort should be made to avoid such commitments, not only because they force government intervention in situations where high production at relatively low prices would still give the fishermen very adequate returns, but also because they deny to consumers (who as taxpayers pay for the support measures) the benefit of a fortuitous increase in support by taking fish off the market. There is also a tendency for minimum prices to become maximum prices in the market. Another weakness of minimum prices, if relatively high, is that they reduce the incentive for the institution of technological improvements and cost reducing practices.

Although such a possibility was probably unthinkable at the time, some 25 years later intervention price supports would contribute to over-fishing and stock depletion by encouraging fishing for compensatory payments which, in many cases, meant bringing fish ashore to be weighed, sprayed with an indelible dye, then back to sea to be dumped.

An alternative compensatory system, McArthur proposed, was through "deficiency payments", compensatory amounts calculated on the weight of landings and returned to the fishing community in the form of infrastructure (such as wharves) on whose construction the fishermen themselves might be employed.

While he accepted the prevailing ethos that the state had a role in fisheries development, McArthur favoured a more enterprising approach and believed that BIM should divest itself of commercial activities, selling or leasing these on to the private sector, which was what happened next.

Reorganisation of BIM was provided for in the 1962 white paper on Sea Fisheries.[27] From 1 January 1963 a new chief executive officer combined that role with chairman – the first full time chairman - and BIM began to withdraw from commercial operations and restructure internally, a process which would take two years to complete. The white paper gave assurance that the government would further support BIM and it also contained proposals to correct perceived deficiencies in BIM's operations.

The most important announcement within the white paper concerned new hire-purchase funding arrangements. As from 1 April 1962, the repayment rate on a new vessel was extended by up to five years, to fifteen. The rate of interest repayable remained at 4% annually. There would be an incentive for hire-purchase loans to be cleared before the prescribed deadline. This would range up to 10% off the agreed price for loans cleared five years before they were due. The capital grant for new boats was increased from 15 to 25%. The cumulative effect of these new terms would facilitate the purchase of a vessel at little more than half of the sale price. New engines would be provided with an initial grant of 25% of their cost and second-hand vessels might be acquired on a down payment of 10%.

BIM boats would have a free inspection service and repair costs would be reasonable, but boats on hire-purchase would have to be maintained as skippers were advised to do.

Boat designs on offer were rationalised. Hitherto, they had been 7.9, 15.2 and 17.1 m oal. Few in the fleet currently exceeded 17.1 m. There was demand for larger vessels and a new design of 19.8 m oal, almost 50 GT, with a draft of 2.6 m and an engine of 169 kW would be provided at a cost of stg£25,000. Preparations were in hand to build this model at the Killybegs yard.

A new shellfish boat design was also proposed. The model would be 9.8 m, with a draft of 1 m equipped with an engine of 24 kW. Each of these vessels would cost stg£3,000 new and would be supplied with a mechanised pot hauler. Crewed by three men, the boat could accommodate 70 lobster pots.

The white paper announced that preparations for major fishery harbours at Killybegs, Castletownbere, Passage East, Howth and Galway were underway. The investment would require stg£1.25 million and would take ten years to complete. Ice would be supplied at each harbour as would facilities for the repair of boats, engines and fish processing. Contributions towards the maintenance of associated services would be levied on fish landings to the harbours and it was anticipated that the infrastructure would attract vessels to berth there.

Minimum prices would be paid for economic quantities of fish landed for conversion to fishmeal by Irish fishing boats registered in the state. Imports of fresh or frozen fish would be allowed under licence. However, importers would be expected to make "voluntary" contributions to the Fishing Industry Development Fund and such monies would be distributed by the minister after consultation with representatives of importers, retailers and fishermen. Suggested uses for these funds included promotional activities to increase fish consumption, and the provision of supports for the cost of ice and transport for landings to a fishmeal factory.

BIM would work with the government department responsible for fisheries, and the state's research vessel, the *Cú Feasa*, would devote itself to a programme of conserving and "rationally exploiting" fish stocks, locating new fishing grounds and improving fishing methods. Any new information would be conveyed to the Irish industry as soon as possible, a pledge of particular relevance to the conduct of herring fishing, where the research vessel was tasked with locating the shoals and reporting their position to the fleet as exploitation took place. Thus, a state agency became a contributor to profit generation by private enterprise.

The agency's next decade saw it grow in influence to the point where it participated in every decision on marine fisheries. Before we explore that, an evaluation of Ireland's introduction to the second great institutional change affecting marine fisheries in the last century, the European Economic Community (EEC), is relevant.

Chapter 7

THE ELABORATE REGIME THAT FAILED

As formalisation of the Geneva Convention (1958) on the boundaries of national marine territories was nearing solution another debate with wide implications was gathering pace. Preparations had commenced to assemble the European Economic Community (EEC) whose embryonic fisheries policy was anticipated, if not actually conceived, in the 1957 Treaty of Rome, article of 38 of which stated:

The function and development of the Common Market in respect to agricultural (fish) products shall be accompanied by the establishment of a common agricultural (fisheries) policy among member states...

The first European community policy to be developed was for agriculture (the Common Agriculture Policy, CAP). When, 25 years later, the Common Fisheries Policy, CFP struggled into existence it was regarded as the more sophisticated of the two. At the time of its birth, each of two policies had to confront its own suite of demons. The CAP was faced with the necessity of rationalising over-production, draining wine lakes and cutting butter mountains down to reasonable size. The CFP was challenged with the need to balance fishing power with available resources. The CFP would be developed over a thirteen year period after 1970.[1]

The first six member states of the EEC had similar fishing limits to Ireland's which, during the 1950s, obliged them to take almost 90% of their landings in international waters. One member state, Luxembourg, was land locked and had no interest in sea fisheries. Following the liberalisation of trade within the EEC area, France and Italy found their fishing industries were economically inefficient and uncompetitive compared with those of Germany, Belgium and the Netherlands. They applied to the Commission (the civil service of the EEC) for assistance to modernise their fleets to improve their financial position.

The Commission took the opportunity to produce a long paper titled "Basic principles for a common fisheries policy" which examined structural matters (fishing fleets), markets, external trade – three of the four policy areas which would subsequently make up the CFP – and social issues. The fourth policy area, conservation, was not part of deliberations until much later.

Before negotiations for accession to the community by any other state could begin, it was vital that a common stance be adopted by the six existing members in order to avoid a situation in which newcomers could destabilise the policy of those that had already joined. Adopting a common front on fisheries matters meant that the agreed fundamentals would have to be accepted by states which acquired membership later; this is in accordance with the accession policy or *acquis communitaire* (community accomplishments) which insists that new members must accept legislation already in force.[2]

The policy agreed by the Six consisted of two separate bodies of legislation. The first, on structural matters, aimed to promote the acquisition of more efficient vessels, technologies and gear. Its purpose was to facilitate the capture of larger quantities of fish and thus reduce imports. The exercise would be funded by subsidy and it led, in due course, to over-capacity - "too many vessels chasing too few fish". When the structural policy was agreed however, there was little scientific expertise within the Commission to warn against over-fishing. However, even if there had been, it was unlikely to have been taken seriously: the structural policy delivered many benefits to the fleet and ancillary industries such as boat-building. Eventually, damage to important fish stocks obliged the Commission to alter its structural policies.[3] At that point a great deal of destruction, much of it possibly irreversible, had taken place.

The second body of legislation agreed by the Six was the markets policy which established market standards, stabilised prices, avoided accumulation of surplus, supported producers' incomes and protected consumers' interests.

In 1969 Denmark, Norway, Ireland and the UK applied for membership of the EEC. All had major fishing resources and ambitions and all, to varying degrees, already shared fishing grounds. Together with the Six that had already joined, they would effectively control many of the rich fishing waters of the north east Atlantic.

In addition to the structural and market policy packages which were ratified in 1970, several other aspects of fisheries management were already in existence. Stemming from the provisions of the Treaty of Rome, exclusion from fishing on the basis of nationality within the Community was prohibited. Thus was born the concept of a "community pond" in which any fishermen from a member state could fish up to the beaches of any other. The concept, it was argued, rather unconvincingly, would benefit both nations that were "fish-rich" because they could exploit the waters of other states, as well as the "fish-poor" who had the obvious advantage of accessing the fishing grounds of the fish-rich. The nations applying for membership in 1969 were all fish-rich and all vehemently opposed the concept of a community pond. Such was their opposition that the Six who had already joined were obliged to sue for compromise which consisted of a ten year derogation from imposition of the "fishing up to the beaches" rule and that was, in due course, extended.

Norway found the negotiations unpalatable and withdrew its application for membership of the EEC.

The arrangements agreed upon allowed nations a variety of access based on traditional usage of the waters in question and they respected the arrangements reached at the 1958 Geneva Convention. The area covered by it would include only Community Atlantic and Baltic waters, not the Mediterranean. Once the maritime limits of the EEZ had been agreed, the need for a CFP became obvious.

The CFP was based on the principle of sharing out available fish in pre-determined quantities among member states, based on their "track record" of landings volumes. In theory, fishing was a traditional activity and, in future, the demands of national member states for fish would be kept in balance with one another, a phenomenon labelled "relative stability". Fishery management theory, at least, was straightforward even if its practice was almost unworkable. The most important shared food fish species (like cod, herring, mackerel) would be scientifically assessed, the tonnage that could be safely harvested without jeopardising the reproductive capacity of the remainder would be estimated as a "total allowable catch" (TAC) which would be subdivided into harvestable amounts (quotas) to each participating nation on the basis of what it had landed over a pre-determined period preceding the introduction of the system.

When the quotas, the percentage shares of a fish stock due to each nation, were agreed, they would be applied, like a key, to the TAC – the harvestable tonnage - of a species proposed in any year. But, once the quota key was decided it had to be adhered to. Even a slight adjustment to any one of them would open a Pandora's box of demands to adjust others. Ireland's industry has still not accepted those allocated to her and has, for years, with perhaps diminishing hope, if not energy, attempted to re-open the debate.

It is difficult to imagine a more practical system of allocating catches than that based on catch history. The CFP could never accommodate ambitions for a larger industry because every nation had them. Fishing is dependent on renewable but finite resources. It is a traditional activity, unusual among modern occupations because it is based on hunting and gathering. For that reason it also stirs atavistic instincts concerned with surrendering birthright, which are more fundamental to a nation than the emotions associated with displacing one contemporary manufacturing or service industry by another. It would have been expected that all nations would want to enlarge their fish landings; it might also have been anticipated that the technology to do this would develop. What might not have been so fully appreciated at the time was that the market for fish, and its health benefits, would vastly increase demand for marine produce in the future.

European waters were, up to 1977, fished by the fleets of the old Soviet Union which scoured the world's oceans in pursuit of cheap protein. Within the Community itself, the most powerful fishing nation was Denmark which had a fleet of 500-600 trawlers to keep in productive employment. Along with Ireland, Denmark was the most truculent of the nations acceding. In 1982 Captain Kent Kirk, representing the Danish industry, publicly attempted to stake a claim inside the UK exclusive zone (within the territorial sea). Denmark stated that, in the absence of unanimous agreement, all member states were free to do what they wished. On the eve of the birth of the CFP, in December 1982, the taoiseach, C.J. Haughey, and the British prime minister, Mrs Margaret Thatcher, combined forces to sort out Danish intransigence. Poul Shlûter, Danish prime minister, was bluntly told to accept the consensus of the other nine or, if he did not agree to do so, he would be excluded from any future arrangements. He acquiesced.

Throughout the stressful years of negotiation and preparation, various initiatives were also pursued by fisheries interests in Ireland to promote their claims and salvage what they could from the encroachment of other states on what was regarded as national territory. The representative body for the industry at the time was the Irish Fishermen's Organisation (IFO) which cleaved to its insistence on a 50 nm national fisheries limit. Supporting its claim, Ireland in 1977 unilaterally attempted to prevent all non–national trawlers of greater than 33.5 m oal from breaching that boundary. The move was obviously in breach of the Treaty of Rome and Ireland was rapidly called to account at the European Court of Justice. The minister holding the fisheries brief at the time, Brian Lenihan (Fianna Fail, Dublin Co, west) shelved the claim.[4]

Did Ireland have any freedom of movement before the CFP was agreed? Critics of the CFP have since argued often and at length that we should not have accepted it. The question is whether Ireland, had she chosen to go it alone, would have been able to maintain an exclusive presence out to the limits of her share of the Community's EEZ? There are reasons to be sceptical.

In 1958, 1971 and 1974 Iceland, attempting to safeguard her cod stock and retain a larger proportion of landings for herself, extended her fishing limits and tried to exclude UK fleets. Three "cod wars" ensued. In the first, support for Britain, whose distant water fleet had been banished from Icelandic waters, came from France, Belgium, Denmark, Germany, the Netherlands and Spain, all of whom became part of the European Community. Iceland put up plucky resistance and shots were fired but any victory she claimed was principally due to international consensus on the extension of national marine boundaries.[5]

The enduring irritant for the Irish industry was the disproportionate proximity of Ireland to the resource, the extent of her sea area (share of the Community's EEZ) and the diminutive size of her quotas. The IFO complained that Ireland had 25% of the European EEZ while Britain claimed 60% of the EEZ waters surrounding it. These figures have often been repeated but they need to be updated. The University of British Colombia published actual areas of EEZ.[6] Ireland has approximately 11% of the EEZ of the north east Atlantic (shared with the other principal participants, Belgium, Denmark, France, Germany, Netherlands, Portugal, Spain and the UK). The UK has 2.5 times as much. Spain and France have large areas of EEZ too but much of theirs is in the Mediterranean, not covered by the CFP. However, in proportion to its EEZ, Ireland's allocation of quota is small (between 3.9 and 7.1% of whitefish and pelagic quotas combined in the period 1990-1998).

It could have been a lot worse. Traditional landings – established on "track record" - registered over a reference period (1973-1978) were the basis for future landing allocations to member states. Ireland, in that time, reported capturing 1.49% of landings for the principal species and that percentage should have been her future share. Foreign minister Garret Fitzgerald, at a fateful meeting in the Hague in 1976, convinced other states that Ireland should be allowed to double her landings as part of her national development programme.[7] The result was that Ireland was permitted to raise its national catch from 75,000 t in 1975 to 150,000 t in 1979. Ireland was the only member state granted this concession.

In addition to the requirement for agreement among those already within the EEC or about to join, pressure was applied by the impending accession to the community of Spain and Portugal which was due to take place in 1986 and would impose considerable additional fishing effort. Spain's fleet was the fourth most powerful in the world and its sea-going personnel 75% of those in the Community.

The CFP came into existence on 1 January 1983. Mandatory recording of landings in the European Communities' logbook on vessels greater than 12 m dates from 1981.[8]

The CFP has been a resounding failure. Persuading nations to accept the CFP was accomplished by promising their industries generous structural funds, thereby over-equipping fleets to compete with one another. Fish resources suffered. Of primary concern to the Commission were quota fish stocks, principally of demersal and pelagic species, which two or more states shared. Today these species account for some 60% of landings to Ireland.

Over the past nine years, the Commission has attempted to monitor an average of 95 fish stocks in the north east Atlantic. The state of almost 60% of them is unknown due to poor data and only one third of the remainder is "safe" in biological terms; two thirds are over-fished.[9]

There is still much resentment about the manner in which Ireland was "gulled into compromise" on its fisheries policy in the years leading up to the formulation of the CFP but much of that is inaccurate and misinformed. Entry to the EEC promised many rewards and advantages and fisheries was just one, albeit highly complex and contentious, brief to be sorted out. A distinguished Irish civil servant, Eamon Gallagher, who later gave sterling service to the Northern Ireland peace process, was its principal architect. Negotiating the fisheries brief, commonly regarded as the most tricky problem of all to be resolved, severely tested his considerable diplomatic skills.

And there were perceived benefits to fisheries arising from Ireland's membership of the EEC. After the declaration of the EEZ in 1977, Ireland was responsible for patrolling its share, amounting to 25% of EEC waters at the time, almost 0.5 million km² (136,000 square n miles). Between that date and Spain's accession in 1986 Ireland was entitled to arrest Spanish hake fishermen who had hitherto fished these waters for centuries, unless they were specifically licensed to fish there; for every one permitted to do so, an estimated one more fished illegally. Fines for first offences ranged from IR£12,000 to IR£100,000, plus confiscation of gear. Second offences merited higher fines and few Spaniards were willing to risk incurring these, although some did. In 1983 alone, naval activity claimed IR£1.5 million from Spanish vessels; arrests of this nationality occurred almost once a week.[10] The following year Spanish fishermen, who were said to be earning IR£200 million on fish from "Irish" waters were reportedly contributing to a pool of funding which was used to pay the fines of those who were convicted of fishing illegally. The penalties were reckoned to comfortably reimburse Irish naval overheads. In one incident in 1984, a Basque trawler, the *Sonia* from the port of Ondorroa, was sunk after a five hour pursuit by the naval vessel LE *Aisling* off the Saltee Islands in the course of which 600 rifle rounds and some machinegun fire were discharged.[11,12] The *Sonia* had allegedly tried to ram the naval vessel. It was the most serious confrontation between an Irish navy boat and an offending prospective – or any - member of the European Community. There was much controversy about the incident but Spain was also anxious to maintain good relations with EEC countries whose CFP she intended to enter in 1986. Later, in 1984, the Spanish navy opened fire on one of its own boats which was fishing in areas reserved for smaller inshore vessels,[13] something that would never have happened in Ireland.

Extending the EEZ boundary to 200 nm had the effect of excluding the Soviet fleet, then the world's mightiest. As the national fishing demand for a fifty-mile exclusive fishing zone, later dropped, raged, Russian and Bulgarian vessels were being arrested off the Irish coast with some frequency.[14]

The CFP was not all bad news for Irish fishermen, although its novelties might now be subsumed by other concerns. One benefit was the opening of markets for species not consumed in Ireland, particularly after the accession of Spain, and we will look at one of these now.

How "Spanish fish" influenced the development of Ireland's industry

As the accession of Spain to the Community approached, the accompanying sense of dread deepened. Of all member nations, Spain was the most demanding and voracious where fish stocks were concerned. On the other hand, there was also an expectation that the market for fish would grow. Anecdotes like the one about the Spanish skipper who landed IR£7,000 worth of fish into Castletownbere which he said would have fetched IR£17,000 in Spain, promised financial advantage to the Irish industry.[15]

Dan Griffin – described as a veteran Co Cork fisherman, and a member of the board of BIM for three years - advised the industry to study Spanish fishing methods. "The fish are in the deep water these days..."[16] which could be regarded as a euphemism for their having been fished out in shallower conditions.

Around the time that Eiranova, a subsidiary company within the multinational Pescanova group, currently the sixth largest fishing company in the world, established a base at Dinish Island, Castletownbere in 1981, the Irish fleet was enlarging its list of target species. Three whitefish species, hake, angler (also known as monk) and megrim, rose to prominence because they commanded high prices on the Spanish market. They became the new "prime species", a halo that had hovered over others at various times in the past. Hake had featured in the Irish diet for as long as known history was available and it had been an important magnet drawing Spaniards towards the Irish coast in earlier times but the other two appeared on the commercial scene quite suddenly.[17] Megrim, also known as "white sole", is abundant in trawl catches, even close inshore, yet I have rarely seen it on retail sale in Ireland. In Vigo it was said to fetch five times the price it would here. The case history I will use to represent the course of commercialisation of the three belongs to angler (monk) fish.

In Ireland's waters, there are two species of angler fish, similar in outward appearance. They consist mostly of head with a wide toothy mouth at the centre of whose upper lip is placed an *illicium* or "fishing rod" which dangles a tiny piece of flesh, serving as a lure, over the gaping jaws. Curious small fish approaching the lure are snapped up. The musculature serving the jaws is powerful and the "face cheeks" can be harvested as meat. Cheek meat is recovered in Spain thought not, to my knowledge, in Ireland. Behind the head the body tapers rapidly to the tail and the trunk also yields edible flesh. The high price of the fish and the low percentage of meat to total weight, has made angler our most costly species of demersal finfish, running close to the price of crustacean (*Nephrops*) flesh, which it has been known to substitute. Yet, before joining the EEC, angler fish was discarded along with other undersized and trash fish from Irish trawls. Perhaps it was something to do with its unprepossessing appearance, a reaction that may not be peculiar to this side of the Atlantic. Taras Grescoe made similar observations on the North American market's similar response to their monkfish which changed only at the end of the 1970s when heavy depletion of other species obliged fishermen to examine the contents of their trawls more discriminatingly.[18]

Angler fish was listed in the Irish official statistics for the first time in 1977 when 56 t, worth stg£26,000, were recorded; less than a decade later its tonnage had risen 35-fold and its value to IR£2.9 million. Spanish demand was in large measure responsible for heavy harvests of angler and associated species in the seas around Ireland which, in turn, encouraged the Irish industry to increase its investment in whitefish vessels.[19]

A distinguishing characteristic of the Spanish fishing industry, in addition to its large size, was its international ramification. Spain sought fish in Community waters but she harvested wherever she could and bought wherever fish were sold worldwide. When, almost thirty years after creating an appetite for the species in Ireland, Spain began to import cheaper angler species from Africa and South America, the Irish market collapsed, demonstrating its dependence on Spanish demand.

In the mid-1980s, surveys of whitefish species off the west coast of Ireland were essentially exploratory: to locate resources but not to assess how they might withstand exploitation. Initially, when the TAC and quota system was set up, Ireland was awarded 2,750 t of angler; Irish fishermen did not distinguish between the two angler species, a fact which complicated assessment of their populations. As with many another species, exploitation reduced numbers of angler fish. In the waters west of Scotland, the TAC rose to peak in 1999 but then declined to half that value. To the south, in the Celtic Seas, the TAC reached its maximum value in 1988 and subsequently declined to between 50 and 80% of that.

Being a fairly sessile species, moving short distances, essentially sitting on the sea bed waiting for its food to swim past, angler is vulnerable to a wide variety of fishing gears. Initially, trawling was the method used to capture it but in the mid-1980s tangle nets targeted the species, and smaller vessels using this static gear became increasingly involved. The range of mobile gears to which angler was susceptible included scallop dredges. It did not take very long for over-fishing to become a fact.

Early in the 1990s Ireland's share of the angler TAC – her quota – was insufficient to support fleet expectations and the fishing effort which had expanded to capture it. For a time Ireland traded some of its quota allocations for other species with member states, mainly France, which transferred her angler quota to Ireland in exchange. In 1993 the IFO, the principal representative organisation of the industry at the time, demanded a quota of 3,000 t for Ireland, 17% more than was her due. It was an argument that would frequently recur in the future and would become known as "socio-economic justification". The industry "needed" more fish to survive, as though fish stocks replicated and grew to order. They do not and the result has been a declining resource base, particularly afflicting stocks of whitefish species. Running an industry on quota swops was far from ideal and in 1998 a shortfall of 1,000 t was needed to balance the books.

One fishermen's representative body, the Irish South and West Fishermen's Organisation (IS&WFO), split from the IFO and set up as a separate representative body in the 1990s on the back of demands for higher angler fish quotas; the IS&WFO would subsequently achieve considerable national prominence.

In order to eke out the shortfall in supply, in 1992, a weekly individual boat quota, allowing a vessel to land only so many tonnes of angler fish, was devised. The industry preferred a monthly one and a series of monthly and bi-monthly Restriction on Fishing orders was obligingly issued by the Department. All co-operatives, the IFO, and other industry bodies agreed to monitor landings of angler fish. Their initiative did not count for much, and in 2006 the Minister, Noel Dempsey, informed the European Commission that he believed the actual landings had been eight times the quota. The announcement was greeted with dismay by the industry which does not appear to have offered any more detailed explanation of what went wrong. In May 2006 Control of Landings Regulations replaced the Restriction on Fishing orders.

Meanwhile, tangle netting which had been introduced by small vessels in the mid-1980s was adopted by much larger ones. It was also a method used by other Community member states. It is difficult to overestimate the consequences of this gear to which angler fish is particularly vulnerable. The netting lies almost in bundles rather than tightly stretched, as would gill netting, and the smaller fish on which angler prey, hide among the meshes. Sooner or later angler fish go in search of them and become enmeshed. Studies have shown that tangle netting which is abandoned or lost can continue operating for up to seven months.

Throughout the 1990s stories circulated of Spanish vessels, fishing out of Vigo, setting literally miles of tangle nets (known as *rascos*) on their way to trawl or long-line off the west coast of Ireland. On their return home, which might be several weeks later, they retrieved the nets, much of whose contents would be too decayed to sell; there were however sufficient freshly captured angler fish to make the exercise worthwhile.

In 2005, estimates, based on research findings, suggested there were almost 9,000 km of static nets soaking off the west coast of Ireland, deposited there by Spanish vessels, and an operation to recover lost gear (the DeepNet survey) confirmed the problem was serious. Ireland complained but probably with little conviction in view of her own efforts to regulate this fishery. To this day we do not know how much tangle netting was soaking in Irish waters in any year. In 1997 there were demands to have it stopped altogether.

The chief executive officer of the IS&WPO (which was set up to demand higher angler quota), Jason Whooley, who took on that role in 2007, blamed the Department for what happened, comparing it to a fire brigade which never anticipates a blaze. The industry which had agreed to regulate landings and failed to do so stayed *schtum*.[20]

Arrangements to locate Eiranova in Castletownbere were finalised at the highest levels, by the Spanish premier, Felipe Gonzales, and taoiseach Jack Lynch. The firm was initially welcomed by the industry in west Cork because it was seen as another buyer and processor for landings in an isolated port enjoying poor infrastructure and communications. Attitudes changed when the desired levels of processing failed to materialise and, instead, freshly landed fish were loaded directly onto lorries and driven to Spain.[21] Eiranova also had six trawlers which joined the Irish register under the Mercantile Marine Act of 1955; crewed by Spanish personnel who were reluctant to share their fishing secrets with Irish fishermen. The initial promise of jobs for Irish people resulting from the introduction of Eiranova to Ireland was not fulfilled. Being registered as Irish vessels, the Eiranova ones were entitled to fly the Irish flag, known by antagonists in the industry as a flag of convenience.[22] On the contrary, using the flag of a member state was in keeping with the thrust of EEC policy which had the ambition of offering opportunities to all citizens of Community nations, without discrimination.

Chapter 8

GOOD NEWS AND SUBSIDIES

The early 1960s witnessed a number of changes at BIM. The appointment of an executive chairman, Brendan O'Kelly, an energetic organiser, put a face on an agency that had, hitherto, presented an anonymous exterior to the world. O'Kelly used the chairman's foreword to introduce the annual report, thus becoming the bearer of good tidings – the annual reports emphasised what was positive and omitted mention of what was inconvenient.

O'Kelly's arrival saw off any commercial pretensions in the reorganised agency whose divisional and sectional titles included "development" or "service" so as to leave no doubt it was not in the business of generating profit. In 1962 BIM's repertoire of activities, which it would retain in one form or another for fifty years, were: a *Market growth division* containing two embryonic branches dealing with home and export markets. It also had the Fish trading section which was being phased out and Ice production, a core activity which we will examine at greater length. *Product development* was about Fish processing. *Advisory services* looked after training and education and also included Experimental and Exploratory fishing and Mussel Transplantation and Purification. A *Technical services division* administered the boatyards and Fleet inspection while an *Operations division* managed the Marine Credit Plan which provided loans to the fishing community. Finally, *Other operations* embraced Gear trading and Fishing operations, managing the small fleets operated by the agency.

But commercial activities had not been totally buried and another division of this kind would mushroom out of the boatyards at the height of their productivity in the future. For the remainder of the 1960s, commercial functions were sloughed off to the private sector and those service functions which remained within BIM assumed greater importance. Divorcing the agency from profit-making had unfortunate long term consequences: policy became a matter of obtaining and dispensing funds without consideration for the consequences which would prove detrimental to the industry over the longer term, and, eventually, to BIM itself.

The various sub-divisions which provided services whose products were more tangible than instruction or information were expected to break even, pay for themselves. Until the end of the 1960s, the Development and Trading Account ran at a loss ranging between 7% and 20% of its expenditure. Among the sub-divisions contained within it, only gear trading – essentially a retailing operation - appears to have made a small profit on occasion. Virtually all of the other departmental areas registered losses year after year.

Fleet-building was accounted for as two operations. Whether the boatyards made a loss or a profit depended on the price realised for their products falling short of or exceeding the investment made in them. The disposal of those products to the fishing community through hire-purchase presented problems of a much greater dimension.

1963 completed the transition from the old to the new regime. The foreword to its annual report announced that policy was intended to achieve the targets set out in the government's second programme for economic expansion and stated

The development programme is being introduced at a time of increasing fishing effort not only for the major fish producing countries of the world but also by many smaller nations which are looking to the sea as a source of protein food.

This particular quotation – or something very like it – is still in vogue when justifying expenditure on sea fisheries development and I have heard it trotted out in almost identical form recently in the context of the latest reform of the CFP. It was not the only reference to the global possibilities for sales of marine produce. BIM's 2003 annual report reminded readers that a European Union of 25 states had 450 million mouths of feed – emphasising the demand for fish and, *inter-alia*, the justification for its own existence - and the annual report for 2010 stated that the Food and Agriculture Organisation of the United Nations estimated that an additional 40 million t of seafood would be required annually by 2030. Ireland's contribution would be by adding value "to our seafood offering". At that point there was not a lot to be gained by claiming we could increase fish landings any further; they had peaked in1995 and were on a downward trend.[1]

BIM in the 1960s identified the obstacles to wealth generation confronting the industry as insufficient processing and marketing; there was no mention of a shortage of the raw material, fish, nor any intimation of a threat to stocks by over-exploiting them. Intensified harvesting of fish stocks was required to ensure expansion of landings, that was all.

The 1963 annual report also contained, for the first time, some census material which hitherto had belonged in the reports of the government department responsible for the industry: the number of vessels and manpower in the fleet (note, the entire national fleet, not the fraction generated or merely managed by BIM). It was not an unreasonable approach in view of the dominance of the industry by the agency at that stage, but its publication by BIM announced the agency was staking a claim to wider territory – the compilation of statistics on the industry had always been a responsibility of the Department. Also included in the report were the quantity and value of fish landings, imports and exports.

The activities of the agency expanded considerably during the 1960s. Two of the principal ones, the growth of the fishing fleet and the search for stocks to supply it will be dealt with separately. Associated back-up programmes of training and education were enlarged to launch and perfect additional skills for virtually every aspect of a complex industry. Processing landings to meet the demands of a fast food market created exports and there were strenuous attempts to increase fish consumption at home.

During a period of fleet expansion, it was hardly surprising that landings records were frequently broken. In 1964 the volume of wet fish landed rose by 15% over the year before while all of the main categories (pelagic, demersal and shellfish) registered increases in value. The following year there was a further rise of 20% in weight of wet fish. Between 1966 and 1967 the total volume of wet fish rose by 10,000 to 42,000 t, an increase of 24%. Landings of wet fish in 1968 were the highest in volume and value since records began. A major contributor was a 42% rise in value of shellfish. This was becoming a feature of the commentary in the later 1960s. A particularly good year's performance by the industry could be attributed to pelagic or demersal or shellfish or to a species among them which would not necessarily register a similar level of performance the following year. A 13% increase in demersal landings in 1968 was due to the improved catches of haddock - a species which displays "spikes" of abundance - sole, brill and ray, while 45% in volume and 53% in value of pelagic landings were attributable to herring, which reached an all-time record in that year. The value of shellfish in 1968 increased by 21% over the previous year. The main contributory species were lobsters, oysters, mussels, *Nephrops* and edible crab; crab landings rose by 400%. The year after that the value of sea fish landings again reached an all-time record, exceeding those of the previous year by 26%. Such figures could well be explained, had there been any curiosity about them, by heavy fishing effort being directed towards a species or group of species in a particular year, towards others the next. Likewise the landings to particular ports displayed considerable growth. In one year landings to Greencastle (Co Donegal) rose by 70%.

If landings were increasing then exports followed. During the 1960s they rose threefold. Imports also rose, by approximately half the value of exports, throughout. Export destinations were France, Britain, Northern Ireland, West Germany and the Netherlands; Spain would feature prominently on this list after her accession to the EEC in 1986. And there were notable successes in opening export markets in Australia and Scandinavia.

The practice became established of exporting fresh landings, in pristine condition, from inshore waters, to the UK in containers. In 1967 the total exports of fresh demersal fish exceeded stg£100,000, an increase of 100% over the previous year. Exports of stg£2.68 million in 1968 were an increase of 16% over those of the previous year. In greater detail they broke into exports worth stg£1.8 million of fresh and processed fish (always excluding salmon which was regarded as a fresh water species and dealt with in the statistical accounts of other agencies) consisting mainly of shellfish and herring. The value of shellfish exports rose by stg£280,000 over those of 1967 to more than stg£1 million. In 1969 the value of exports rose once more, by 33% to stg£3.6 million. Exports of fresh and processed fish were worth stg£2.3 million, an increase of 29% over the previous year. Shellfish exports rose by 24% to stg£1.26 million. Herring exports (fresh and processed) increased by 52% from stg£577,000 to £877,000, with processed herring accounting for stg£583,000 of the total.

Efforts to increase home consumption of fish intensified during the 1960s. A special year to celebrate was 1967 when *per capita* consumption rose by 11% while imports of fish and fishery products declined by 4%. Sales efforts to persuade the public to eat fish included radio commercials, a housewife's recipe and information service and sales drives in retail stores. Cookery advice was extended to the hospitality trade and institutional cooking. School cookery competitions and promotions organised with the Irish Country Women's Association and defence forces became fixtures in BIM's calendar. Advice was even made available on the design of retail outlets.

In order to ascertain quality standards it was proposed in the mid-1960s that exporters of fish be licensed. BIM set up a product research laboratory. Agency personnel worked with staff of the Department responsible for fisheries and various state agencies to achieve these ends. The agency also consulted with the planning authorities to anticipate industry infrastructure in coastal areas. The point was close, if it had not already been reached, where BIM would be referred to on any and everything to do with the sea.

Before the 1960s concluded, BIM announced diversification in its funding arrangements. In future grants would be made available under four headings, for New fishing vessels, Capital improvements, Incentive and Investment.

The early 1970s proffered new opportunities in an atmosphere of considerable unease as the prospect of Ireland's entry to the EEC approached. In Brussels BIM personnel secured positions on an Advisory Committee of Fisheries and three working groups of the joint social committee for education and training. Application to the European Social Fund secured finance for its training programmes while the European Agricultural Guidance and Guarantee Fund (FEOGA) provided funding for industrial plant and vessels. In the future, until 2005 when the taps were switched off, those and other European Community monies would supply up to half of BIM's annual income.[2]

BIM would also become involved in the implementation of European Community policy, and would play a central role in the formation of producer organisations. In 1971 it became clear that EEC policy would favour certain nations – Belgium, Germany, France, the Netherlands and Britain all enjoying traditional fishing rights within designated parts of the six –

twelve nm band around the Irish coast through the London agreement. Potential new arrangements were concentrating minds and BIM's policy was now expressed through four divisions: Fisheries development, Market development and Investment development; the fourth, Commercial boatbuilding operated under a separate management structure.

It will be remembered that in 1963, BIM renounced all commercial activity but, in keeping with its practice, if not its renunciation of money-making, a venture that showed promise was re-conferred with commercial status. For several years the boatyards were so busy turning out vessels that they earned this accolade. In 1968 BIM's boatyards employed just fewer than 100 souls; seven years later they were staffed by more than 300. It was very unusual for BIM to give details of staff numbers in their annual reports and doing so was a statement of confidence by the organisation. But 1975 was a peak from which boatbuilding fell away; 1975 was a watershed in many respects. Following two further years in which employee numbers in the boatyards dwindled, that statistic was no longer reported.

BIM never appeared to rush into anything but it became gradually involved in ever more activities. Various lines of endeavour, apparently independent and innocent of one another's existence, at times coalesced and another divisional reshuffle saw the birth of a topical section. In the early 1970s "Resource development" began to survey locations for fish farms and, by the end of the decade, BIM was providing financial assistance to those who operated them.

While BIM was recklessly increasing fishing pressure on stocks and remorselessly looking for more to keep the fleet financially afloat, it could not be oblivious to the way things where heading. The Chairman's foreword states clearly the concern which was widespread at the time. This is from 1973:

The problems of fishery development are complex and varied. The fact that the production process is based on hunting, rather than farming the basic raw material, introduces elements of uncertainty which affect the distributive and processing sectors of the industry. On the other hand over-intensive fishing of any resource can deplete stocks to a level which will necessitate the introduction of conservation measures to preserve the species.

Current policy is aimed at expanding the catching sector of our fishing fleet so that it can more effectively fish the resources of our continental shelf. In implementing this policy regard must be had to the economics of any such operation particularly in those waters outside our fishery limits where the greatest competition exists. While we must aim at increasing volume landings it is also imperative that the financial income of fishermen should be maintained at a level which will ensure an adequate return on capital thereby encouraging a satisfactory level of investment in this high risk area. Increasing our presence in these waters is not merely a question of investment in larger boats. We must also, through technical education and training, develop a corps of fishermen who are capable of exploiting those waters more intensively than heretofore.

It is a very confused mixture of messages. The vulnerability of the resource was acknowledged but it would be fished just the same and BIM had no intention of making any concession towards stock management. The following year's foreword acknowledged that conservation measures for herring were required − BIM had, for several years been involved with intensifying fishing effort by extending the herring season,

Such measures now being taken at international level are designed to introduce more rational exploitation of stocks through the introduction of quota systems and it is hoped that, by better management of the resource, the serious situation which is now arising in this area of development will be averted in future.

BIM accepted no responsibility for what had befallen herring but it readily recognised a problem which had to be tackled. Perhaps BIM was anxious not to fall too far out of line with the EEC which had become its new source of funding!

Fishing activities by community trawler fleets must be tailored to the maximum sustainable yields...if there is to be rational development which will ensure prosperous living standards for all those engaged in European fisheries.

Only a revised CFP can restore long-term stability and confidence to European fisheries after the difficulties of recent years. This policy should be geared to rebuilding fish stocks and not just to conserving them at current levels. The proposed quota arrangements are by themselves inadequate for this purpose as they ignore the vital role of the coastal state in any resource management programme.

Over the years Ireland has maintained continual vigilance in stock conservation in its coastal waters. Given an exclusive fishery zone our industry will be in a position to manage stocks more effectively, by ensuring that investment and effort accurately relate to the resource levels of such a zone.

The last paragraph was blatantly untrue. Ireland had behaved and was behaving irresponsibly and BIM was the principal agent through which the damage was inflicted. Furthermore, there was no intention of changing behaviour. Arguments that fishery management would be more effectively undertaken in exclusive national waters are still expressed now and they still are, as they were then, simply a rationale for maintaining exclusivity.

In the early 1970s, any concern for the fate of depleting fish stocks was expressed outside Ireland. At home, landings records continued to be broken. In 1971 the volume of demersal landings rose by 35% and wetfish landings by 25%. Haddock landings doubled and those of cod and whiting by a quarter in a single year; pelagic landings fell by a quarter. The following year wetfish landings again increased by one quarter and achieved first sales of stg£5.3 million, the highest to date. In 1973 landings rose by a further 42%; cod and whiting both increased by more than 60% over the previous year.[3] This kind of expansion could not last indefinitely and, in 1974, things began to go wrong.

Initially there was not too much to worry about. The volume of wetfish landings fell from 85,700 to 84,600 t but its unit value rose. The year after, 1975, saw a decrease in wet fish landings to 75,800 t. Haddock, now on the falling arm of one of its spikes of abundance, collapsed from 2,400 to 1,000 t between 1974 and 1975. Among pelagic species, only mackerel showed an increase. Pelagic landings, particularly of mackerel, would greatly increase in the future, peaking in 1995. Shellfisheries, especially by boats targeting prawns, *Nephrops*, would not register their highest landings until 2001, but valuable demersal species were close to their maximum and non-sustainable output. Two additional sources of wet fish were now added to the annual table of landings in BIM's annual report.

The first came from the realisation - perhaps it reflected a new way or doing business or it might have been overlooked in previous years - that Irish boats landed a proportion of their catches abroad. These would, in future, be included in the national statistics. Foreign landings could also be registered as exports so they were of additional value in demonstrating the revenue-generating potential of the industry.

The second was the mention, for the first time, of salmon drift net catches in BIM's account of sea fish landings. It is a detail but a highly significant one. Atlantic salmon spends approximately one third of its life in marine waters and it was always, to some extent, harvested there. However, traditionally, the species was regarded as a freshwater one, a legislative legacy of the nineteenth century.[4] Several half-hearted attempts were made, to my knowledge, to alter the administrative system so that salmon would be treated as a marine species but, so far, they have not been successful and salmon has been treated by separate sections of the department responsible for fisheries generally and by associated state agencies as a species belonging in freshwater; at the present time sea fisheries and salmon are administered by separate ministries. It must be said there are practical reasons why this distinction should be maintained into the future.

A rise of 6% in tonnage of wet fish landed in 1976 represented a slight recovery from the previous year. Prices for shellfish increased by 41%. As was hitherto the case, there was something to celebrate. The 1977 foreword to BIM's annual report had the following:

In October 1976 recognition of the need to expand the Irish fishing industry was endorsed at the council of ministers' meeting in the Hague, when a commitment was given to a doubling of the Irish catch from a figure of 75,000 t to 150,000 t by 1979. The attainment of this objective is however dependent on the establishment of a common fisheries policy which will recognize the importance of conservation and the role of the coastal state in managing its own resources within an exclusive zone.

The enlarging EEC required a common fisheries policy and, in exchange for agreement on one, Ireland, the only aspirant nation to be permitted this concession, would be allowed to double its landings of quota species.

During the early 1970s Irish fishermen demanded an exclusive band of fifty nm width excluding all other community nationals. It proved to be one of the most intractable demands of Ireland's accession to the EEC. On 22 February 1978 the IFO dropped its demand after being told by Finn Gundelach, the EEC Agriculture Commissioner, that it was not possible to "dine *a la carte* at the Community table". Benefits from joining the EEC required sacrifices. If the fifty nm claim were dropped, he assured the industry, it would be given the best terms of any country in opportunities to expand its fishing industry. He pointed out that a stg£90 million proposal for the expansion of artisanal (inshore) fishing had been put to the council of ministers by the Commission and that Ireland could claim a "very significant portion" of the sum. He also promised that stg£30 million would be made available to upgrade Ireland's naval service, affording more effective protection of its resources.[5]

Things begin to fall apart

Despite occasional sparks of heavy landings, such as an increase of 77% in sole and 138% in sprat in 1976, the prevailing uncertainty about the consequences for the industry entering the EEC, allied with the increasing scarcity of many sea fish species, was a dark cloud spooking the industry. BIM continued to put a brave complexion on developments but a more defensive and subdued note crept into its utterances. For the first time, the 1977 annual report revealed the number of its employees, 403; a year later that had reduced to 375. (These numbers may not have been comprehensive however and it was more than a decade later that the breakdown of staff into salaried and contract numbers was provided.) When the boatyards were sold as going concerns, employee numbers fell to fewer than 160. BIM retained responsibility for its former boat-building employees and, when the independent yards began to shed staff, the agency paid redundancy monies to them.

It might have been coincidental but the colour appeared to have drained out of its jazzily illustrated annual reports.

In 1981 Brendan O'Kelly, who had done so much to inject personality into BIM's operations, retired. But, if BIM's staff complement was eroding, new organisation charts presented a robust line-up: *Market development* was subdivided to deal with home (with eight sections) and export (six) branches; *Fisheries development* (eight sections), *Investment development* (nine) and *Accounts*. The 1981 annual report promised

A sustained drive for greater productivity of the fleet will be a priority for the board

Hardly a reassuring statement for anyone concerned with the condition of the resources expected to generate the raw material. The following year saw the announcement that reorganisation of top management would place the main emphasis on marketing and export development.

Within divisions, there was also a subtle change of emphasis as the 1980s advanced. Grant assistance for pilot fish farming projects became a section within *Fisheries development*. By mid-decade there was a national mariculture grants scheme. BIM, in 1983, reminded its readers

The board is the state agency with primary responsibility for the overall development of the fishing industry

although its understanding of what constituted the industry was now very different from that envisaged in its founding legislation of more than thirty years before.

The number of salaried staff at BIM decreased further; in some years at the end of the 1980s, the declared number approached one hundred, although there are reasons to suppose this was not the full complement; for the first time in 2008 numbers of salaried and contract staff was provided. There were 97 salaried personnel and 78 ice plant operators and contractors. In the course of perusing 59 annual reports on the agency, I do not recall encountering that breakdown on any other occasion.

But, if staff numbers at BIM were slowly decreasing, the agency continued to gather responsibilities and influence. Virtually any committee involved with fisheries or the marine environment in Ireland had at least one member from BIM, and it supplied the secretariat for a number of policy-formulating committees drafting policy papers on the conduct of fisheries in Ireland.

In 1991 Dr T.K. Whitaker[6], an economist by training, and Ireland's best known civil servant, credited with, among many other achievements, mobilising opinion within the civil service and government in favour of a change of emphasis from agriculture towards industry and services which became the first programme for economic expansion, was asked to chair an advisory group to review the CFP. Whitaker was a polymath with extensive interests, of which salmon angling was one. Another of several celebrity chairman heading up BIM committees was Padraic White[7], formerly chairman of the

Industrial Development Authority. He performed prominent roles in a number of other public and private corporations, including the utility company Northern Ireland Water. The third to make a significant contribution was Dr Noel Cawley[8] who had previously served as chairman of the Irish Dairy Board and the Irish Horse Board. All were men of solid achievement and integrity whose record and history lent authority and gravitas to the reports issued under their names; their primary interests were not, however, in marine or fisheries matters. BIM, as secretariat, supplied them with specific details about the catching sector. With a history of forty plus years behind it, the agency, together with the more influential members of the industry, had become an "establishment". The various committees organised by it churned out reports which supported the *status quo*, perpetuating the demand for more investment and for greater self-governance by the catching sector. On the contrary, what was badly needed, was more restrictive management of exploitation by an over-capitalised industry.

Fleet formation had run into a brick wall. The EEC assumed responsibility for major funding and was unwilling to continue with the construction of more vessels. But there were chinks in that armour and BIM found other aspects of fleet enlargement to supply. The Community's preoccupation was with shared or "quota" species and BIM declared a drive to exploit species which were not regulated in this way. It offered assistance with improvements for safety measures for boats already part of the fleet and for "modernisation" of existing vessels. Both would bring additional fishing pressure to bear on the resource. BIM had been in existence for some forty years and the vessels built during its earlier years and by its predecessor, the ISFA, were unable to compete with more modern designs and needed to be replaced. And the European Community had the money; all BIM required was a pretext to gain access to it.

In 1985 prospects for the industry rose again. Falling inflation and oil costs, a rise in first sale prices of fish, improved export earnings and rising home consumption combined to brighten the outlook. There was "considerable scope for further growth".

As the 1980s drew to a close BIM's view of its priorities was summarised under six headings[9]: fishing vessels would be constructed and modernised, aquaculture developed, exploratory fishing voyages undertaken, certain vessels withdrawn from the fleet, port facilities would be provided and markets found for "under-fished species" whose existence was yet to be established. Against a background of resource depletion and over-fishing it was an agenda for a prospecting rather than a resource management agency. In 1989 BIM's annual report noted the hardly surprising news that whitefish were "scarce on inshore grounds"; the public was becoming reconciled to the fact that fish were disappearing and there was no public outcry. In which case, who would have responsibility for management? Within Ireland no agency or government department was prepared to stand up to BIM so those responsibilities gravitated to the European Commission which would be demonised for its efforts to impose regulation on an industry that was now proportionally too large for the stocks on which it depended.

A statement in 1993 recognised that overfishing had occurred and that stocks must be managed and conservation methods introduced. Who better to prepare grants for these activities than BIM? The following year a decommissioning scheme, to remove some of the excess fish catching capacity was announced. That too would be administered by BIM.

Towards the end of the decade a slight change in emphasis made its way into BIM's perception of its purpose:

To promote the sustainable development of the Irish seafish and aquaculture industry both at sea and ashore and the diversification of the coastal economy so as to enhance the employment, income and welfare of people in coastal regions and their contribution to the national economy.[10]

Sea fishing and aquaculture rated equal mention while "diversification of the coastal economy" would introduce another line of grant-dispensing possibilities. How much things had altered in the marine fisheries-based economy can be gleaned by glancing at the agency's published output: from the late 1960s into the next decade its reports carried a list of resource assessments – essentially descriptions of exploitable fisheries such as shellfish beds. In the late 1990s, the list of published reports switched to marketing opportunities. BIM had put a lot of work into seafood markets but it had also amassed a vast quantity of detail on those participating in the industry to whom large quantities of grants had being distributed. A proportion of the reports listed in the 1990s were marked for limited circulation: they had been privately commissioned. The fact that an agency with a fund of private data at its disposal would have restricted some of its market appraisals to a few individuals must have raised the possibility of a conflict of interest. I have no inkling that it did but the facts raise the question as to whether anyone cared at that stage, or whether those in the industry werc in such awc of BIM that thcy thought better of asking awkward questions.

There was another line of appeal for public support. Ireland was experiencing unprecedented economic growth which was scheduled to continue. It was unevenly distributed, geographically and by economic sector. The fishing industry had the ability to distribute wealth more evenly. In short, the nation needed BIM:

...I urge the key partners, both government and the private sector, to take the initiative to increase investment, employment and earnings in the industry. Encouragement can be taken from the whitefish initiative...[11]

The whitefish initiative referred to would prove an unmitigated disaster.

Until the late 1970s expenditure through BIM was to enlarge the fishing fleet and keep it solvent. A breakdown of the accounts between 1952 and 1976 attributed 19% (approximately) of its income to administration, 63% to the fleet, the balancing 14% was expended on processing and marketing.

An overview of investment through the agency for the period 1977 - 2010 inclusive indicates a shift in priorities. Administration accounted for 19% and aquaculture 16%. The largest absorber of funds remained the fleet and exploratory fishing (56%). "Others", including processing and marketing were funded with 10%. Of one fact there seemed to be little doubt: the world of marine harvests was changing, however brave a mask was put on the fact; expenditure had moved away from capture fisheries towards aquaculture.

The change in fortunes of fish stocks demanded that different projects were required to substitute for them and several were rapidly improvised. A plethora of imaginative grant-devouring suggestions made themselves obvious; their variety burgeoned between 1995 and 2000 when the EU supplied almost half of the total income of the agency. Several sources of European money – including a Peace and Reconciliation fund – contributed to a bewildering array of projects.

There was something in this last great burst of European grant-giving largesse which overflowed into the inshore fleet of small vessels. They would have an extended list of species to supply them too. Mussel and scallop would provide for their needs and grants were proffered to engage the community in developing the resource. Business and health and safety courses were tailored to enhance the wealth-generating abilities of small scale fleets.

Training and instruction became something of a growth industry. BIM ran two centres, in Greencastle and Castletownbere. In 1997 alone they had 1,900 trainees who ranged from school leavers to fishermen and fish farmers. In addition to the ubiquitous health and safety courses, students were schooled in a variety of sea-going skills and methods of handling a good quality product - "caring for the catch"; fish farmers were instructed in a number of techniques including scuba diving.

Processing was encouraged and there were grants for upgrading premises and an inquiry into establishing a tuna cannery. Exports were incentivised.

The market potential of turbot, halibut and lemon sole was examined. At home the funds were directed at launching a major inquiry into the fish eating habits of the Irish. There were promotions to integrate sea food and tourism. Schools' cookery competitions, one of the longest running endeavours of the agency, were now funded by the European Commission as was another campaign titled "make friends with fish". European funds supported various sea food festival promotions along with a drive to encourage the consumption of seafood in pubs.

European-sponsored publications rained down like confetti: they included, among many others, a practical guide to technical conservation measures for the fishing industry, recipe leaflets for the consumer and a millennium sea food calendar for everyone.

Rope mussel farms were susceptible to toxic algal blooms which damaged their product; the culture of Japanese oysters had been adversely affected by hot weather. A hardship fund was devised to compensate farmers of both. A bereavement support service was set up for the survivors of those lost at sea.

But old habits die hard and there was more than a residual desire to winkle out whatever fish remained. By the late 1990s the shallow inshore waters had yielded most of their secrets although the search for undiscovered mussel reefs would continue indefinitely. Deep waters were the last frontier and grants were available to plumb them. Boats were encouraged, with financial assistance, to fish for blue whiting, a new species for the Irish industry. Deep water long-lining was grant-aided too. Efforts to share the albacore tuna fishery with traditional exploiters of it, particularly Spain, manifested themselves in pair trawling trials. Inshore, a new approach to capturing whitefish using hook and line automatically "jigged" up and down was also sponsored by BIM. Additional species investigated for harvest included squat lobster bluefin tuna and brown squid.

Trials were undertaken with trawls which would, supposedly, fish in a more selective manner, doing less damage to non-target species; the experiments had started in the early years of the 1990s and continue today, more than twenty years later.

And the technology of harvesting more fin and shell fish proved irresistible. Grants were dispensed to develop artificial baits for shellfish traps. An improved hydraulic crab pot hauler was developed.

Scientists and environmentalists were becoming more loquacious and alarmed about the fate of the marine environment and fish stocks exposed to unrelenting pressure in spite of which the build-up of fishing effort continued. In 1995 the PESCA programme of funding was introduced by the EU and, as had become customary, BIM administered that too. PESCA, according to its publicity was, primarily concerned with the socio-economic diversification of coastal areas, in other words re-positioning personnel who could no longer be accommodated in capture fisheries. The existence of the PESCA funding initiative was an admission that a major problem with fish stocks existed. The stated purpose of the fund was to alleviate problems associated with fleet reduction (a recognition that the fleet was too large) and consequent stock depletion. Funding would provide a new beginning in marine tourism for those hitherto employed in sea fishing. One of the consequences of BIM's efforts in enlarging the fleet had been the removal of trophy-sized fish on which a recreational tourism industry largely depended. Ironically, the agency now became involved in dispensing grants for angling vessels to relieve some of the congestion among commercial fishermen.

Although the PESCA initiative was devised to find alternative occupations for those engaged in fishing and it was a frank recognition that the capture fish industry was over-subscribed, its arrival did little, if anything, to dampen the enthusiasm of BIM for further expansion of sea fishing effort. If such was the sentiment of the initiative however, the letter of its formulation allowed a more fluid interpretation, and PESCA co-existed with other contemporaneous funding mechanisms which set about intensifying the problem it was supposed to be relieving. Vessels were decommissioned to remove congestion on the fishing grounds while other grant schemes modernised fishing craft and improved safety on board. Both allowed boats to stay at sea for longer, effectively increasing fishing effort. Once more, what was required was a basic rethink of where the industry was heading.

The need to reduce fishing effort and get personnel involved doing something else prompted a number of diversionary endeavours such as lobster management by *V*-notching, marine tourism and aqua and mariculture projects. The list of candidate species considered for culture included abalone, sea urchins, Manila clams grown under mesh, Canadian and Icelandic charr, ornamental fish, marine algae, fresh water eel and perch, a wholly fresh water species.

Investment of IR£171 million (€217 million) in EU and exchequer funding would be pumped into the industry in the course of a new national development programme, 2000 - 2006, the largest proportion, €76 million, to fishery harbours, sea fisheries second with €57 million and aquaculture €32 million, third.

Refurbishing an already overgrown fleet was uppermost in the list of priorities. Grant assistance would be provided

...for the construction of new vessels, modernisation and provision of safety equipment on existing vessels and the purchase of second hand vessels...

The programme was intended to make the entire fishing enterprise more "sustainable", an aspirational term which had no demonstrable application or visible outcome.

The annual report for 2008 stated:

The marine environment, the sustainability of fish stocks and the protection of marine biodiversity were critical themes for fisheries development during 2008...

By the end of the twentieth century and into the new millennium, BIM had developed a schizophrenic persona. On the one hand the need to do something to allow stock recovery was obvious, on the other its compulsive commitment to funnelling money into the industry was an unbreakable habit. The fleet was too large and technology creep was making it more deadly every year. Clearly it could not go on indefinitely and the strain was beginning to tell. In 2003, BIM set out a vision of itself, tinged with desperation:

One of the key drivers of BIM's success in supporting the development of the seafood industry is its capacity to operate as a one stop shop with adequately funded integrated development programmes from the sea to the table. This uniquely and effectively promotes sustainable and viable development in the most remote and deprived coastal/rural areas of Ireland – areas well beyond the reach of conventional industrial development. It is vital that BIM continues to provide this integrated, market led development support, working with the industry to maximise output and deliver a high quality, value added and healthy product to consumers in Ireland and abroad...

The self-justifying tone was obvious in a number of aspects of the annual report. Every page of the accounts now carried three signatures, more details of the board's remuneration were set out and on one occasion there was even an actuarial report on the life expectancy of retirees!

BIM's efforts had been counter-productive. It had created a monster fleet, out of proportion with the marine resource and dependent on a constant revenue stream from the taxpayer to keep it afloat. Instead, the agency administered a plethora of projects under a variety of headings, little, if anything, of practical benefit being created as a result. This excerpt from the

2009 annual report demonstrates how procedural jargon was taking over from rational thought:

In addition to nine private sector projects, the National Development Programme 2000-2006 also provided funding for the continuing implementation of an Environmental and Quality Programme. Delivered through a regional officer programme, this Bay Carrying Capacity project to assess the sustainable levels of shellfish production was implemented under the aegis of the Aquaculture Initiative EEIG as part of a Cross-border Aquaculture Development Project.

Once again, there was a failure to understand, or an aversion to admitting, where earlier mistakes had been made and how the mess might be rectified.

As a dispenser of grants and facilitator of loans BIM would not have had a function if a limit to expansion by the industry had been acknowledged. Accordingly, shortly after finance from the European Union dried up in 2005, the survival of the agency came into question. BIM was the only semi-state body recommended for disbandment in a review of the public service in 2009.[12]

Chapter 9

GUIDANCE, GUESSTIMATES AND THE FIRST CASUALTY OF MANY

In economic terms the industry in the twentieth century in Ireland was evaluated by quantifying four descriptive elements. Landings of fin and shellfish were its primary product. The money they generated at first sale paid for the mortgage on the fleet, the wages of crew and for fuel, gear and other consumables used in the course of operations. The amount spent purchasing vessels represented annual investment in the fleet; factory capacity was a third element of capital growth and the additional value conferred on the raw product after processing, expressed in the value of exports, a fourth.

Of the four, the first, landings, was the keystone. Without adequate landings there was insufficient revenue to operate a fishing fleet, no incentive to build more, no justification for capital investment in factories and insufficient material to process or export.

The combination of the four was not a complete portrait of the industry but each conveyed a measurement of economic activity and they were consistently used to monitor progress. The Central Statistics Office would have collected export and other data but BIM was in an excellent position to furnish insightful interpretations of all four because, after its first decade in existence, it was intimately associated with all aspects of productivity and, through its facilitation of exchequer subsidies, with the business arrangements of perhaps the majority who participated in sea fisheries.

Although the early years of the decade were not promising, the 1960s proved a significant decade in the development of the modern industry. The second programme for economic expansion set targets which were monitored and reported annually.[1]

Between 1960 and 1963, landings declined in value from stg£1.6 to stg£1.4 million. Consequently, exports also fell short of expectation. There were explanations: a dispute at Dunmore East disrupted the herring fishery and prices for demersal species fell. Landings had been expected to double between 1960 and 1970 but in 1964 this prognosis was deemed to have been "ambitious".

In spite of obvious drawbacks there was a general consensus within the establishment on how the industry was evolving although there may also have been doubts about its rate of progress. Another consultancy report, this one by an American survey team[2], supported what was happening but had some cautionary observations. Its enquiry was initiated during a conversation between taoiseach Sean Lemass and president of the United States of America, John F. Kennedy. The team consisted of four individuals, an economist, an exploratory fishing and gear specialist, a technologist and a biologist. Like McArthur,[3] their sojourn in the country was short, just two months, but they travelled widely in that time and met representatives of the industry, BIM and the Department of Lands which was currently overseeing it.

The team made 29 recommendations. These were in keeping with the direction in which policy was trending. They were optimistic for the prospects of a larger sea fishing industry in Ireland. They recommended that landings be increased and fish processing intensified, fish quality raised and standardised, and marketing streamlined; to further guarantee quality of product export licences should be required for all fish products sent abroad. Exploratory and experimental fishing should be part of BIM's range of responsibilities and research was required to anticipate problems in shellfish management.

Like McArthur,   the American team also had views on the way the industry should be assisted by government. They too favoured encouraging private enterprise rather than state management wherever possible, liberalised trade policies and the investment of foreign capital.

Because Ireland was a small area in close proximity to fishing grounds and the majority of boats made brief trips of one or at most two days' duration, the team advised against any programme of subsidies for landings.

...even temporary subsidies to make initial sales because they have a way of becoming a permanent part of the price structure and thus render product or cost improvements less necessary...subsidies have the appeal of a simplified solution to an involved problem. However, they have major disadvantages. Our principal objection is that, by protecting less efficient units, they remove the onus for carrying out changes that would remove the need for the subsidy...

The hire-purchase method of fleet building attracted criticism because it kept fishermen financially and psychologically dependent on government. The explanation for this is not immediately obvious unless one considers the mounting backlog of arrears, despite which some hirers managed to hold onto their vessels.

There is an apparent tendency for at least a portion of the hire-purchaser owners to feel that their welfare if not their prosperity is the responsibility of the government. Vessels owners, as private businessmen, should be given complete responsibility for their own livelihood.

If, the team appeared to be recommending, there were a high initial casualty rate in hire-purchase arrangements followed by immediate repossession, this would encourage others to become more independent and conduct their business in a more pragmatic fashion.

The American team recommended that the hire-purchase scheme should be extended to vessels of 22.9 - 30.5 m oal and that it should be a prerequisite for obtaining a boat of this kind that a prospective skipper would have served on foreign offshore vessels. The team was confident that within seven to ten years Irish skippers would be capable to fishing waters as distant as the Arctic.

Although the American team leaned in favour of private enterprise managing the industry, they could not altogether exclude state participation. Services which would be provided, simply because there was insufficient profit for private enterprise to do so, would be slipways for larger vessels, ice production and cold storage facilities at ports. The largest fish market in the country was in Dublin and a case could be made for this to be managed by the state or local authority.

The most efficient form of vessel management was by skipper-owners. However, such work was all-consuming and the landings were invariably passed on to some other agency for sale so the catcher did not get best reward for his efforts. Co-operative selling was more beneficial for the boat owner, and government might usefully take a hand in the training of co-operative managers who would be incentivised by some kind of bonus arrangement to obtain best value for their members. Some shellfish resources might also be managed co-operatively or by private entities.

The American team anticipated mariculture developments, among them centralised hatchery and spat collection arrangements for oyster. The transplantation of seed mussel beds to on-growing conditions was favoured and fishermen were advised to seek unexploited beds of clams, cockles and palourde.

The introduction and culture of Manila clams was considered worthwhile.

In keeping with current preoccupations, the objective of the American team was to recommend how the industry should be put on its feet in the most practical manner. Little consideration was given to the capacity of the resource to sustain exploitation but occasional comments indicated that harvesting had already had a discernible impact on inshore fish stocks. *Nephrops* appeared to be fully exploited in the Irish Sea and the reduction in size of the prawns suggested a larger mesh size should be used to catch them. In spite of the fact that the average size of lobsters had declined, the team recommended more productive harvesting methods. Whitefish stocks may also have been perceived as eroded because a search for under-fished populations inshore was proposed. Paradoxically, the team also recommended extending fishing effort on cod, haddock, hake and pollock. Studies should be undertaken as a basis for predicting herring abundance and larger vessels might be employed to harvest this species while mackerel fishing should be conducted between 20 and 30 nm offshore to compete with foreign vessels.

The need for better statistical information was axknowledged and the desirability of compulsory reporting of landings per trip was alluded to. The licensing of all sea fishing vessels was also recommended; at the time, only boats above a certain size were required to be so authorised. The activities of BIM confirmed that progress in industrial build-up was taking place and that was mentioned by the American survey team.

BIM's first decade had drawn it into a variety of enterprises but the discipline associated with the second programme for economic expansion would test its competence; were its efforts measured and directed or was the agency merely responding spontaneously to demands?

Following the publication of the second programme for economic expansion in 1963, reports on its progress were issued in 1964 and 1965. BIM's annual report for 1966, in an ambitious format, set out targets for the value of landings and investment in the industry and the contribution of the catching sector to GNP from 1964 to 1970. The presentation, consisted of histogram columns bearing target numbers, provided with space for the addition of columns recording achievement beside them in succeeding years. However, the target figures appeared to remain as they were set whereas the achieved results were in current values, and the annual inflation rate averaged 5.4% between 1964 and 1970. In the three compared years, landings fell short of expectation by between 10 and 17%. Taking inflation into account, the shortfall in value climbed to 37% in 1966. Forecasting output in marine fisheries is notoriously problematical because so many natural variables can influence outcome. Predicting investment should have been more straightforward consisting as it did of orders for boats and plant, both of which have a lead-in time. However, predictions of investment were even wider of the mark than landings. The results which were actually achieved diverged from the predicted outcome by between 11 and 31%; taking inflation into account, the divergence in 1965 rose to 73%. The histograms were not reproduced in succeeding annual reports and this attempt to measure performance was shelved.

Another version of the exercise was introduced in the annual report for 1967, employing four performance indicators: landings and exports (which could be described as "outputs") and investment in vessels and processing ("inputs"). To be on the safe side, the target selected for the value of landings (stg£2.45 million) was less than forecast (stg£2.8 million) in the earlier exercise but in an attempt to make up for a lower start, the predicted rate of increase was higher. By 1969 the achieved figures were wide of the mark for the outputs; slightly more successful for the inputs where targets were exceeded in 1968 (investment in vessels) and 1969 (investment in processing), but the shortfall otherwise fluctuated dramatically, suggesting that prediction was hopeless. This exercise, like its predecessor, was discontinued.

Perhaps BIM was not the only organisation struggling to make sense of the future. The third programme for economic and social development[4] covered the period 1969-1972. Its section 23 stated that, in retrospect, the working out of the national growth targets for the second programme was

...unduly elaborate, and created a widespread preoccupation with detailed figures to the detriment of a more appropriate concern with broad policies for change and growth. Partly for this reason and partly because of the government's wish, in the limited time available for the preparation of the present programme, to concentrate on policies relevant to the elements in a strategy for achieving full employment...the programme is selective in its approach...much background information has been omitted.

The section on sea fisheries opened with a review of landings between 1963 and 1967 which had, it claimed, increased by an average of 11% a year. Because earnings had improved, more vessels had been persuaded to join the industry and more recent additions were becoming larger. Employment in the catching sector was 5,370, 27% of whom were full time, and the proportion of those working part time was falling. BIM's advisory and training facilities were being expanded. Five fishery harbours would be provided before the end of 1971. Research was continuing with two fishery vessels, *Cú Feasa* and *Cú na Mara*. BIM would continue its marketing efforts abroad and increase the consumption of fish at home. Greater processing of landings would be another objective and heavier reliance would be placed on the sales of processed fish.

The abandonment of targets allowed the comparative histograms to be dropped but the statistics involved continued to be expressed in another way. In current values, four criteria: investment in the fleet, investment in shore processing facilities (inputs), landings and exports (outputs) were combined to place a value on the industry which could be expressed as its contribution to Gross National Product (GNP). On their own, at times of high inflation, all showed spectacular year on year increases set out in large graphs in BIM's annual reports.

Released from the tyranny of targets, the four criteria describing the industry remained a part of BIM presentations, and, between 1967 and 1975, their combined value contributed an increasing, then levelling percentage of national income. In 1975 the outputs of the GNP totals were re-defined. Salmon made its way onto the marine catch statistics and, for the first time, landings of Irish vessels into foreign ports were included in the national totals. These data were also expressed, in part or entirely, as export figures which bolstered a favourable outlook.

The early examples of target setting were lamentable. Guessing the future of the fishing industry was hit and miss and suggested, if anyone was in doubt about it, that there was no substance to a development strategy. For the remainder of its history BIM usually, though not always, avoided following up firm predictions and made the most of positive results when they occurred; when they did not, as little mention of the fact was made as was decently possible.

In the Chairman's foreword to the annual reports and statement of accounts in 1994, it was announced that BIM would play a key role in the development of aquaculture, fisheries, processing and marketing in the Operational Programme, 1994-1999. On this occasion the targets were defined in terms of monies to be invested – a total of IR£36 million, IR£11 million from the EU, IR£2 million from the exchequer and IR£23 million from the private sector – and jobs created, after the investment has been fully made, in the hope that if you pump money into something jobs come out the other end, another assumption that does not necessarily hold. Investment in the fleet was intended to

...take up unused quotas, increase catches of non-quota species and bring the fleet into better balance with available resources...

which could, alternatively, amount to increasing fishing effort by seeking hitherto elusive quota species, thereby causing more damage to already depleted fish stocks and alienating the fleet even further from the environment that supported it.

One of the few predictions I came across in frequent references to this programme was the value (output) of aquaculture which was projected to increase from IR£40 million in 1991 to IR£101 million in 1999.[5] The prediction having been made in 1994, the projected outcome must be enlarged by 8% to take account of inflation. Rephrasing the proposal to take this into account, aquaculture output of IR£40 million in 1991 was predicted to rise to IR£109 million at the end of the operational programme. Its current value in 1999 was IR£67.5 million, a shortfall of 38%.

Predicting anything gives a hostage to fortune and BIM had not got any better at it in its forty plus years of existence. If the agency could not offer guaranteed outcomes, it could reassure everyone that best efforts had been made to avoid disaster:

...Implementation of the (1994-1999) programme to be monitored by representatives from the EU Commission, Department of the Marine, BIM, the industry and other relevant government departments...

There was another investment programme which would have profound consequences for the sea fishing industry and in that case BIM peered into a crystal globe once again with a level of success similar to its earlier efforts. A whitefish renewal scheme was the outcome, and that deserves more detailed consideration in a chapter of its own.[6]

The erosion of the biological resource that supported the industry occurred over a number of years. Concerns were expressed both within and outside the industry about what was happening but they made no difference. As some species declined, the commercial significance of others, originating in larger catches, increased, and the list of exploited species lengthened. But the problem was not entirely one of resources melting away. As the century progressed public attitudes acquired a harder, less sympathetic edge. There was less concern and less patience for the fauna underwriting wealth generation. The phenomenon was clearly demonstrated by the rapid disimprovement in the circumstances of a small, seashore-dwelling mollusc, the palourde.

The brief clam boom

Peadar O'Donnell was a native of Co Donegal where he was born on a small farm in the Rosses in 1893. He trained as a teacher and had numerous other achievements in a busy life of 93 years. He was a member of the Irish Republican Army during the war of independence, a political activist, a newspaper editor; he fought in the Spanish civil war and he wrote a number of novels. After the Second World War, he went on to edit the literary journal *The Bell*. He began to write *Islanders*, published in 1927, when incarcerated in Mountjoy Jail. The novel about poor fishing families was critically acclaimed and ran to many editions on both sides of the Atlantic. When asked, O'Donnell replied that it was not politically inspired, merely a recital of stories he had heard in his native area. Mac Laughlin, who has written authoritatively about fishermen throughout Ireland's history praised *Islanders* as a realistic portrayal of many poor fishing communities in post-Famine coastal Ireland.[7]

The novel describes the "heroism of human beings who can force a living from the rocks" to whom "two bucketfuls of winkles may be a considerable addition to the wealth of the home". O'Donnell was writing about gatherers of shore food whose foraging abilities did not reach far below the tide line. For them, herring was a major bonanza but only at such times as the shoals ventured sufficiently close to become susceptible to their primitive fishing gear.

As might be expected, O'Donnell had views about the way the fishing industry should be organised. At a symposium in 1956, he considered the dilemma of accommodating small operators in a world of bigger boats. While it did not make sense that the seasonal herring shoals should go by untouched, larger boats would mean bigger catches and they would depress market demand for the occasional landings of fisher-farmers.

O'Donnell had thought about it and, in tune with the growing consensus, threw his weight behind modernisation; on the perceived primitive state of the fleet he expressed impatience:

...instead of pottering about with tiny, yearly allocations in the estimates...scrap all out-of-date boats and gear, place a fair, even generous, value on them and recognise it as a down payment on new boats and gear...[8]

A species which would have been known to O'Donnell's islanders is the palourde, also known as the native clam and locally, in south west Ireland, as the *circín*. The species grows up to 7 cm in length and it ranges between the tide marks, occurring so high on the shore that some beds have been obliterated in recent years by road building. The clam frequents a very mixed sediment consisting of sand, mud and gravel in which it is the dominant bivalve species. Often these sediment beds are associated with fresh water inflows. Palourde has a high tolerance of fresh water and, indeed, of pollution caused by algal blooms. The animal does not lie deep beneath the surface and it is possible to excavate individuals, whose presence can be observed with the naked eye, using a pocket knife.

Palourde is a species which, on a geological time scale, is in retreat. It was far more extensively distributed and abundant in Pliocene times, between two and five million years ago. It currently ranges between Bergen, Norway (61°N) and Senegal, West Africa (12°N). It occurs in the Mediterranean and Red Seas, having invaded the latter following the construction of the Suez Canal in 1869.[9]

In the northern part of its range, palourde has retreated to a western fringe, no longer occurring in the North Sea. In historic times there were some records of palourde on the east coast of Ireland but its favoured habitat is within the bays and *rías* (fjords) of the Atlantic coastline. It occurs in several locations in

Co Donegal, O'Donnell's county, favouring sheltered shorelines with a mixed sediment of gravel and mud.

Commercial production of palourde today takes place in France (Golfe de Morbihan) and the Mediterranean coastal lagoons (*étangs*). In Italy it is harvested in the vicinity of Trieste; smaller landings are made in Taranto, Venice and Molfetta. The Algarve in Portugal produces the majority of that country's landings. In Spain most palourde are gathered within the *rías*, along with the Gulf of Cadiz and Cartagena Bay. Wherever it is exploited throughout its range, palourde is highly regarded for its culinary qualities.

Before examining what happened to palourde in Ireland, it is necessary to say something about the biology of this species. The life strategy of the palourde belongs to what is known as *K*, as opposed to *r-selected* animals. Species like cod pursue an *r* life strategy; they produce enormous numbers of eggs, suffer high mortalities at all stages of their growth and development and, in circumstances where their populations are strong, yield few adults (relative to the number of eggs in the first instance). *K* species are better adjusted to their circumstances, biological and environmental. Mortality throughout their development from egg to adult takes place at a more subdued rate and a higher percentage of relatively fewer eggs develops to mature adults. Among the fishes, sharks, with few young, late maturity and long gestation time are *K-strategy* species. So are deep water species which will feature later in our story.[10]

Visualise a population of *r* strategy organisms. It would consist of high numbers of eggs but declining numbers of subsequent development stages. However, while those numbers decline, the surviving individuals grow. At maturity, the surviving adults outweigh by several times the quantity of eggs that produced them. *K* strategy populations are made up at any time of fewer eggs relative to the number of adults. The majority of the biomass resides in the adult stage.

Palourde was gathered probably throughout the period of human occupation of this island; it could not be described as an abundant species because the locations in which it thrives are of limited extent although undisturbed populations could reach high densities and the species is the dominant bivalve where it occurs. In the early twentieth century there appears to have been very little commercial interest in palourde in Ireland. The species was gathered in the south of England where it was sold as "queen cockle" and in France where it was known as *l'eclovisse* but there are no earlier statistics quantifying its collection here. Our first indication of palourde exports dates from the 1930s. They became more consistent in the 1960s. Most extraction of the shellfish took place in the inner Kerry and Cork *rías* and, until 1973, exports (there appears to have been little interest in the purchase of palourde within Ireland) of between three and eleven tonnes occurred annually. In 1974, at the instigation of a shellfish buyer, attention focused on palourde beds around Galway Bay. News spread rapidly and depletion of the recently re-discovered beds was quickly accomplished.

Although palourde beds were few and of limited extent, the species fell beneath the radar of the agency anxious to establish a shellfish export trade. Within BIM, two officers were deputed to visit newly discovered populations of the species and encourage local people to dig them out. Many of the individuals who did so may have had little or no connection with the fishing industry; I can recollect university students in Galway earning much needed cash by digging in the strand, equipped with no more than a garden fork and a bucket.

Prior to the mid-1970s, when palourde became the focus of intense exploitation, it was gathered locally in small quantities which did not jeopardise its populations. This we can surmise because, when commercial interest focused on the species, those populations occurred at high abundance which was drastically and very rapidly reduced.

In spite of the fact that marine biology was becoming a popular subject in the 1970s no effort was made to anticipate any problems that might arise from digging out readily accessible populations of palourde. There was no precautionary appraisal of its ability to survive a fishery of the wild stocks. In 1974 a mere 20 t were captured but the species fetched high prices and heavier landings were anticipated. In 1975 more than 100 t were registered, and the species was separately listed as one of 42 varieties (mainly species) of marine landings in 1975, the first year of heavy catches, in anticipation of its contribution to inshore fishing income in the future. Despite its miniscule annual yield, palourde had been ear-marked as a species worthy of additional exploration a decade earlier.[11] More than 100 "varieties" are currently registered among the landings but palourde is no longer among them. In 1976 landings were less than 200 t and the following year only 30 t were harvested.

The early years of the "clam boom" were well documented but not for anything to do with the management of the species in the wild. Palourde was so valuable that investigations were undertaken to explore the possibilities of artificially cultivating it in mariculture. Its ability to survive in contaminated waters would have been beneficial but its slow growth rate rendered it uneconomic. The introduced species, Manila clam, proved a more amenable alternative.

Palourde were dug out with spades and forks – a crude process if one compares it with the technique of removing the animals with a dessert spoon, practised in France; that method was also reportedly employed in Murrisk (Co Mayo) in the recent past. In the *rías* of north west Spain, palourde, known as *almeja fina*, is "cultivated" in shellfish beds, much in the manner that garden crops are raised.[12] Karl Partridge provided some indications of the rewards of harvesting it in his thesis on the clam: allowing for inflation, calculated using the Consumer Price Index (1977 - 2008), part time digging could yield €61.5 per hour and that would have been tax free.[13]

The *Sunday Independent* of 4 April 1976 in an article titled "Kilcolgan is digging for gold" reported up to 300 people simultaneously doing so on the shore at Clarinbridge, Co

Galway. Partridge himself recalled 160 people digging at the head of Killybegs Bay. He did not anticipate the stocks could sustain this kind of pressure and, by 1977, landings were in steep decline.

What Partridge discovered, and later investigations confirmed, is that palourde at our latitude, although not necessarily further south, is a *K-strategy* species. Animals which conform to this life strategy live for a long time and reproduce slowly. Their populations consist mainly of adults with very few juveniles interspersed among them. Once a *K*-selected species becomes depleted by removal of those adults it cannot regain its former abundance unless considerable effort is made to protect and restore it.

The collapse of Ireland's palourde fisheries was dramatic and, 35 years later, populations of the species have still not recovered.[14]

Chapter 10

MEANWHILE, BACK IN THE TERRITORIAL SEA

While the fleet was growing in size, and European institutional changes with enormous implications for enterprise were underway, the associations representing fishermen were in a state of flux. Mirroring the fleet arrangements outlined in chapter 3, two types of association would branch from what was originally a single stem.

It is self-evident that smaller vessels fish close to the coast – contemporary terminology describing their operation refers to them as small scale coastal fisheries; fishing deeper water, further from shore must be undertaken by larger boats which carry more gear and whose size makes them more weather-worthy. Many nations make distinctions between their "inshore" and "offshore" fleets, reserving areas close to the coast for smaller craft and forbidding certain activities, like trawling, in shallow depths. The definition of what constitutes an inshore boat varies from one nation to another. In the UK, 10 m oal is the critical dimension, the EU favours 12 m and Ireland has traditionally referred to vessels of up to 15 m as belonging to this group.[1]

Ireland's fishing fleet in August 2010 was made up of 2,133 vessels. Almost three quarters of them were under 10 m and almost 90% were less than 15 m oal. At first glance these statistics suggest an "inshore" rather than a "high seas" fleet and this definition is the official one. As recently as September 2011, I listened to a senior official in the Department of Agriculture Food and the Marine[2] say that Ireland does not have an offshore fleet. Yet, if two other vessel characteristics are examined, that statement is questionable. Taking the EU definition of inshore boats being less than 12 m, the number in this group is 86% of the Irish fleet but their tonnage amounts to only 10% of the total, so boats greater than 12 m oal account for 90% of GT; the percentage of kW, engine power, belonging to smaller vessels is a little higher, 30%, but 70% resides among the larger ones. It is not possible to give a precise breakdown of employment between inshore and offshore vessels but it is probably around fifty percent each.

The illusion that Ireland has only an inshore fleet is a cultivated myth which is at variance with administrative policy catering for larger vessels. There is a history explaining it. Initially, the demand within the industry was for an adequately equipped inshore fleet, in keeping with the logic that, if the inshore fleet of the UK were capable of harvesting seven times the quantity of whitefish consumed in Ireland, something similar here would satisfy requirements.[3]

Very soon afterwards the wind changed – though, it must be said, not initially in any definite direction. It vacillated but was settling into a more offshore tendency. As signs of fish depletion became more numerous (in 1954, for instance, there were reports of over-fishing by trawling in Dingle Bay[4]) and the limitations of the shallow shelf seas became more obvious, deeper waters began to exert a more powerful attraction. Despite the increase in numbers of vessels (hardly a month went by but "a new 50 footer goes plunging off the launching slips..."), greater efficiency in fishing technique and the installation of radio telephones and echo sounders, landings in 1953 were lower than in 1948.[5] Annual landings, indeed landings recorded in shorter time spans, were scrutinised and debated as an augury of progress. In August 1954 a headline alarmingly announced "The industry is going astern…" The report which followed stated "The fact that a larger, more up-to-date fleet has landed less fish is gravely perturbing." Despite all the money borrowed, the industry in 1954 had not progressed beyond where it had been in 1952,[6] hardly a reason to be critical of an activity where so many environmental, technical, market and human behavioural factors interact to influence productivity, but such was the impatience of the time.

A letter from the Valentia fishermen's Guild of Muintir na Tire crystallised their frustration in dramatic terms:

…The only fishing worth while today…is deep sea fishing, not inshore which can only scratch the surface…Inshore fishing must be nursed and highly protected…Fish costs nothing only the gear to catch it…If the Irish race is to be saved, something must be done…[7]

At least the letter recognised the necessity to manage inshore fishing; unfortunately that concern was trampled in the subsequent desperate stampede for, first, wealth and, later, solvency.

The political establishment was chary about large boats if, for no other reason, than they would compete with smaller ones, its current constituency. Joseph Brennan (Fianna Fail, Donegal west) expressed the realities more prosaically when he observed in 1955 that there were 17 – 20 "large" fishing boats in Killybegs, a development to be welcomed but

…we are inclined to forget we still have with us the inshore fisherman with a small bit of land who goes to sea whenever the conditions are favourable…he feels (the larger boats) should be kept off our shore for some distance and that he should at least have an undisturbed portion of water to himself…[8]

And that is the kernel of the argument justifying the zonation of fishing activities.

Like all politicians with a flair for longevity, the minister, Oliver J. Flanagan (Fine Gael, Laois-Offaly), debating the landings statistics of 1955, tried to have it every way:

It is government policy that the fishing industry should be based to the greatest possible extent on the inshore fishermen. The word inshore *I may say very often gives rise to the misconception in the mind of the public. It is often thought the word implies fishing very close to the shore and that the outer waters are left untouched by our fishermen...That is not so as with the type of boat, ranging usually between 56 and 60 feet overall with first class equipment, now being used by our fishermen, fishing activities are not confined to the grounds close to the shore...*

An editorial in the trade press in 1959 warned of the perils of investing in offshore fisheries, based on experience elsewhere:

In Ireland we must appreciate that an entry into middle-water trawling is an entry into a phase of the European fisheries in which over-fishing keeps the industry in a chronic state of crisis...[9]

It was possibly the most prescient statement ever issued by the Association.

As is often the case in Ireland, it was left to an outsider to clarify a particular policy. The American team which visited in 1964 did just that; it recognized the contentious and inevitable extension of fishing operations towards the shelf edge. To facilitate such a development, inshore waters should be accessible by boats of all sizes. Not to do so would discourage entrepreneurs from taking a leap into the unknown – the Association's point above - with larger boats on untested grounds without the comfort of more familiar territory to fall back on.

In legal terms, two fishing sub-fleets had been in existence since 1952 when legislation specified that no vessel greater than 10.7 m should fish without a licence (meaning that smaller boats could). The same legislation allowed the minister to impose restrictions which could have meant excluding boats of certain sizes and fishing methods from near-shore waters.[10] However, although I have attempted to ascertain whether that was ever done in practice, I have been unable to. It may well be another - and there are many examples - putative regulation which was never enforced. That said, there is a belief that such conditions were placed on certain vessels at times.[11] I cannot see the point of imposing this kind of restriction on a private and confidential basis; for one thing, nobody would know whether the law was being broken unless the offence was common knowledge.

The existence of restrictions of this kind of regulation on larger vessels might have survived from the more remote past. Bye-laws excluding steam trawlers from coastal waters had been enacted in the late nineteenth century and survived into the following one after vessels of that kind had ceased to operate. Restrictions on trawling in inshore waters and bays had been introduced at about the same time in order to promote the orderly development of fisheries, largely by preventing vessels towing mobile gears across gear set by herring drift netters. They were later deemed to be unenforceable and unnecessary and, if they ever had been applied, had fallen into disuse; hence their revocation was recommended.[12]

The concession to Northern Ireland boats in 1960 allowed those of less than 22.9 m (75 feet oal) to fish inside the Republic's waters while boats from the UK were not permitted to cross the six nm exclusive national boundary. The length limit was regarded in the industry as some kind of recognition that vessels of larger size were, or would be, prevented from working close to shore in the Republic. So much in Ireland has been surmised from rumour and wishful thinking, coloured by self-interest and this was a good example of all of them. Those who already had boats did not want competition from larger ones so they felt secure in having exclusive national waters to themselves. At least they would be safe from the depredations of larger vessels belonging to other nationalities. It did not take long before that illusion was shattered.

Before its stocks were run down and mackerel displaced it as the most valuable marine fish species, herring landings were locally and temporarily enormous – described as gluts - and very valuable. The efforts of the small Irish fleet were supplemented by those of other nationalities who fished alongside and this was welcomed by the industry in Ireland because large catches attracted foreign buyers. Of course, not all landings were absorbed and passed on for human consumption. In fact, large quantities of herring were abandoned, sometimes rotting on the quayside. The most lucrative destination for these was fishmeal. So typical were these glut fisheries that the provision of fishmeal plant was regarded as an indispensible adjunct to harvesting herring.

The establishment of fishmeal plants was a part of BIM's agenda which was actively embraced; a fishmeal plant was among the assets it inherited from its predecessor. But there was a problem. Herring glut fisheries were neither regular nor predictable and fishmeal factories ran to a time table. If they were to be maintained, they had to be constantly supplied.

In 1957 a pilot fishmeal plant employing 28 people was built at Killybegs at a cost of stg£132,000. The product was bedevilled by Peruvian imports and the factory was taken over by a German concern, Atlantic Fisheries Ltd. The plant had a daily capacity of 100 t and herring could not be relied on to satisfy that demand. Supplies had to be sourced elsewhere and an Icelandic purse seiner, *Harfon*, was chartered for that purpose. The activities of the *Harfon* were controversial and it was unsuccessful and, following representations by the industry to the Department the vessel was instructed not to fish within Irish national limits. Two years later two trawlers from the Netherlands, *Gronland* and *Mercator,* were brought to Killybegs by an Irish registered fishing company to make up for the shortfall in factory feed. They were equally unproductive.

When ownership of the Killybegs plant then passed into Dutch hands, very generous arrangements were made to maintain its fish supplies. An agreement was formalised between its principals and the minister allowing them to operate two trawlers for twelve months within the territorial sea between Inishowen and Ballycotton. If this were insufficient, ten further vessels would be allowed to operate inside national limits when they were targeting sprat, herring and pilchard. Despite this, and the fact that it was prepared to use fish waste from other vessels, the Killybegs factory never got sufficient material to keep it occupied for more than a couple of hours a day.

The activities of these fishmeal vessels exacted a discernible toll on fish stocks in the north west of the country. A local fishermen's port group, the Killybegs Fisheries Association became agitated and framed a number of proposals which could well have resulted in inshore waters becoming a distinctive zone in terms of the vessels working within it and the methods of fishing used. Its advice was however ignored.[13] The National Fishermen's Organisation (an umbrella group for port associations, formed in 1958) took up the call to zone national waters and exclude vessels greater than 21.3 m (note, 70 rather than 75 feet oal) from exclusive national waters.[14] Their petition also fell of deaf ears. There are several general presumptions underlying fisheries management in Ireland. One states that once a job provider has been established on the basis of fish stocks, its survival will take precedence over theirs even if that means that in the medium term all collapse!

The fishmeal experience was repeated in 1968 when another plant was constructed at Mornington, at the mouth of the Boyne River in Co Meath. The capacity of the plant was intended to be 350 t per day, an optimistic target in light of the Killybegs experience. Before the fishmeal plant arrived on the scene, Mornington was known only for its salmon draft net and mussel rake fisheries. There were no commercial sea fish landings at that time. However, there was always a need for another source of employment and the plant would create a new port facility. One of the mussel fishermen told the *Irish Skipper*, the new industry journal: "We need the factory…there's nothing to do here for six months of the year only walk about".[15]

The plant was welcomed by trade journal. In an interview with its editor, Arthur Reynolds, an executive of Salvesens, the company which owned the operation, observed that BIM was striving to expand Irish fisheries and there was no reason to suppose that expansion would not continue. The plant would be a considerable outlet for trash and industrial fish and offal. Salvesens had at an earlier time been a whaling company but they were now concerned to make money from sprat, claimed Reynolds's headline;[16] indeed, working down the food chain was a fitting metaphor for the way the industry was heading.

Throughout its history, controversy simmered about whether vessels supplying the Mornington plant should be restricted to offshore waters. Support from the trade press may have prolonged its brief life. Another supposition masquerading as a rule saw the light of day, although it was never implemented either: that trawlers greater than 100 GT and 294 kW must remain outside the territorial sea altogether – that is outside the twelve nm limit.[17] The unfounded presumption that fisheries were being prudently managed was Mornington's saviour, and supplies to the plant from inshore waters were maintained, at what long term damage to finfish stocks, particularly herring, we can only guess.

Like its predecessor in Killybegs, the Mornington plant had considerable difficulties obtaining fish supplies to keep it in business. By December 1969 it was receiving only 200 t of fish per day. Those catches it managed to get hold of included significant quantities of juvenile clupeids (sprat and herring) which might otherwise have grown and gone to human consumption. It never exceeded half its capacity and in April 1976 its closure was announced.

If rules and regulations are to be introduced to any fishing enterprise it is a good idea to do so at an early stage before bad but temporarily profitable behaviour becomes the norm. The passage of time, the growth of vessel numbers, individual size and power were accompanied by administrative and political change; political and business power and influence drifted away from the small coastal fishermen, if they ever possessed any, to the owners of larger boats. What might have been justified in the mid-1960s when an argument was made for the use of inshore waters as a kindergarten for offshore operators, was hardly valid a decade later when vessels were routinely venturing further afield; but the debate had faded, ceased to exist. In the late 1970s, the highly productive inshore, shallow waters, vital as nursery and spawning areas, were accessible by vessels of any size, using any fishing methods they pleased.

And the rancour between the two elements of the fleet, the inshore and industrial vessels, lived on and showed its teeth from time to time. As late as 1986, small boat fishermen drift netting for salmon were complaining about much larger vessels crowding in on them for the same purpose. An editorial comment in the trade press remarked:

...Inshore men have been saying for years that their main problem is not intervention by foreign vessels but by trawlers many times bigger than themselves which often hail from nearby ports...[18]

The road taken by the industry, hand in hand with the development agency and the government department collaborating with it, has been one of creating larger and progressively more modern functional units and making resources available to those. Yet the ethos of the industry and particularly of the agency – BIM – charged with developing it, was ostensibly about something quite different – promoting peripheral, underdeveloped regions. The argument that grants should be dispensed where jobs are needed and where there is no other way of supporting them other than through fisheries has frequently been rehearsed to justify the work undertaken by BIM. The chairman's statement in the 2010 annual report did it again:

...the real significance of these (fisheries) jobs is that they are mainly located in our remote coastal communities – Ireland's

seafood is what keeps many of these communities alive and well in the absence of economic alternatives...

So, what is the official attitude to the small-scale coastal fleets which employ, at relatively lower energy cost and less environmental impact, some 50% of those engaged in capture fisheries?

Late in 1979 fishermen's organisations approached the minister, Brian Lenihan (Fianna Fail, Dublin Co, west) and requested the establishment of an exclusive zone for fishermen working close to the coast. For the associations in question the protection of smaller vessels was a major mandate. A typical political response greeted their petition: the minister would be delighted to comply but he had to have agreement within the industry first.[19] At that stage, in an atmosphere of high tension and a deficit in supplies of fish, no sector of the fleet would have willingly foregone any fishing rights and the proposal got no further.

In the course of negotiations on the CFP in 1980, the European Commission refused to allow exclusive national fishing zones other than those which were already in existence under the provisions for the territorial sea. A scenario was "vaguely" proposed that for "certain particularly dependent areas protective measures may be adopted for local fishermen". These should include the designation of fishing zones exclusively for the boats operating from a local area and might also place limits on the size of boats allowed to fish certain areas.[20] An explanation for the uncertainty may have been the putative location of these arrangements within the territorial sea which was, strictly speaking, under national control and outside the jurisdiction of the Commission, hence its reluctance to dictate solutions.

When, in the 1980s, European fishing fleets had expanded beyond the point where the resource could support them, the Commission introduced a series of plans – Multi-Annual Guidance Programmes (MAGP) to stem their growth.[21] At variance with the uncaring attitude of the administration in

Ireland, the Commission was sensitive to the consequences for peripheral communities of reducing the number of small-scale vessels. An alternative was available: vessels less than 12 m in length could be "ring fenced" and omitted from the MAGP reduction targets. This would have meant that the GT and kW of small boats would be frozen into a new "inshore segment" and would not be available to transfer into larger vessels. In 1994 the Department informed the industry that this alternative was not acceptable. Surplus GT and kW should remain available to be cannibalised into larger catching units should the occasion demand it. Small boats were seen to get in the way of "proper" fleet development; smaller numbers of larger vessels were more conveniently regulated.[22] Thenceforth the activities of the small-scale coastal fleet fell beneath the regulatory radar and those vessels were left to their own devices. The result was that the area of some 27,300 km^2 of marine and transitional (estuarine) waters from the Irish coast out to six nm beyond baselines, the only area under Ireland's exclusive control, would be less regulated than the sea area under Community jurisdiction outside it.

I am reluctant to attribute responsibility for what occurred to any agency, industry association or government department, other than to say that the department charged with looking after the industry was ultimately to blame, probably without adequate power and certainly without motivation to do anything about it. Once again, situations evolved and parties interacted and policy was created by default.

The evolution of relevant representation

The Irish Fishermen's Association, founded in 1925, was the first of a number of representative groups serving the industry in the twentieth century. Membership was open to "all who are engaged in fishing for a livelihood in Irish inland or coastal waters" as well as to "non-fishermen approved of by the local branch committee and willing to subscribe stg£1 per year". The aims of the association were to protect and promote the interests of fishermen and to encourage improved methods of fishing and marketing landings.[23]

The multiplicity of representative groups which formed in succeeding years was attributed by de Courcy Ireland to the difficulties of organising a body of men who had become disillusioned, suspicious and uncooperative. In which case, the many crises of the burgeoning industry after the 1960s, together with its necessary adjustments to the EEC context would have been expected to spawn a succession of them to cope with rapidly changing circumstances.

The first flashpoint bearing portents of a species wipe-out was the marine fishery for Atlantic salmon which gathered momentum during and following the 1960s. Salmon was, administratively, a fresh water species but some had always been harvested in the sea. The use of drift nets to capture it greatly increased the sea harvest and the synthetic materials with which they were made from the 1960s confronted the fish on their return to spawn with lighter, more portable, easily handled and deadly walls of mesh set across their migration route.

Today most people are aware of the deteriorating plight of a number of marine species, cod and bluefin tuna are two with high profiles. In the 1970s inroads into marine fish stocks had not impinged on the public consciousness but salmon had a supportive constituency. For one thing, the species had an alternative commercial use - for recreational and tourist anglers - which was, arguably, of greater value to the national economy than its price, dead, on a fish-monger's slab. Hoteliers, riparian owners and scientists urged the government to curb the expanding marine fishery. The issues were highly charged and very emotional and they appeared, to both sides, to be clear-cut, each protagonist interpreting the facts differently. Government responses to one or other were fashioned with political advantage in mind.

In May 1973, an advert in the trade press requested subscriptions to fund a campaign to counter a proposed reduction in the number of salmon drift net licences. The name of the group was the National Fishermen's Defence Association (NFDA). It went on to organise parades, television and press conferences as a result of which the legislation in question was amended. The parliamentary secretary, Michael Pat Murphy (Labour, Cork west) conceded the statutory instrument had been too restrictive and he backed down.[24] In July of that year, the group formed a national executive and prepared for a fight.

John A. Mulcahy, an Irish American millionaire living in Waterville, Co Kerry, of which he was a considerable benefactor, took up the gauntlet. He stated he was not against drift netting but he felt too many people were engaged in it. He offered to personally remunerate fishermen with the value of their last three year's landings and buy them out. The NFDA retorted that vested interests were using conservation as an excuse to further their own ambitions and it countered with a proposal that the state should acquire all fishing rights which would be leased to local angling clubs and that former owners should be precluded from getting them back.

The following year the Salmon and Trout Conservation Council of Ireland (STCCI) was formed by hoteliers and private commercial fishery owners to elaborate the conservation arguments. The lack of appropriate data made for confusion. Natural fluctuations in salmon numbers, environmental change, freshwater pollution and partisan interpretations of the facts confused the issues. Neither side could be certain about its case but that did not mean each was not convinced by it, and an unbiased observer might be swayed by either. At one point, a university lecturer was called upon to adjudicate and found in favour of the NFDA.

Over a longer timescale, the changes wrought by over-fishing, and possibly accelerated by climate change, clarified matters. Drift netting for salmon off the Irish coast was banned in 2007 under EU legislation because this "interceptory" fishery was netting fish returning to other member states. At that stage, the returning runs of maturing fish were considerably depleted.

In its second year the NFDA revealed ambitions to place its supporters on state and local authorities administering fisheries matters. Through its work defending salmon drift netters, the NFDA had happened on a much larger proportion of the fishing industry which was crying out for representation. The following year, 1975, it demanded that some of the investment in big fishing craft be diverted towards inshore boats: "...the scramble to build bigger trawlers had overshadowed the traditional "small" fisherman..." This led on, naturally, to demanding an exclusive zone for small boat operators, secure from the depredations of trawler activity. In November 1975 the NFDA changed its name to the National Salmon and Inshore Fishermen's Association (NSIFA). True to its provenance, it organised within terrestrial rather than marine administrative boundaries: one representative was elected from each of the seventeen fishery districts outlined by river catchment boundaries, reflecting the management arrangements for salmon.

The late 1970s was a time of turmoil; declining fish stocks intensified the struggle for advantage and pitted one group of fishermen against another. The CFP was blundering its way into shape and the Commission needed agencies to administer its policies. One of these was the market policy of intervention: establishing a floor below which fish prices could not fall. The mechanism to achieve it was to purchase fish from fishermen at a guaranteed price, using European funds. The Commission's agents were to be Producers' Organisations (POs). All fishermen's representative organisations wanted to secure this lucrative role. The NSIFA inquired how doing so might further their members' interests. They were told that the produce targeted by their members, salmon and shellfish, were always in high demand and would never be put into intervention which was primarily intended for other marine finfish species.

During the closing years of the 1970s the NSIFA pursued the campaign for which it was established and widened its demands to asking for change within BIM: a more representative board, marketing assistance for their produce and the introduction of a ban on trawling in inshore areas (specifically to update the defunct prohibitions on steam trawling). They also demanded that lobster, crawfish and crab fishing be licensed.

In 1979 funds were made available for the NSIFA from Europe through the Combat Poverty agency which pigeon-holed their status in what had become a highly acquisitive pecking order. The drift net controversy rumbled on, claim and counter-claim becoming more strident. The NSIFA tried to maintain pressure to secure exclusive inshore zonation, inquiring in October 1979 whether owners of trawlers of 37 m were permitted to fish up to the shoreline. They complained that their members were not being consulted where aquaculture projects were undertaken. They also argued that small boat fishermen should be permitted to drift net for herring in the Celtic Sea when similar activity by larger vessels was forbidden for conservation reasons. Many of their queries were rhetorical, most of what they said was ignored. Some of their demands, for regulation of shell fisheries for example, were appropriate. But they were never seriously considered and, with the passage of time and the growth in fishing effort, their implementation became progressively more problematical. A variant of another of their suggestions, the creation of a quality mark for wild caught salmon and sea trout, would become a major initiative for BIM in later years. They also proposed that legally caught salmon be distinctively tagged to discourage poaching and that was, in due course, adopted.

In 1980 it was announced that a single body, the Irish National Fisheries Council (INFC), would be created to represent the interests of those working in and supplying the catching sector; its purpose was to forge an overall strategy for the development of the entire industry.[25] It was officially recognised as advisory to the Department and the NSIFA was included among its members. The same year a 10% levy on the first sale of salmon, payable by the purchaser, was introduced.[26] The NSIFA announced its members would not pay it. The issue was taken up by the INFC which also demanded it be withdrawn and, in 1983, the levy was abolished.

Membership of the INFC diluted the NSIFA's claims in a more authoritative agenda, which may have unduly favoured suppliers of services to the industry over the catching sector, and contributed to the emasculation of the inshore group. Concessions were made to the NSIFA on salmon drift netting and restrictions on this form of fishing were eased. In February 1981 Combat Poverty funding ceased and the NSIFA moved to temporary premises. A note in the trade press in 1983 observed that it had been "moribund for some time". The comment continued "one difficulty in keeping it together was that the fishery bye-laws vary from district to district (this referring to salmon fisheries districts whose local regulations varied between one and the next) and the men had only a few fundamental issues".[27] It was true that the NSIFA had a brief shopping list and some of its demands were profoundly wrong but others were eminently sensible. Most significantly, it represented a neglected section of the fishing community which would, if the establishment consensus were taken to the extreme, have been relegated to oblivion at an earlier time than actually happened.

In December 1983 salmon fishermen in Co Donegal applied to join another representative group, the Irish Fishermen's Organisation (IFO) which picked up the baton from the NSIFA.

The IFO itself had come into existence the year after Ireland acceded to the EEC at a speed which reflected the momentum of the time. The Federation of Irish Fishing Co-operatives (FIFC) had recently added "and Associations" to its title to embrace non-trading groups of fishermen before assuming a new title altogether, the Irish Fishermen's Organisation.[28] Its first mandate was to agitate for minimum whitefish prices and that was rapidly joined by other issues. Within four months of its formation the IFO had complained that Ireland's delegation to the UN law of the sea conference, which would culminate in the 200 nm EEZ, did not have fishermen on it. Minimum whitefish prices were symptomatic of a much longer list of grievances concerning the cost of fishing operations and the low level or lack of profits from it. To this was added the concern about poaching by foreign fishing boats, a number of which belonged to fellow EEC states.

The NFDA had bolstered its reputation by confronting officialdom soon after its formation and the IFO got a similar opportunity in 1975. Departmental estimates were cut by 5% which translated into a much greater reduction in the fisheries budget: funding for new vessels, exploratory fishing and harbour development would all be curtailed. The IFO protested together with the Federation of Marine Industries.[29] The government reversed its decision to cut loans for boat purchase.[30]

In April, the IFO organised a blockade of four main ports to highlight the plight of fishermen. Each member of the organisation was asked for a donation of stg£10 to make good any damage to vessels incurred in the course of the action.[31]Thus far, the issues selected by the new organisation represented a different interest group to the NFDA. The IFO campaign was about fleet expansion and, in the same month, its chairman demanded the fleet be funded to expand into middle waters, to coincide with enlargement of the EEZ.

The first annual general meeting of the IFO on 19 July 1975 was a very confident affair. The morning session was given over to policy, the afternoon to fisheries development plans. The list of invited observers included guests from related organisations in Ireland and the UK.

The IFO demanded a separate minister for fisheries, an extension to fifty nm of exclusive Irish fishing limits, a ban on third country fish imports to the EEC, adequate port facilities for handling and storing fish landings, short term financial assistance and adequate representation of fishermen on all state bodies of relevance.[32]

However representative of their members, meddlesome organisations do not win friends in government. We have seen the way in which the NSIFA was neutered: its list of demands was absorbed and smothered in the agenda of a larger group appointed by the Department. Now the government had to decide on a body to represent fishermen in an advisory capacity. The

Irish Fish Producers' Organisation (IFPO), a much more docile model was chosen. The IFO's chairman, Joey Murrin of Killybegs, was outraged when the minister, Michael Pat Murphy referred to his members as "just another pressure group". Murrin had no doubt that the April blockade was the reason the IFO had been over looked. The decision would be a crucial policy marker.

In January 1976 the IFO was joined by three other fisheries representative groups, the Irish Fish Processors and Exporters Association, the Irish Federation of Marine Industries and the Dublin Retail Fish Merchants' Association, who, together, presented a long list of demands:

There should be a separate minister for fisheries to whom responsibilities for harbour development would be transferred. There was a need for improvement in training facilities for the industry and education in co-operative management. A major reorganisation of BIM was demanded along the lines of the British White Fish Authority; seats on its board should be allocated to the four groups making the representations. The functions of BIM's chairman and chief executive should be separated and the chairman elected annually. Boatbuilding should be removed from the agency and left to private enterprise and there should be a reduced deposit on boat purchase. Retail purchasing facilities needed reorganisation. The Dublin fish market should be transferred to another site and there should be grants of 50% to establish fish shops outside the capital. At sea, fishery limits should be extended and baselines redrawn and poachers should be more heavily fined. The list given here is not complete but it reflected the general trend of thought within the industry at the time. On one further matter, introducing a licensing system for shell fishing, the IFO was in agreement with the "inshore" shopping list of the NSIFA.[33] On another item they differed. IFO chairman Murrin held that until such time as there was an effective marketing structure and extended fishing limits, the speed of inshore vessel production should be decelerated. The government had substantially increased grants for vessel purchase.[34] It was a matter he would return to two years later. In the meantime, the Irish Transport and General Workers' Union greeted his remarks with considerable unease, concerned for the prospects of their members in the boatyards.

In addition to badgering government for institutional change and greater funding for the industry, the IFO was practically involved on the water and in the ports, mediating in local disputes between fleet sectors where one displaced another from its traditional grounds as a result of intensifying competition for fish. It was constantly immersed in its banner campaigns, urging tactical change in the talks which would lead to the setting up of the EEZ and the proposed fifty nm exclusive fishing limit, on which it was supported by UK fishing organisations and later, the Angling Council of Ireland. Imaginative approaches were devised to keep the issues fresh in the public mind: a subscription list in support of the exclusive fisheries limit was opened and published, then a second one.

At its 1976 annual general meeting in July the IFO made a bid to comprehensively represent the marine catching sector. Membership was opened to all boats for an annual fee of stg£5 plus 0.5% of gross landings value. A decision was taken to form and operate through port committees. By November the organisation had issued a statement to the effect that it represented the majority of trawler owners in the country. It continued to work in association with the three other marine bodies, none of them representing the catching sector.

The January issue of the *Irish Skipper* in 1977 carried a highly emotive comment column titled "The new imperialism". It spoke of the chaotic uncertainty of the future: falling catches and constant harassment from foreign fishermen. But, if the nation were poor in assets, it still had some advantages: "strongest of these is this island's very survival after 400 years of war, persecution and economic oppression". The IFO championed the cause "relentlessly". For the first time the general public had supported the industry "which politicians at home or abroad will not ignore or cheat."

If, at this stage the IFO had burned its boats with government, it was highly popular with the industry.

Two months later, to appease the deepening despair, the minister, Paddy Donegan (Fine Gael, Louth) issued a fishery order that no boat greater than 33 m oal or with engine power exceeding 809 kW should fish within 50 nm of the Irish coast. It had to be non-discriminatory under EEC law and it was - Ireland had no vessels of those characteristics – so, presumably, it was legal. The measure was said to be temporary and in support of the new exclusive zone. Britain however, opposed any unilateral move by Ireland which preceded a fully resolved EEC policy.[35]

Writing in the *Sunday Independent*, a journalist, Hugh Munro, fretted about the costs of patrolling an exclusive fifty nm zone. Frank Doyle, secretary of the IFO, replied in the trade press that an exclusive zone of that size was required to stop the reduction in fish stocks. Nobody to whom it might have occurred, publicly inquired whether the policies pursued by the state, in agreement with the IFO, were not having precisely that effect.

In May 1977, Joey Murrin, reassured that the government was committed to unilateral conservation measures, retired as chairman of the IFO to concentrate on local issues in Killybegs where he served as chairman of the Killybegs Fishermen's Association.

At that point the minister for fisheries changed, along with the government, and Brian Lenihan (Fianna Fail, Dublin Co, west) assumed the role. By July 1977 there were concerns that he was not so committed to his predecessor's pledge on the fifty nm exclusive zone. A new organisation came into existence in Killybegs, the largest fishing port in the country and Murrin's heartland, to support the IFO claim,[36] and keep pressure on the minister who was believed to be vulnerable to the farming lobby, the larger proportion of his ministry, to conclude a deal with Europe. At a meeting of fishermen from the Republic, Northern Ireland and mainland Britain in November, Lenihan stated he was seeking an exclusive zone of "up to fifty nm"; pressed, he hastily corrected his statement to "fifty nm" but those assembled realised their claim would, at best, be watered down.[37]

Early the following year the IFO abandoned its claim to the fifty nm exclusive zone. Other items had moved up the agenda, notably sorting out quota allocations to member states, and the commissioner Finn Gundelach had told Ireland in no uncertain terms that is was not possible to dine *a la carte* at the Community table. If Ireland wanted the benefit of high prices for fish and farm produce it could not arrogate a large portion of Community waters to itself. The trade press complimented the IFO's graceful acceptance of the inevitable; it had done its utmost and fought convincingly. Persistence in the face of insurmountable reality would have risked disunity.[38] But acceptance did not last long; the IFO's chairman, Joey Murrin (he had returned to the post) was over-ruled by its executive committee which pledged itself to continue the campaign for a fifty nm exclusive fisheries zone.[39]

Murrin was one of, if not the most, politically adept leaders to arise from the industry; a fisherman for twenty years, he had a genuine flair for administration and occupied high rank, sometimes simultaneously, in several national and notable local fishermen's associations. He offered leadership which was slightly ahead, but not too far out of reach, of his membership. In 1978 he welcomed landings by other Community nations to Ireland at a time when the mood within the industry was to forbid them. On several occasions he spoke against BIM's policy of continually building vessels when the fleet was already too big for the resource. When Murrin retired for a second time as chairman of IFO in 1979 his resignation was interpreted as likely to weaken the organisation.[40]

The IFO resumed its campaigns. The matter of the exclusive fishery zone having been, for the moment at least, parked, another issue assumed prominence on the agenda. In 1978 the IFO expressed a view on salmon drift netting, a campaign which was already being pursued by the NSIFA:

The IFO can neither condone the poaching of salmon nor accept the logic of turning the might of the Irish navy against the genuine fisherman who, through no fault of his own, has been deprived of a licence to fish legally.[41]

The IFO emphasised what it identified as the inappropriate awarding, as a result of political interference, of licences to people other than traditional fishermen who were entitled to them. Both organisations, the IFO and the NSIFA, were now focused on several similar issues; licensing shellfisheries was another of them. A third was the designation of an inshore area for the use of small craft.

In September 1978 the IFO moved from the Merrion Square offices of the Irish Agricultural Organisation Society Ltd, which had organised fishing co-operatives – whose federation (the FIFC) had morphed into the IFO - in Ireland, to its own premises in Nassau Street, Dublin.

A ban on herring fishing in the Celtic Sea was introduced for conservation reasons and the IFO announced its boats would fish in defiance of it because the minister was ignoring draft proposals on conservation within Ireland's sector of the EEZ. The trade press thought this action very unwise because it could only inflict damage on the herring stocks; the IFO should work within the law.[42] This was the second occasion in a short space of time on which the trade press, usually highly supportive of fishermen's organisations, took issue with IFO tactics.

During the remaining years of the 1970s and into the new decade, the IFO campaigned on a large number of matters of concern to its members. Ever present in the background was the looming encroachment of European policy and fishing fleets.

The IFO's demands included the appointment of 300 fishery officers as observers on vessels from third countries greater than 24 m oal operating in Community waters, an end to fish imports from third countries to the Community and an instruction that purse seines be banned within Ireland's sector of the EEZ. The policy of registering, or "flagging", the boats of other member states with Irish identity to avail of Irish quotas was challenged, as was fishing for hake off the south coast of Ireland by UK-flagged Spanish-owned boats. In June 1983 the IFO protested at the proposed location of a support base for Spanish trawlers fishing Ireland's EEZ at Cappa, Co Clare. The prospect of Spain's accession to the Community caused considerable stress to fishing organisations in all nations already embraced by the developing CFP, not least to the IFO, which wanted Spaniards to remain outside fifty nm for ten years and the territorial sea forever. In 1982 the IFO stated that nothing less than a twelve nm exclusive fisheries zone would satisfy its members. It was a climb-down from the fifty nm demand but equally unrealistic because the arrangements for the territorial sea were more than twenty years in existence and had been generally accepted by member states of the Community.

These issues were highly emotionally charged but they also evaded some realities. The TAC and quota method of distributing landings among member states was deplored on the grounds that nations other than Ireland were depleting fish stocks and there was no justification for "sharing out the misery". A plea was directed to the minister to ensure that the countries which caused the problems bear the brunt of the loss. The industry in Ireland was in denial that its activities were contributing to the deepening crisis. Instead, Frank Doyle told his supporters and the establishment: "The Irish should not be penalised for the mistakes or greed of others".

Unfortunately the IFO also opposed a number of technical conservation measures introduced to manage fish stocks more sustainably. Its obduracy on herring in the Celtic Sea was a case in point; its refusal to comply with fishery closure was intended to be a demand for a wider regulatory regime. It also opposed the introduction of a 70 mm net mesh for *Nephrops*. Then in 1981, the IFO demanded a "sea prosecutor" be appointed because too many law breakers were getting away with poaching offences.[43]Paradoxical as this demand would turn out to be, it was in keeping with an ethos that foreigners breached regulations and the Irish industry would never dream of doing anything of the sort.

Meanwhile, the industry continued to be caught in the pincer of rising costs and declining supplies of fish, both exacerbated by growing fleet numbers and technology creep. In April 1980 the IFO protested at the rise in fuel prices borne by fishing, which had become ever more capital intensive and fuel-demanding, while agriculture could avail of lower prices; subsidies were demanded for fisheries in the interests of fairness. Later that year the IFO complained its members could not sell their catches. In a thinly veiled threat, the organisation announced that boat owners were faced with the alternatives of going broke or breaking the law. Obviously a breathing space was urgently required: fish stocks needed to recover so that they could become more abundant and valuable and the costs of catching them would decrease so that lower wholesale prices would yield proportionally greater benefits to boat owners. But the frantic treadmill onto which the industry had got itself did not allow the luxury of that kind of reflection; boat mortgages had to be paid. Racing to hell in a fish basket is an appropriate variation of the often quoted descriptive shorthand.

These demands were essentially populist, without any coherent policy running through them. Thus, while the fleet continued to expand, with consequent problems, the IFO insisted on the rights of its members to purchase new vessels abroad where keener prices were available, and insisted that BIM should put second-hand vessels on the market to facilitate the entry of younger skippers to the fleet.

Confronting irrationality in the fishing industry was never straightforward. The IFO blockaded ports and it was given to breaking out in placards and picketing the ministry and parliament in Kildare Street, Dublin. In September 1980 it organised a well received free give-away of fish to draw public attention to the plight of its members.

The IFO sought official mediator status with BIM where vessels were re-possessed. The problems continued to fester into 1984 when the fleet was acknowledged to be in negative equity.[44] The IFO issued proposals to sort it out by revaluing all vessels and recalculating and extending periods of repayment accordingly.

As the new producers'organisations made ground, they attracted larger vessels to their membership. In that respect the IFO lost prestige and must have suffered a reduction in income. When, in 1983, the Donegal salmon drift net fishermen sought representation, the IFO agreed to provide it and resumed at the point at which the NSIFA disappeared from view. Constant badgering is effective; there is an attrition of resistance and, in 1984, the minister Paddy O'Toole (Fine Gael, Mayo east) responded with a classic unworkable Irish compromise. The use of highly effective monofilament nets had been banned but now their use to catch other species would be permitted while possession and use of these gears to catch salmon would remain illegal.[45]

The IFO remained steadfast in its opposition to any interference with drift netting; it had opposed the short-lived 1980 levy on first sale of salmon. In March 1984 proposals emanating from the IFO for the management of salmon fisheries recommended that greater numbers of fishermen should be accommodated in a fishery which was widely regarded by scientists and the public as out of control.[46] Four years previously, a member of the naval service had reported frequently encountering illegal salmon drift nets of five and six nm in length.[47] Now the IFO recommended that permitted lengths of drift net should be increased and they should be constructed with deadly plastic mono- and multi-filament materials. Riparian owners should be obliged to implement restocking programmes and there should be greater penalties for river poaching. In short, those with responsibilities for the juvenile habitat would generate larger numbers of sea-run fish to supply the members of the IFO. Two years later, the IFO was protesting at the draconian penalties for infringements of salmon fishing regulations and urging that money spent on protection should be re-directed into restocking.[48] When a Helvic fishermen was jailed later the same year, a year that was reputed to enjoy the best salmon runs for years, the IFO proclaimed:

Legally or illegally (monofilament drift) nets have been in use by fishermen for more than a decade and salmon stocks are no nearer in danger of extinction now than then…

Another of the policies championed by the IFO was an inshore fishing zone restricted to small boats on which it made official representations. Later in 1986 an intriguing conflict of interest might have arisen when salmon fishermen (whom it represented) complained about incursions onto their grounds by trawlers whose cause the IFO had supported, although currently to a decreasing extent.

Like the NSIFA before it, the IFO was gradually being eclipsed by events and an industry which was given to fracturing spontaneously was being drawn apart by more topical leadership. The decline was gradual and never as conclusive as that of the NSIFA.

When, in 1980, a National Fisheries Advisory Council was appointed, four major national organisations were nominated to participate in it: the NSIFA, the Irish Fish Producers' Organisation, the Irish Fish Processors and Exporters' Association and the Irish Federation of Marine Industries. The IFO was not among them.[49] Three years later, its chairman, Joey Maddock, appealed for greater support from the industry[50] and just over a year after that the IFO, which had raised finance through subscription and a levy on catches, asked its members for an annual subscription instead.[51] In March 1985 it was announced that another former chairman of the IFO, trawler skipper Brian Crummey, was taking up fish farming, and the following September the IFO moved offices from Nassau St to less fashionable Fenian St in Dublin whence it has since continued to defend the industry at a more subdued intensity.

Producers' Organisations

When it first saw the light of day, the embryonic CFP had only two of its four policy legs: one dealing with structural funds and financial provision for fleets, the other with markets, although the latter was more an ambition than a fledged reality. The market policy would represent, in the words of the official user's guide to the CFP

...that large part of the industry which in fact represents the majority of the value chain...

Obviously, becoming a crucial link in the chain would be a prize worth fighting for. Producers' Organisations (POs) operate as agents of the Commission in Brussels. They function by compensating fishermen when prices for their produce fall below a certain level – in other words, they operate an intervention system. POs also "bring together fishers to manage the take-up possibilities (of quota species) over the course of a year so as to stabilise first-sale prices" in order to distil the highest available value out of the landings and to avoid gluts and shortages.[52]

Occasional mention was made of POs in the trade press before and immediately following Ireland's entry to the EEC, notably by Jim O'Connor who, as Advisory Services Manager and then Fisheries Development Manager in BIM, had a monthly column which kept his agency in touch with the industry. Early articles introducing the subject to fishermen were written by the editor of the *Irish Skipper*, Arthur Reynolds and by Frank Doyle, secretary for many years to the IFO. Doyle's first appointment as organiser of fisheries co-operatives was made by the Irish Agricultural Organisation Society (IAOS), and his efforts later created the IFO.

In 1975 the Irish Fish Producers' Organisation (IFPO) was nominated the first of the new administrative bodies.[53] It was expected to become operational on 1 February 1976.[54] Jim O'Connor of BIM was appointed its chief executive and secretary.[55]

Before the IFPO came into existence, a bruised ego or two had to be salved. On 24 September 1974 a meeting of fishermen's representatives was held at BIM's headquarters to hammer out rules for setting up an Irish fish producers' organisation, and Doyle attended. The atmosphere was cordial and there was apparently unanimous agreement on the way ahead.[56] The IFO would have had adequate credentials and high expectations to assume the role but was not offered it. At a meeting to launch the new PO, the parliamentary secretary Michael Pat Murphy made at least three inflammatory remarks which caused Joey Murrin, chairman of the IFO, to take umbrage, leave the meeting and write to the press about it. Murphy declared that Irish fishermen had no organisation to represent them, that the IFO was "just another pressure group" and that the IFPO would be recognised by the government as the representative voice of the fishermen. So much for democracy.

Another letter – from the secretary of NSIFA – sat alongside Murrin's in the trade press. NSIFA had been informed that its produce would never be subject to intervention because its value was too high and hence, that POs were of no relevance to its members. O'Connor, according to the letter, had also told them at an early stage in the formation of the first PO, that it was envisaged that membership would be confined to vessels of above 15 m and this turned out to be substantially if not absolutely correct. It is not possible to easily recognise which vessels belong to POs at the present time – membership lists are not published – but these organisations are known to be the preserve of larger boats although their organisers say they admit smaller ones.[57]

The IFPO was intended to serve all of the Republic, but three years later a second one came into existence: the Killybegs Fishermen's Organisation (KFO) whose secretary and manager was one, Joey Murrin. The start of his new occupation coincided with two other events: his resignation as chairman of the IFO and his appointment to the board of BIM. Murrin had stood as Fine Gael candidate in the Connaught Ulster constituency for election to the European parliament – unsuccessfully, although he acquitted himself well.[58] In time he would become BIM's chairman too and a member of the board of the Marine Institute.

A letter to the trade press from O'Connor some eighteen months after the formation of the IFPO claimed that 80% of trawler owners were shareholders in it.[59] That must have drained some blood out of the IFO which had been the representative organisation for vessels towing mobile gears.

The Irish Fish Producers' Organisation (IFPO), the first of its ilk in Ireland, had a chief executive of relatively quiet disposition, although he was a fluent communicator; Jim O'Connor had worked in BIM for a number of years before taking up the post. In contrast, the second, the Killybegs Fishermen's Organisation whose business would be pelagic landings – particularly those of the lucrative mackerel fishery - had a leader who was no stranger to publicly and loquaciously expressed his members' interests, in whose cause, he was far-sighted.

Early in 1993 a new organisation of fishermen splintered from the IFO in south west Ireland, for the stated reasons that quotas for angler fish and hake were inadequate and that BIM funding for fleet modernisation should be greater.[60] The Irish South and West Fish Producers' Organisation also agitated in favour of drift netting for albacore tuna, an activity which was vehemently and effectively opposed by Spain.[61] On 19 November 1994, the first meeting of the IS&WFPO took place in Skibbereen, Co Cork. It was the third PO to be formed.[62] Its creation was another wound inflicted on the IFO.

The South and East Coast Fishermen's Co-operative Society Ltd was formed in 1962 to represent the interests of herring fishermen. Based in Dunmore East, it was responsible for organising auctions and achieved much in uniting the fishermen of the south coast; it also arranged and administered local conservation measures.[63] In 2002 a newly formed South and East Fishermen's Organisation assumed a wider brief.[64] The following April it announced it had received recognition from the Department and was on an equal footing with other representative organisations; some twelve or thirteen vessels were "about to join".[65] In late 2004 the IS&EFO received official notification that it had become a PO.[66] In January 2006 the IS&EFO was reported to have 60 members. Although a PO it had not adopted that title.[67]

In 2006, the four POs, IS&WFPO, IS&EFPO, KFO and IFPO united under the umbrella title Federation of Irish Fishermen (FIF). They claimed to represent more than 90% of full time fishing vessels over 12 m in Ireland.[68] The FIF and its constituent organisations, in association with the government department and BIM would become a "fisheries establishment" , dominating policy in the future.

POs will play a prominent and privileged role in the remainder of this story. Their *modus operandi* would further enhance the status of larger vessels and add considerably to the problems of the smaller and more numerous components of the national fleet.

Chapter 11

FISH PRODUCERS' ORGANISATIONS TAKE POSSESSION

Producers' Organisations (POs) were devised to grease the wheels of commerce. They guaranteed that the producer – the fisherman - would obtain a reasonable income by intervening in the market when prices had fallen to a level at or below which fish would be withdrawn from sale. The PO would then compensate the fisherman at an agreed price, from EEC funds, and the fish would be sold for fishmeal manufacture or dumped. In order to participate in this, individual boats (not fishermen's representative groups) had to purchase shares in a PO or become members.

POs were crucially about products, not locations, and the development of the Community had emphasised that bias time and time again. In business one could not discriminate against another member state on the grounds of nationality. Community institutions kept local problems at arm's length; solutions were applied through Community, not national, mechanisms, in accordance with the principles underlying the Treaty of Rome. It followed that EEC policy viewed local, inshore fisheries organisations askance because they were essentially restricted to small geographical areas. This did not mean that the Community was unsympathetic to inshore fisheries. As we have seen on several occasions[1] Brussels identified a procedure allowing provision to be made for them but there simply was no mechanism available to do so within the Commission's armoury.

POs were expected to prepare fishing plans that ensured the annual harvest took place at an organised pace which allowed the market to absorb what was caught rather than making erratic landings resulting in gluts and shortages. POs were intended to benefit the consumer by keeping wholesale prices low. They were supposed to trade across national boundaries and to deal with large quantities of marine produce to supply Community demand. It was to be expected that their members or suppliers would be larger vessels making substantial landings.

POs were also an acknowledged vehicle to petition the Commission on a variety of Community matters of concern to their members.[2] It did not take long for their ambitions to extend beyond distributing quotas, to demanding heavier ones.

Within general rules of procedure, there was latitude for national and regional variation. A review in 2001 listed some: Denmark's POs were not concerned with fisheries management; POs in the Netherlands tended to embrace boats targeting a variety of species, rather than those regulated by TAC and quota, but there were differences in the way they treated fisheries for cockles and for mussels. In the UK, as in Ireland, smaller vessels were not members whereas in France some artisanal boats were included.[3]

Among fishermen's representative organisations of their time, POs conferred enormous wealth. In the UK they were reckoned to handle 95% of landings of quota species in 2001.

By 2001, the IFO had been relegated from representing the trawling fleet – the principal fishing method in use – to defending the interests of dispossessed inshore fishermen.[4] The organisation represented the interests of salmon drift net fishermen and the operators of smaller craft. On several occasions the IFO attempted to secure zonation favouring smaller vessels, displacing competition from larger craft to further offshore. In 1975 Murrin was part-time secretary and public relations officer of the Killybegs Fishermen's Association which made what were regarded as important proposals to regulate the activities of vessels of different sizes and provide protection for smaller ones from the activities of larger boats in inshore waters. Murrin's move to head up the Killybegs Fishermen's Organisation (KFO), a PO, coincided with a change of attitude within the industry.[5]

Killala Bay, sitting between Cos Mayo and Sligo, opens to Killybegs to the north-north-east of it. In August 1990 large RSW trawlers (refrigerated sea water boats, equipped with tanks in which fish were stored), members of the KFO, up to nine at a time, invaded the bay, which was inside the baselines, to fish. These vessels were among the 3% of largest boats in the national fleet. The local inshore fleet numbering some 35 boats saw their fish scooped up and removed and they also lost their gear – gill nets and pots in the same way. Murrin, speaking on behalf of the KFO, commented that if the reports were true, he could not condone such activities - "We can't defend the indefensible".[6]

Within months, the same problem manifested itself closer to Murrin's home base. As the crow flies, Inver Bay is less than 10 km to the east of Killybegs. In the midst of the Killala incidents, local fishermen working in the bay had similar experiences. On occasion the navy was asked to intervene to impose order. Murrin volubly opposed the use of the navy in these circumstances and told a local newspaper that "millions of pounds worth of navy boats were anchored in the area while Spanish vessels were poaching off the west coast." The same report stated that the KFO was prepared to contest a supposed "law" which excluded its trawlers from Inver Bay.[7] Such a regulation did indeed exist. Enacted on 16 February 1857, the bye-law was, in the opinion of a reviewer of the legislation in the late 1970s, devised to protect the interests of those engaged in using static gears (long lines and trammel nets).[8] The commercial interests of the largest pelagic boats, clients of the KFO, had replaced the inshore causes championed by the Killybegs Fishermen's Association and the IFO. What had happened in this case was another metaphor for the industry: owners of the largest boats wielded the greatest influence in the way policy was shaped. That influence would be exercised in a progressively more acquisitive manner.

Adverse reports of the activities of RSW vessels continued to accrue. When, a year later, these boats were said to be rampaging through static gear set for crab along the Mayo coast Murrin was defensive, replying that other fishermen should be more careful to conspicuously mark their gear:

Blanket allegations of indiscriminate destruction of static gear by the Killybegs boats along the Mayo coast will be hotly rejected in the future and unfounded allegations will be contested legally.[9]

Incidentally, the protesters argued that the boats were fishing illegally so close to shore but they were unable to seek legal redress for their problem. This illuminates the regulatory confusion surrounding the operations of large vessels in the territorial sea and demonstrates how ineffective supposed, dormant or simply non-existent regulations were to protect their livelihoods.

Other incidents demonstrated the rising perspective of the ambitious leader of a PO. In June, 1991, the government declared Irish waters to be a whale and dolphin sanctuary. The gesture was strongly personally associated with Charles J. Haughey who was taoiseach at the time and had a genuine interest in natural history. The announcement came from his office, not the Department of the Marine which, at that time, was responsible for fisheries. It was the kind of environmental gesture which had become characteristic of the nation: it looked and sounded good and it did not (apparently) discomfort any vested interests. In fact, the sanctuary was "not a legal entity…it did not have the legal status of a reserve or refuge". Whatever legislative protection could be afforded to whales and dolphins already existed in established law.[10] And that was poor enough. The UK Whale and Dolphin Conservation Society described the declaration as meaningless as long as the Irish government allowed its fishermen to drift net.[11] Murrin anticipated that any conservation measure might obstruct the use of certain gears, particularly gill nets for albacore tuna, to which dolphins are vulnerable, and he described the lack of prior consultation as "irresponsible and high handed".[12]

As time passed, the use of marine resources, decided after consultation and participation by "stakeholders", became the new norm. Fishermen's organisations, principally POs in Ireland, would become the only stakeholders who mattered and they would, progressively, take whatever decisions they were able to on their own, without reference to anyone else, not even their fellow fishermen in other representative groups within the catching sector. When, in 2002, a committee in the Department of Communications Marine and Natural Resources was set up to review vessel licensing policy the only representatives of the industry were drawn from the POs; the inshore vessels, accounting for some 80% of the total were not represented on it.[13] Unfortunately, many of the utterances of the POs opposed measures introduced by Brussels on the basis of scientific advice to curtail fishing effort in the longer term interests of the industry generally.

Scientific investigation and stock assessment of resources had been undertaken since the foundation of the state by government inspectorate rather than semi-state agency. In my view, that is the appropriate way to provide dispassionate and objective scientific advice, at a remove from vested influences. But the inspectorate in question was unproductive. In 1992 the government department responsible for fisheries set up a Marine Institute which would undertake all scientific investigations in the future, to which the staff of the inspectorate were transferred. The institute was expected to generate a proportion of its funds through consultancy rather than rely on government hand-outs for everything, and that concentrated minds; the institute has had a considerably greater work output than its predecessor. The prospectus for the Marine Institute announced it would be concerned with a range of marine activities including fisheries, aquaculture, amenities, recreation, engineering and electronics. Murrin was quick out of the traps complaining that its board did not include a representative of the fishing industry. An apt simile was to hand to capture the deficiency, describing it as "...like building a boat worth IR£8 million without a rudder to steer it".[14] Murrin had the vision that fishing matters would steer the Marine Institute, which meant putting POs, in which he was then the driving personality, behind the wheel. And so it has been since; the Marine Institute has had up to two representatives of the POs on its board of eight whose number includes the chairman. Nobody from the majority of vessels in the fishing fleet, the inshore boats, has so far been a member.

One of a minority of leaders of POs to date, Murrin had experience of working at sea as a fisherman. A growing majority of his successors, men, and recently women, have had none. Increasingly they have recruited from the new establishment made up of the agencies charged with administering aspects of the industry and the department responsible for fisheries – one incumbent was employed both there and in BIM. Another executive moved the other way, back from a PO to head up BIM. The naval service provided a third. Their background has varied from administration within the industry itself to the professions: economics, business organisation, engineering and law; they have served on the boards and committees of other state agencies associated with the fishing industry, BIM itself, the Marine Institute, which prepares scientific assessments of fishing opportunities, and the Sea Fisheries Protection Authority, which regulates the activities of the catching sector. Skilled lobbyists, the leaders of the POs, formed strong political links, particularly with Fianna Fail which was in power for the majority of years in which industrialisation of the fleet took place. The links have been particularly strong in the constituency of Donegal west where the largest fishing port in the country, Killybegs, is situated. The local champion of the industry, Pat the Cope Gallagher (Cope deriving from Co-operative, a family association with the fishing industry) served as minister in the Dail and member of the European parliament where he is the acknowledged leader of the industry in Ireland. However, the

Fine Gael dominated coalition's minister of Agriculture Food and the Marine, who came into office in 2011, formed immediate and strong bonds with the Federation of Irish Fishermen, the umbrella group within which three of the four POs currently cluster.

Long gone are the raucous days of brandished placards. These have been supplanted by lobbying, their upmarket equivalent. Members of the KFO, the most prosperous of the four, include the "mackerel millionaires", who own some of the largest fishing vessels in the country; their product is large volume but low unit value. A disaffected founder member described it in 2004 as having become a "pinstripe suit and bow tie operation".[15] There is a massive gulf between the way this and other POs function and the lives and work of the majority of fishermen, whose employment in peripheral coastal regions is used to justify the existence of BIM.

POs operate by representing their members to the establishment, and they deliver within the constraints of European regulations. Leadership is said to be stressful. Some skippers are reputed to be alpha males, insisting their demands require immediate solutions; the industry is not characterised by its longer term perspective.

Interactions between POs and the other elements of the establishment are frequent but one of their established rituals, which is profoundly misunderstood by the public, largely because its purpose is misrepresented, is the annual divvying up of the Community harvest of species regulated by TAC and quota. The process operates in the following way:

The species in question are the most important of our food fishes, mainly finfish. The list of exploited species has lengthened with the passage of time as more, seen to be vulnerable to over-exploitation and in need of regulation, have been added to it. In the share-out which took place in December 2011, Ireland had interests in some 25 finfish (mostly single species) and one crustacean, *Nephrops*, varieties distributed among 64 "stocks" in the north east Atlantic. The combination of a species and statistical sea area (composed of one or more statistical divisions) is described as a "stock unit".

All of these stock units are shared among two or more member states whose scientists assess them. The analyses are pooled to ascertain what catch a stock might yield without impairing its ability to reproduce and provide landings into the future, so that the fisheries remain sustainable. The analyses are peer-reviewed by a number of committees associated with the International Council for the Exploration of the Sea (ICES) and the Commission in Brussels. In early autumn each year the Commission publishes a list of "fishing opportunities", setting out what removable tonnage it recommends from each stock. The figure is the proposed Total Allowable Catch (TAC) which is then subdivided among participating nations in pre-determined percentages (quotas). Despite the assessment work undertaken on stock units – it can be expensive, involving as it does research cruises, laboratory analyses and workshops – predicting the future is never straightforward; the natural environment is subject to too many variable influences, many of them incompletely understood. For this reason and because there are insufficient resources to examine every stock unit at a desirable level of detail, the CFP recognises the "precautionary principle" which broadly states that if one is uncertain about the status of a fish stock, one should err on the conservative side rather than suppose that lack of information means the stock is in robust good health, and act accordingly.

Lack of information tends to be interpreted in another way by the POs who assume that shortage of data means there is no reason to exercise restraint where landings are concerned. In addition to the fact that it is the human condition to toil to attain greater wealth, there is the deepening indebtedness of the fleet, widespread throughout European nations, and further afield, and this is frequently cited as the excuse for heavier landings than recommended by the Commission. The argument is cited as "socio-economic" which, broadly stated, goes: "we cannot survive on that TAC, we need more". Taking more means eating into the brood stock which replicates future generations of the species. The inevitable result is a spiral of decline and the evidence of that is to be seen all around us.

After publication by the Commission of the advice on fishing opportunities, the minister is lobbied by POs to get a better deal. Lobbyists vie with scientific advice and the strategy adopted by the minister is invariably a compromise. Whatever is agreed, the scientists advising the minister are expected to row in behind him and support his stance, even if it runs contrary to the results of their deliberations, undertaken at considerable public expense.

Matters come to a head at the Fisheries Council meeting in Brussels in late December. The minister does not go to Brussels alone; he is accompanied by an entourage of supportive scientists and departmental officials while a delegation of PO officials and members provides encouragement, in the nature of a claque, urging the national representative to get them a better deal. A flavour of the decision-taking all-night meeting was captured on film for the KFO :

...viewers can see the...tiredness and the tension as the fishing industry representatives await their fate for the coming year...[16]

The pressure on a minister not to disappoint his constituents' expectations might also be imagined. Final decisions may not be taken until the facts are debated at a tripartite confrontation between the minister responsible for fisheries, the fisheries commissioner and the appropriate minister of the country which currently holds the presidency of the European Union. The size of TACs can owe much to personal and political interactions among these three.

The December council meeting is a high point in the fisheries calendar. The publicity build-up to it, faithfully documented by the print media, presents the event as a trial of strength in which the national David wrests a greater allocation of fish from the European Goliath. When the minister returns to Dublin, should he have breached the recommendations which were based on scientific enquiry, his trip is hailed as a success.

The misunderstanding of the process lies in the fact that Brussels does not have any fish to withhold. Securing a larger quota for Ireland means doing the same for any other nation which is party to exploiting a stock unit. Quotas follow automatically from TACs and they cannot be independently justified and fixed for any state. And raising the TAC means stealing fish from the future and reducing the size of the resource in the years ahead. The extent of the damage done by the process wherein the Commission proposes but national, short sighted politics, dispose, is quantifiable and published. Between 2003 and 2010 for example, political squabbling inflated TACs by an average of 47% above levels recommended on the basis of scientific assessment. In 2005 the figure was 56%.[17]

There can be no denying the fact that the CFP was a disaster. The first problem, corrected far too late, was the over-generous allocation of subsidies which allowed fleets and infrastructure to expand beyond the capacity of the resource to support them. The mechanism for managing and allocating landings might have worked but the procedure allowing politics driven by irrational demand and greed to take precedence over science is its ultimate flaw.

The CFP is a creature of the EEC and its successor treaties. It is their original sin. It is however, interpreted by the industry as a convenient scapegoat for all of the problems associated with a complex industry which takes no responsibility for what has occurred. In other, puny, ways the Commission has tried to rectify its errors. It has argued for rational exploitation of resources but does not have the power to enforce those recommendations. It has advocated more sustainable inshore fishing methods but, as a prisoner of its founding legislation, is unable to do more to impose them. Finally, under other directives, such as the Habitats Directive,[18] it has applied the brakes to the certain rush to extinction of Atlantic salmon as a result of over-fishing by drift netting. A major question is what the national government and the industry in Ireland have done to conserve or sustainably manage any wild fishery: the answer is precious little.

Intervention

Not long after the first PO became operational in 1976, the long anticipated market intervention system was activated. The procedure was straightforward enough: if the landings of certain species could not be disposed of at a market price above a specified threshold through normal commercial channels, it was withdrawn from sale, its captors compensated and the fish reduced to meal, the lowest valued product. If, however, factory capacity were not available, the fish was sprayed with gentian violet, a dye, and dumped back into the sea. The purpose of the dye was to prevent their being scooped up by other enterprising fishermen and sold at withdrawal price on subsequent occasions.

Intervention then, could be said to have opened another market opportunity. Before long however, public unease was being voiced about the fate of mackerel, dumped because the market could not absorb them.[19] In 1979, the Commission expressed its disapproval that large landings of this species were being brought ashore, not to supply the market for human consumption, because it was known in advance that the commercial market could not absorb the fish, but rather, for intervention.[20] It was then made clear that no funds would be made available for intervention in the case of species whose quotas had been exhausted.[21] However, quotas were being set at higher levels than recommended on the basis of scientific assessment as a result of the political pressures described earlier.

In November 1981, the European Parliament adopted a resolution deploring the fact that 100,000 t of fish had been destroyed through intervention throughout the Community in 1980.[22] Later in 1981, a meeting of fifty POs demanded higher withdrawal prices: France wanted an increase of 60%, Denmark 20% and the Commission offered 6%.[23] Problems were anticipated, should withdrawal prices increase; wholesale market prices were expected to move in the same direction and that could well result in greater quantities, unaffordable to the public, going to intervention instead.

In March 1982 almost 5,000 t of herring were withdrawn from sale due to lack of demand and almost all were dumped. The bidding at auction opened at IR£7 per box, the minimum allowed under intervention rules, and when that threshold was not reached, the fish were condemned.[24]

As a policy, intervention did not find favour with the general public. When, in 1982, 150 t of herring were dyed and dumped into the sea off Howth because the market could not absorb them there was adverse comment by both fishermen and the press. The event took place on the instruction of the IFPO because the acceptable threshold price had been reached. The IFPO paid IR£5 per box for the fish which, had they sold, would have obtained IR£8. The incident coincided with the jailing of five fishermen on charges of illegally fishing for herring; five others had been released on bail for a similar offence.[25]

Responding to the spreading concern, the Commission announced it would apply withdrawal compensation on a sliding scale: the heavier the landings, the lower the rate of remuneration.[26] That was not welcomed by the industry. Withdrawal of mackerel had caused most concern because of the large quantities involved, but the dumping of herring in the Celtic Sea, the rebuilding of whose stock – an exceptional event in the history of Irish fisheries - had required a great deal of sacrifice by fishermen, provoked uncharacteristic statements of unhappiness from the invariably docile scientific community.[27]

Difficulties in distributing fish to the home market were an explanation for so much going to intervention. The industry was becoming sensitive to public disquiet at the destruction of human food. Almost four years after being set up, the IFPO, claiming to represent 85% of trawlers, decided to establish a retail outlet in the salubrious suburb of Sandymount to sell some of the landings which would otherwise be condemned.[28] It was a welcome initiative but an inadequate measure. The IFPO reported that in 1981, 5,000 t had been destroyed for which IR£550,000 had been paid to fishermen, some of whom where accused by the organisation of targeting fish for this purpose. (A later report on 1981 amended the earlier 5,000 t to 6,000 t of fish withdrawn from market sale at a cost of IR£550,000.[29]) In some ports, 20% of the landings had been withdrawn and the IFPO remonstrated that the amount should not exceed half that percentage.[30] The problem was that rising fuel prices were again imposing stress on the industry which needed intervention payments to remain in business. Indeed BIM reckoned that intervention would keep only some heads above water.[31]

Harvesting food fishes which would subsequently be condemned had other consequences too. If a skipper knew his catch was destined for intervention he took less care with boxing and icing it and this fostered a careless approach to handling wet fish which would result in a less valuable product when it was presented to the open market.[32] BIM at the time was endeavouring to improve the way fish were handled at sea and intervention was encouraging bad practice.

Despite strong reservations in the public mind and in the Commission about the use and effects of intervention, the IFO in April 1981 proposed, in association with the new KFO and the IFPO, a "top up" on intervention payments by the Irish government. The case for additional money was made in terms which could not be ignored: conditions were so dire in the industry that boat owners had two options, going broke or breaking the law. Whatever about going broke, no politician wanted to endure the embarrassment of tolerating law breaking while the malefactors flaunted their offences. In June a subsidy package was launched. A total of IR£1.6 million was earmarked for distribution to the industry through the newly formed POs or *via* fisheries co-operatives. IR£1.25 million would support whitefish and mackerel boats, IR£200,000 prawn (*Nephrops*) vessels. The amount would be additional to intervention payments made through the European Commission.[33]If it had not been the case hitherto, the industry was now irrevocably wedded to high subsidy levels.

The Celtic Sea herring fishery was the second largest in the country yielding landings which fluctuated dramatically. The stock was fished intensively by Irish and continental fleets from the 1950s to the mid-1970s. Landings peaked at 44,000 t in 1969 and then declined. The fishery was closed in 1977 and did not reopen for five years, although it was never adequately monitored in the interim. When it resumed, landings failed to regain their previous peak and demand for herring slumped.[34]Loss of market, once a species has been removed from the fish monger's slab, poses a serious risk to its commercial viability. Its absence clears space for other tastes and preferences to become established. Closure of the herring fishery had given mackerel an opportunity to gain a foothold in consumer affections. But faster swimming mackerel were not so readily landed and herring continued to be brought ashore in large quantities.

Accusations that fishing was taking place purely for the purpose of getting intervention payments persisted. Herring landings were particularly associated with the practice which continued into the 1990s.[35]In 1992 greater landings of herring from the western Celtic Sea depressed market prices. The industry sought higher compensatory payments at the same time as demanding higher quotas thereby creating the impression it was capturing fish only to bring them ashore, claim compensation, and then dump them. Ireland could not be seen to seek higher quotas and intervention at the same time.[36]

Harvesting marine produce to obtain subsistence money while at the same time destroying valuable human food which the market could not absorb, thereby jeopardising future fishing prospects, was symptomatic of the desperation of the early 1980s. There simply was not enough fish to support all of the vessels involved, so some boats would have to be sacrificed. The smaller and less politically powerful members of the national fleet had already been earmarked. In the years ahead the centres for the larger vessels would relocate to Killybegs, Co Donegal (pelagic fleet) and Castletownbere, Co Cork (demersal) and the other fishery harbours at Dunmore East, Co Waterford, and Rossaveal, Co Galway. Howth, Co Dublin, the principal fishery harbour on the Irish Sea, would be gradually forsaken to become a quaint tourist attraction, its fish shops and restaurants along the pier supplied with produce from abroad delivered, not by boat, but by plane through Dublin airport, a greatly reduced fleet of mainly *Nephrops* fishing vessels providing a suitable backdrop for relaxed dining.

Towards the close of the millennium, POs were urged to use intervention mechanisms sparingly and only as an emergency measure, without distorting the operations of the market.[37] Once the realisation of stock collapse was accepted, the EU proposed suspending the intervention system for any species subject to a recovery plan.[38] However, not all whitefish species were candidates for rehabilitation. In the spring of 2002, favourable weather conditions led to large whitefish landings in the south west. In one week in early April of that year, 50 t went to intervention. A local fisherman complained that large boats, targeting whiting, were capable of taking "a two-month quota in one day's fishing".[39]

The fate of whiting, once the staple of the Irish industry, was the inevitable consequence of over-fishing.

Whiting: "the most popular... easily caught... apparently inexhaustible species"

It would be useful at this stage to examine how a species of particular interest to the POs fared throughout the history of industrialisation up to and through the introduction of the new intervention system.

Whiting is another iconic species which illustrates the development of the industry from the 1950s: a species which was abundant in the north east Atlantic, rare in the warmer waters further south; a member of the same family but considerably smaller than cod, it reaches 70 cm in length and three kg in weight. Whiting formed a considerable proportion of the landings from the Irish Sea. In the North Sea its landings increased after the Second World War and exceeded 90,000 t annually.[40] The species is piscivorous, preying on the juveniles of other fish species and it also consumes crustaceans.

A succession of whitefish have, at one time or another, dominated the consciousness of the industry and the purchasing public. Today our preoccupation is with angler and cod but, in 1954, the most popular demersal species was whiting; herring was its pelagic equivalent in the public taste. Herring was in such great demand that when it was not landed in sufficient quantity to satisfy the market, it was imported. There was also a practical link between the fisheries for herring and whiting. A successful whiting fishery in the summer provided the finance to pursue herring in the winter months. Whiting was the most abundant whitefish species, making up 37% of the landings and it was also cheap, 30% the price of cod.

Reflecting its importance, one of the first Irish whitefish studies was of whiting which was captured mainly by seine nets. The males were reported to be sexually mature at 20.3 cm, females at 24.1 cm. The latter was the size limit fixed in 1951. The catches consisted of small fish, averaging less than 200 g in weight and the majority of the landings were approximately two years of age.

By 1955 the boat building programme organized by BIM was well underway. In that year the price of whiting rose and the trade press reported it occasionally exceeded that of cod. Whiting was

...the most popular fish of all. Even at periods of dearth there is no big swing to other varieties...

It is a salutary reminder – and there are others – of the necessity of managing fisheries towards stable output in order to sustain markets for the species. Public taste is fickle and a dearth of supply opens opportunities for other species and flavours.

The early years of fleet industrialisation were accompanied by shortage of investment (in the stridently expressed opinions of those who favoured it) and frequently acrimonious debate. The quality of inshore-caught fish could be maintained because it was captured within easy reach of markets. Impossible to grasp now, freshness was an elusive quality in the 1950s.

Whiting was the barometer of inshore fleet success. In 1957 the price of whiting fell from stg£1-15 shillings per 50 kg to stg£1- 5 shillings (29% reduction) and in June reached stg£0-3 shillings (a further reduction of 88%). By August the bottom had fallen out of the market for cod, and whiting was being thrown overboard. Within the industry the mood swung around. What was required, it was suggested, was more prime and less seine-caught fish; more fish in winter, less in summer. In April 1958, the problem for the whiting fishery was reduced to easily understandable terms in the trade press. In order to survive, the industry had to sell twenty boxes of the species in each county every Friday!

Despite the frustrations and vicissitudes, the possibilities of a fishing industry proved attractive. Erskine Childers (Fianna Fail), minister for Lands, gave a speech in 1958 warning that an export trade in fish was essential if Ireland were not to become a tourist paradise and cattle ranch in fifty years' time. We were, he recognized, the third lowest consumers of fish in Europe.

Where there is a vacuum there will be a plan to buy one's way out of it. In 1958 a scenario was advanced requiring stg£10 million investment over ten years (approximately ten times the current fisheries budget). This, it was forecast, would give full time employment to 10,000 men, rather than the 1,500 at that time. The fleet would expand to between 4,000 and 5,000 vessels. The prospects for whiting landings were extolled as sufficient to generate momentum for an export trade; the species obtained 50% higher first value sales in the UK, and in France fetched three times the price.

Late the following year, whiting again returned to favour when it was described in the trade press as the most unappreciated of all species and still the most valuable after herring. The species was

...edible, easily caught, abundant, in constant steady supply and apparently inexhaustible...

However, heavy landings led to gluts and small size made for poor quality, and the landings were fetching £stg0-7 shillings per 50 kg, even when there was little else on offer. As early as 1957 it had been noted that the average size of whiting landed was getting smaller. The industry began to recognize that the quality of landings should be improved.

In April 1960 the trade press contained a report by an inspirational driving force behind the fishing industry, John K. Clear. Cod and haddock, he reported, provided good fishing from time to time but whiting was the only whitefish which was abundant in Irish waters. Some 100,000 t were harvested from the north east Atlantic of which Ireland took 3%. Whiting was the species generating the third heaviest landings to France and the sixth most valuable, twice as valuable as the total fin and shellfish landings to Ireland. In 1959 France's landings of whiting exceeded by 50% Ireland's total marine landings of all species.

By mid-1961 questions began to be raised about over-fishing. The trade press recognized there were problems with cod, plaice, haddock and hake and mentioned the possibility that whiting, "the backbone of the industry", had suffered the same fate. Had there even been a "gross mismanagement of fishery affairs?" it inquired.

The second investigation of whiting, carried out in the mid-1960s,[41] provided a comprehensive account of the species and some insight on its fate. At that time mobile gears were changing over from seining to trawling. Seining, the author observed, targeted dense shoals of demersal fish and those of whiting were made up of young individuals averaging one year old. Trawl-caught whiting were of better quality. Such was their quality that fillets from the largest individuals were occasionally sold as haddock.

Then, in the late 1960s fishing commenced supplying the Mornington fishmeal plant in Co Meath. The fate of whiting, a relatively low-value species (it had stabilised at about half the price of cod) which topped a food chain consisting of sprat and sand eel, among other species, was under consideration; perhaps these fodder fish species might be put to better use if they were consumed by a predator of greater value than whiting? Uninformed fantasising of this kind is not uncommon among fishermen and sometimes scientists who should know better. A new legal regulation reduced the minimum length of whiting captured for fishmeal to 20 cm; the existing size limit (24.1 cm) was retained for whiting for human consumption. At the same time, a trawl fishery for another species of much higher value, *Nephrops*, had begun to gather pace and whiting became a by-caught casualty of that.

The industrialisation of the fleet intensified in due course, reaching full bloom after Ireland's entry to the EEC. Landings of whiting to Ireland peaked at 15,643 t in 1981. It had much in common with other whitefish species, registering heavy yields until the TAC and quota system of management was introduced. Then we had the sad spectacle of quotas, set unrealistically out of reach of landings, both declining year by year. Ireland's total landings of this species in 2009 were less than 2,600 t. The last refuge of the species is the Celtic Sea. In the Irish Sea, once its great producer, official landings to all nations fell from 11,700 t in 1987 to 100 t in 2004.

The fate of whiting is not dissimilar to that of cod but its landings during the twenty years after 1950 were much greater.[42]

Mention of whiting as a candidate for intervention payments was made in the trade press in 1972. Other likely species were herring, cod, saithe, haddock, mackerel and plaice.[43] They would be joined by hake.

As whitefish species declined, the fauna on which they fed were relieved of predatory pressure. Populations of fodder organisms like *Nephrops* (prawns) and marine snails (whelk), previously held in check by species like cod, saithe and whiting, grew and multiplied without hindrance. In due course they became fisheries in their own right. Were they managed any better?

The decline of top predators

The removal of finfish provided a respite for the invertebrates on which they fed. In the absence of predators, shellfish stocks expanded. Some of these were the so-called "new" fisheries marked by BIM for development; in fact they were new only in the sense that they had recently joined the list of species to be ruthlessly exploited without any management constraints.

The government department for fisheries was ultimately responsible for rationally harvesting marine produce. It had its hands full reconciling the larger boats to the Community regime. There was little to be done for smaller craft which were left to their own devices. Inshore regulations were not enforced; had they been there might well have been greater unrest to add to the problems of the industry and a thorn in the side for politicians; a large proportion of employment in the catching sector resided among vessels of less than 12 m. Many of these craft were fully licensed and entitled to fish using a variety of methods. These entitlements would have required them to purchase, on the open market, had they not already been in existence when the CFP was set up, tonnage and kW. Those unfortunate boat owners now had to work alongside others who fished without legal entitlement but with impunity, simply because the authorities were not interested in enforcing the law.

The harbour of Arklow is situated 67 km due south of Howth. It is one of the east coast's larger harbours with an established marine tradition. In the early twentieth century, only Arklow and Howth on the east coast of what would become the Republic had fleets of motor-propelled boats. Despite being disadvantaged as a fishing centre – Arklow had no fish market and was subject to siltation – its vessels were noted for the versatility of their methods and the ease with which they moved "according to disposition, over the waters of the entire southern half of Ireland".[44]

By the late 1980s, the fishing port of Arklow was a shadow of its former self. The older wooden boats of characteristic design with pointed ends, bow and stern, constructed by the ISFA and BIM, had retired to the harbour, smaller numbers to its neighbours, Courtown and Wicklow, some of them to rot there until they were finally broken up early in the new millennium. The vessels ranged in size between 6 and 19 m oal. The majority had, at their launch, been intended to fish trawls, but the combination of more modern and powerful craft with a lower density of fish, had eliminated them from dragging mobile nets. Instead, the relict fleet had retreated to using static gears, gill nets and pots or traps. Their major target was a marine snail, whelk.

Whelk is a large gastropod, growing up to 10 cm in length. It occurs at different densities in all waters around the Irish coast. It is a fodder species for whitefish but is also consumed by large crustaceans, lobsters and edible crab, particularly the latter. Among high concentrations of crustaceans its prospects of being eaten increase, and the snail has the ability to grow a thicker shell in these circumstances. Apparently, the development of heavier armour diverts energy from other life processes and whelk, which adopt this survival strategy, mature more slowly than those that have a thinner shell. The snail does not have a planktonic stage and it occurs in self-replicating "patches" or "stocklets", unevenly scattered over the seabed of the south west Irish Sea. Once thick-shelled whelk patches are harvested, they require a long time to recover and rebuild; these populations perform rather like K-strategy species.[45] The Irish Sea has small numbers of edible crab – less than 1% of landings from the seas around the Republic are sourced there; it is no coincidence that a large biomass (a maximum of almost 40,000 t was estimated) of thin-shelled whelk occurs there too, at times. The antithesis of the thick-shelled form, lightly armoured whelk grows and matures rapidly and is fecund. This fishery can withstand heavy exploitation and recover quickly afterwards.[46]

We have records of a small whelk fishery in the Irish Sea exporting product to British markets in the 1960s but its landings probably never exceeded 100 t annually. Later, the removal of predatory whitefish species released a population expansion coinciding with a collapse of whelk populations in parts of the Pacific Ocean which had hitherto been harvested to supply south east Asian markets. South Korea in particular, has a large appetite for marine gastropods, known locally as "bai tops", which are reputed to be an aphrodisiac. A whelk fishery to supply this market was developed by fish processors in south east Ireland with the assistance of BIM. Landings between 1995 and 2006 fluctuated between 3,000 and 9,000 t annually.

The whelk fishery was not a costly one for fishermen to enter. Traps were constructed from used industrial polythene containers and the most expensive element of the equipment was probably the rope used to anchor and retrieve them. It was the ideal fishing method for vessels that had passed their earning prime but the fishery required regulation if it were to perform at an optimal level. A mechanism to do so was readily available.

It is an old and trusted maxim of managing fish that an individual should be allowed to reproduce once before being harvested. Whelk in the south west Irish Sea matured at an average of five years old and at a fairly uniform size. The obvious way of conserving stock was to release back to the sea those which did not reach the critical dimension. It was not a difficult objective to secure. The animals could be sieved through automatic graders as the catches were brought aboard, and the undersized ones discharged to the ground from which they had been captured.

In the early days of the export fishery whelk were dealt with manually in the factories which did not welcome small individuals because they were uneconomic to process. Later, machinery was developed to cope with all sizes and the inclusion of small individuals ceased to be a problem. However, whelk size was a significant instrument of conservation. The EU Commission specified a dimension and the Department in Ireland transposed it into national regulation.

The regulation was never enforced. Nor were licensing rules applied to the Irish Sea fleet of relict boats. At some ports, more than half the individual whelk brought ashore were beneath the critical size dimension. The activity pattern of the fleet increased as the biomass of this, the only remaining species it had to pursue, rose and fell as the abundance of its prey fluctuated. The Department appeared to take the view that the fishermen should regulate themselves, which was a convenient way of shrugging off responsibility for it. Unfortunately, few fishermen burden themselves with the labour of complying with regulations involving the loss of sellable product unless their fellows are obliged to do likewise. Over twelve years of monitoring this fishery, the number of undersized whelk averaged 30%, their weight 16% of the landings. Thus, a fishery which could so easily have been put on an even and productive keel, was allowed to lurch from boom to bust. It was typical of the way business was conducted within the six nm exclusive limit.[47] It was also eloquent of the attitude to rational fisheries management within the fisheries establishment.

Chapter 12

A FLEET FOR ALL SEASONS

Despite its influence and control on the way the industry developed over some sixty years, BIM had not displayed much flair for the business of fishing. Before withdrawing from commercial activities at the end of its first decade, it had briefly, and unsuccessfully, attempted to profitably manage three fishing operations: the "deep sea" cutters, the Gaeltacht fleet and the inshore boats operated directly by the agency[1]. In the future, BIM became the conduit for subsidies issuing from the exchequer and, later, the Commission in Brussels, channelling funds through a number of state bodies to the fleet. The dependency of the industry on a finite fish resource should have been an obstacle to fleet expansion but it was the business of the agency to promote it. So, instead, BIM, in association with government, contrived easier repayment terms. The result of these was a bloated and uneconomic fleet, hooked on subsidy, and, inevitably, exploiting stocks beyond their limits of endurance.

The most important announcement within the fisheries white paper[2] produced in 1962 revised hire-purchase funding arrangements. From 1 April 1962, the repayment rate on a new vessel was extended by up to five years, to fifteen. The rate of interest repayable was reduced to 4% annually. While desirable from the purchasers' standpoint, a lower interest rate obliterated any profit margin to cover risks to BIM arising from future hire-purchase repayment shortfalls. The new arrangements offered an incentive to clear loans before the prescribed deadline. It would range up to 10% off the agreed price for loans that were cleared five years early. The capital grant for new boats was increased from 15 to 25%. The cumulative effect of these terms would facilitate the purchase of a vessel at little more than half the pre-1962 costs.

New engines would be provided with an initial grant of 25% of the cost price and second-hand vessels might be acquired on a down-payment of 10%. BIM boats would have a free inspection service and repair costs would be reasonable, but boats on hire purchase would have to be maintained as skippers were instructed to do.

Boat designs were rationalised. Those on offer hitherto had been 7.9, 15.2 and 17.1 m oal. Few in the fleet currently exceeded 17 m. There was demand for larger vessels and a new design of 19.8 m, almost 50 GT, with a draft of 2.6 m and an engine of 169 kW was available at a cost of stg£25,000; preparations were in hand to build it at the Killybegs boatyard.

A new shellfish boat design was also advertised. The model would be 9.8 m oal with a draft of 1 m, equipped with an engine of 24 kW. One of these vessels cost stg£3,000 new and came equipped with a pot hauler. Crewed by three men, the boat could accommodate 70 lobster pots.

The white paper announced that preparations for major fishery harbours at Killybegs, Castletownbere, Passage (Dunmore) East, Howth and Galway were underway. The investment would require stg£1.25 million and would take ten years to complete. Ice would be supplied at each, as would facilities for the repair of boats, engines and fish processing.

The introduction of new, easier hire-purchase arrangements ushered in a phase of greater compliance with repayment arrangements but those hirers already in arrears made up 50% of the total. In March 1967 hire-purchase arrears of almost stg£1 million had accumulated. The annual report for 1966 announced:

Because the board is now primarily a development authority rather than a commercial organisation it cannot expect to eliminate this deficit from further operations. It intends to make proposals to the department of agriculture and fisheries to have the position rationalised.

When the American team submitted their observations on the industry in 1964,[3] they observed that BIM's vessel building programme was currently suspended because the fleet was sufficiently large and recommended it should not be reactivated until the need became obvious. Their observation did not count for much; once the temporary hiatus of indebtedness had been got around, boat-building surged ahead. The size of the fleet at the time the American team visited was something in the vicinity of 13,000 GT.[4] When, in the late 1970s, depletion of fish stocks became undeniable, it had expanded to more than twice that size. Eventually, the fleet would peak at 86,000 GT in 2004, nearly seven times its displacement in 1964.

The American team recommended the hire-purchase scheme should be extended to vessels of 22.9 - 30.5 m and it should be a prerequisite for obtaining a boat of these dimensions that a prospective skipper would have served on foreign offshore vessels. The team was confident that within seven to ten years Irish skippers would be capable of fishing waters as distant as the Arctic. Their advice was assiduously grasped; BIM favoured improving marine credit to facilitate any Irish company wishing to invest capital in middle water boats in the mid-1960s. While increasing success was drawing more investment into the industry, still more was required to be placed in bigger boats.[5] Gradually, such vessels made their debut.

In 1966 it was announced that initial capital grants of 25% for new vessels would cover boats of varying sizes, including two vessels of 33.5 m. Credit lines were established between BIM and finance houses in France, the Netherlands, Norway and Poland, and eleven vessels were ordered under the new arrangements. BIM boatyards at Killybegs and Baltimore were extended and upgraded for the construction of larger boats.

But that was not all. Forty new fishing vessels of various sizes were commissioned under BIM's marine credit plan. There was also a drive to encourage industry outsiders to take up fishing, especially in western areas. Advertisements assured small farmers it was relatively inexpensive to become involved in its lobster boat scheme. In 1967 a total of 73 vessels joined the fleet, eighteen of them second-hand and 49 less than 15.2 m. The estimated capital value of the fleet was stg£500,000. In 1967 the Killybegs and Baltimore boatyards were modernised in anticipation of rising demand for home-built boats. A new slipway was installed in Killybegs to service those up to 27.4 m. A similar facility was planned for Baltimore.

Fleet commissioning continued the following year with ten vessels of greater than 15.2 m and 40 smaller ones, mostly exceeding 9.1 m for shell fisheries. In addition, 21 second-hand boats were acquired, nine of them above 15.2 m oal. Many of the smaller vessels were constructed in fifteen small boatyards scattered around the coast; BIM assisted with the expansion of their services. The mid-1960s saw a massive growth in fishing effort and a small loss recorded in BIM's boatbuilding account was attributed to disruption in the course of expanding the boatyards. After 1968 boats of up to 27.4 m could be accommodated in the yards at Baltimore and Killybegs.

The closing years of the 1960s witnessed another increase in fleet numbers. A further fourteen vessels of greater than 15.2 m joined in 1969 along with forty smaller new shellfish boats. More than half of the larger new vessels had been constructed in Irish yards, the remainder purchased abroad. Six second-hand boats were also introduced from overseas.

The most dramatic and sustained increase in the growth of the Irish fleet was now underway. At the time it may have looked as though it would last forever. For almost a decade the value of annual output in the boatyards increased from stg£0.26 million in 1968 to a peak of stg£2.75 million in 1976. Staff numbers engaged in boatbuilding grew from 99 in 1968 to 318 in 1975 and the activity was re-designated "commercial", despite the agency's disavowal of the adjective almost a decade before.

1970 saw the first vessel of 22.9 m launched and the keel for a second one laid. Such was the increase in demand for boats that the yards would work to full capacity until 1973. The early 1970s appeared to have put the hire-purchase crisis well behind them. In 1972 all repayments were up to date, for the first time in four years. Everyone began to breathe easily once more. Then, in 1974, only four fifths of the hire-purchase monies due were collected.

The growth in numbers of larger vessels was changing patterns of fishing: an increasing proportion of the fleet was making several-day "trip" operations farther from shore, rather than single day ones, particularly on the east coast. More distant grounds were opening up and heavier landings were expected.

In 1971 35 boats over 50 m joined the fleet; others between eight and 14 m were also grant-assisted. Newcomers included a steel boat of 26.2 m, marking the beginning of an influx of vessels of this approximate size and stern trawler design that had been constructed abroad. During the next few years steel hulled vessels would attract a growing proportion of new fleet investment. Orders were spreading beyond Irish yards to Sweden, France, Gt Britain and the Netherlands and in 1974 there were seven steel hulled boats of 26 – 38.4 m on order from these. However, demand for smaller craft of less than 15 m had not slackened and in 1974 orders for 35, worth stg£0.25 million were placed with Irish yards. In addition, modernisation grants, tailored to improve the quality of electronic equipment and other instrumentation, such as winches, were being used with greater frequency to upgrade vessels.

The following year the structural funds, one of the four sub-policies of which the CFP was made up, came into existence.[6] This category of grant from the Commission under FEOGA (European Agricultural Guidance and Guarantee Fund), made available to construct new vessels and, in time, upgrade existing ones, would, eventually, transform the fleet. Business seemed to be booming. BIM handled 460 applications to grant aid marine equipment and vessels from the fishing community in 1975, 329 of which were approved. More than 100 vessels ranging in size from less than 15 m to 27 m – the largest of them steel hulled trawlers – were either purchased or put on order in that year. The total order book was worth stg£6 million, of which sum the state would contribute one quarter. In 1976 the number of boats delivered and on order fell to below eighty but the value of the order book at year's end was stg£8 million of which one quarter was state grant. In spite of the value of investment in that year, a reduction in boatbuilding was about to materialise. In 1976 the output of the yards reached its peak of stg£2.75 million but personnel employed in the boatyards fell, from 318 in 1975 to 295.

When a business is operating at full capacity, a slight disturbance can spook its success. What was occurring now would be described as a "severe downturn". In 1977 output at the BIM boatyards fell from stg£2.75 to stg£2.47 million and personnel declined by a further ten. That was the last year in which the success of the boatyards, expressed as numbers employed, was presented in tabular form in the agency's annual report.

Surrendering the boatyards did not come without resistance. Although fewer vessels were launched in 1976 and the yards did not operate to full capacity, two new designs, 19.8 and 26.2 m oal, were added to the available range. They were more comfortable to work on and better designed. In 1977, 73 vessels, whose grants had been arranged by BIM, were added to the fleet. Smaller boats (62 of them less than 15.2 m) were Irish-built while a proportion of larger ones (11 ranging between 18.3 and 27.1 m) were constructed abroad.

In 1977 indebtedness of the fishing community to BIM resurged and reached stg£1.5 million. FEOGA grants were used to reduce it. And vessels continued to be added to the fleet. In 1978, 63 were commissioned, nine of them above 19.8 m, 54 approximately 15.2 m oal. Three trawlers were constructed, speculatively, in the hope that the market for wooden boats would recover and in order to keep some momentum alive in the yards. The strategy did not work and when delivery dates were exceeded without sales having taken place, substantial losses were incurred.

In 1979 the boatyards were sold as going concerns. Once again the agency had stepped away from a commercial enterprise. It had also retreated two steps from the coalface of the industry; in the first instance the managment of fishing vessels was abandoned and, in this one, their construction was ceded to an intermediary. BIM would continue to be associated with boat-building, offering technical expertise, and the activity would remain on its organization chart. It would also retain responsibility for the staff who would be paid redundancy money by the agency when the boatyards eventually closed.

An astonishing aspect of endless fleet enlargement was the equanimity with which it was accepted by the industry. One man, who performed in a number of nationally significant leadership roles and later became part of the establishment as chairman of BIM, Joey Murrin of Killybegs, was the exception which proved the otherwise general rule of acquiescence. As chairman of the IFO, he delivered a speech at its fourth annual general meeting in 1978 which stated the blindingly obvious:

The continual expansion of our present fleet in its present form – i.e. geared mainly to fish herring, is economic madness and common sense will tell us that you cannot continue to build boats indiscriminately. Fish stocks of all species are at such a low level that I feel there should be stabilisation of our fleet. What expansion amounts to in the type of boat that we are allowed to purchase at the moment, is a smaller slice of the cake for each fisherman in the future...

However, there may have been a tinge of self-interest in his words which continued:

It is against my nature to suggest that there should be any curbs on our expansion but I would be acting irresponsibly to suggest anything else in light of the unavailability of the raw material. The view is not a contradiction of my proposal for bigger boats, in fact they are two different suggestions.[7]

Murrin's career took him to head up the KFO, a PO catering for the largest vessels in the fleet.

The evolution of the fleet still had a long way to go. As the 1970s drew to a close, the major advance was the construction of five 40.5 m vessels whose launch was scheduled for 1980. The boatyards were also occupied with a number of modernisation assignments, fitting out vessels with a variety of equipment: power blocks, deck cranes, steering gear, pot-haulers, radar, sonar, echo sounders and refrigerated salt water tanks, all major elements of technology creep which would accentuate fishing pressure on the marine environment.

In 1980 four steel vessels greater than 27.4 m, eleven boats between 19.5 and 27.4 m and forty boats of 15.2 m all began fishing. The following year FEOGA grants to modernise fishing boats came on stream; before that they had been for vessel purchase only.

Grant applications under BIM's marine credit plan ran at between 400 and 500 annually during the early 1980s. As the early decade advanced a credit squeeze led to another crisis in loan repayments, in spite of which interest in boat purchase did not decline. In 1982 money owed to BIM by fishermen increased threefold to nearly IR£3.6 million, despite a special government aid package in 1981. The explanation from BIM was that fish prices were low and operating costs had increased. But the landings had risen considerably; in 1982 they were valued at IR£47 million, a 21% increase over the previous year, and exports, at IR£72 million, were 33% higher and 21% greater in volume than the year before.[8]

The same year, the minister, Brendan Daly (Fianna Fail, Clare) introduced the Sea Fisheries Amendment Bill, 1981 which enabled BIM to increase its borrowings from the exchequer above its current limit of IR£15 to IR£40 million.[9] For all that, there was a shortage of funds to maintain existing vessels and there were mutterings that the fleet required a total refurbishment. To rectify the deficiencies BIM implemented stricter rules for those applying for grants and carried out more critical analyses to assess the viability of fishing projects.

The early 1980s also witnessed the slow crystallisation of the CFP. There were still high hopes that larger quotas would be allocated to Ireland and, at the time, there was scant regard for observing quota limits. In the case of mackerel, a warning was issued that larger catches

...(would) lead to the continued rapid depletion of the stocks, with the danger of its total collapse...

The industry itself readily admitted that Irish fishermen caught over 30,000 t more of this species than permitted in 1981.[10]

As we have seen, forging a common fisheries policy was not an easy negotiation; financial handouts and unrealistic commitments by the Commission in Brussels were a means to achieving general acceptance by member nations. The promise of new fleets was part of the bargain and no nation would look a gift horse in the teeth; all wanted to avail of grants and subsidies, indeed all desperately needed the assistance. Once heavy and systematic depletion of fish stocks got underway, a decreasing minority of fishermen would be able to make a living only by accelerating the rate of technology creep to keep ahead of the others. None but the decreasing minority with the largest boats and most advanced technology would be able to stay in the race.

In 1983 FEOGA approved 73 projects valued at IR£3.24 million which included 68 boat builds (17 exceeding 12 m and 51 less than12 m). It was the largest approval to date. Since the instigation of FEOGA, 224 Irish fishing vessels had been approved for grant aid of almost IR£15 million.[11]

At the end of December 1983, arrears on repayments of IR£6 million languished on BIM's books. The agency put a brave face on it:

Despite the difficulties being experienced by the industry 48% of borrowers were not in serious arrears in the loan repayments on 31 Dec 1983 ...

Which meant that 52% were in serious trouble and an unquantified number were simply "in trouble". Something had to be done to curtail the haemorrhage. A reserve of IR£7.6 million was created against bad debts. The following year when 21 new vessels valued at IR£5.2 million were approved, two thirds receiving grants and loans and one third grants only. The shutters came down on loans at the end of December 1983. Arrears had then risen to IR£8 million. New vessels were excluded from future finance packages.

The emphasis for exchequer and European Commission financial assistance switched to modernisation grants, essentially improving the catching abilities of existing vessels. Some new boats, whose purchase was approved at an earlier time, continued to trickle into the fleet, 21 valued at IR£5.2 million in 1984.

Repayments continued to trouble the agency which, once more, took an optimistic line:

Approximately 40% of the loan accounts were less than 6 months in arrears on 31 December 1984 and in a seasonal business such as fishing, arrears of six months, while undesirable, are nevertheless not a cause for serious concern...[12]

According to credit arrangements with foreign lenders, BIM underwrote the loans to fishermen who made purchases through the agency, and the agency was in trouble too. The year 1985 was marked by a fall in fish prices and a rise in fuel costs which combined to create a cost-price squeeze. It amounted to a slump from which the industry was rescued by a government aid package worth IR£3 million.[13] In July a government assistance package was brought on stream and the loan crisis ameliorated.

...over 80% of the bank loan accounts which have been guaranteed by BIM and which represented approximately 50% of (BIM's) total outstanding loan liability were fully up to date on 31 Dec 1985...

This might be rephrased as "50% of outstanding loan liability was not up to date..." and, if so, it was not a great improvement

A corresponding figure prior to the introduction of the aid package would have been 33%...[14]

The marine credit plan remained open to applications and vessels which had been approved earlier continued to join the fleet. The annual report for 1985 was the last to carry the tabular analysis of fleet composition. BIM no longer wished to highlight an enterprise which was obviously reaching saturation. In 1989 BIM issued the first of its fish farming newsletters, a diversification of its activities and a diversion from the increasingly more troubled catching sector which the agency had been set up to serve. Details of employment at sea were henceforth omitted from its annual reports and, instead, tables showing the number at work in the industry referred to shore-based employment, including those working at aquaculture.

The emphasis, in the immediate future, moved to restructuring and rolling-over existing loans. Some small vessels (less than 15 m) continued to enter the fleet. After 1985 no new application for boats over 15 m was entertained. FEOGA assistance remained available but it was expressed as "modernisation" or "improvements" which, in 1985, enlarged three vessels from 27.4 to 36.6 m in length.

The power of the European fishing fleet, measured in kW, expanded fourfold between 1970 and 1996.[15] As early as 1970, the need for a mechanism to balance fleet size and resource was recognised. It was some time before it materialised and when it did it would be known as a Multi-Annual Guidance Programme (MAGP). The mechanism would, ideally, limit the amount of fishing capacity on stocks that were over-exploited. Initially, it was a weak governor of fleet size but MAGP was the only one available and it required time to accumulate credibility. An obstacle, at the outset, was the perception that fish stocks were abundant. Another widespread conviction was that a member state must at least hold onto its share of marine produce or, better, increase it.

The first MAGP programme (MAGP 1) ran for three years from 1983 to 1986. Its objective was to maintain capacity levels at the 1983 level. It was ineffective. In 1986 only two of twelve member states had done what was required.

In 1986 loans to fishermen stood at IR£21 million; FEOGA provided assistance to 342 applications. The following year new structural funds became available for the next ten years from the Commission under whose terms large vessels (greater than 33 m) became eligible; there was a special grant of 40% for first time fisherman-owners and the minimum expenditure available for modernisation was raised.[16] In 1988 the Commission ruled that FEOGA grant aid was not available to increase engine power or GT. The fleet had been capped, or, rather, an attempt had been made to curtail its further expansion.

MAGP 2 (1986-1991) was more ambitious than its unsuccessful predecessor. This time, five states complied but legal obstacles were cited as a reason for a less general enforcement of the rules. Some states were obliged to reduce their distant water fleets (the

UK being one of these) as a result of the introduction of EEZ bands of 200 nm by third country host nations, but a reduction in their fishing activities some distance from their coasts port did not necessarily benefit fish stocks closer to home.

In 1990, in association with MAGP 2, the Irish government introduced a stipulation that any new vessels entering the fleet (at that point there was a backlog of 260 applications for new whitefish vessels)[17] must first remove kW and GT equivalent to the new vessel from the registered fleet; were a vessel to be "modernised" by, for example, enlargement, the same condition would apply. Informally, within the industry, the rule would be known as the "buy two, fish one" regulation.[18] It created an informal market in which notional kW – "kiloWattage" – and GRT –"tonnage" – could be traded and it also raised the cost for anyone wishing to introduce a new vessel to the fleet. Such a person would be obliged to buy the necessary units at the open market rate whose value fluctuated in accordance with the availability of kiloWatts and tonnage and the demand for them.

In the run-up to MAGP 2, some complications arose. First, Ireland discovered she had miscalculated the size of its fleet in 1983, the beginning of MAGP 1, by almost 8,000 GT above the agreed target at the time. Second, Ireland introduced three new pelagic vessels of almost 8,000 GT to the fleet. In order to comply with MAGP 2 therefore, Ireland was required to remove approximately 16,000 GT or 27.1% of GT from the catching sector. Inevitably, Ireland exceeded the targets agreed in 1986 although more than 19% of capacity was taken out – specifically from the demersal section of the fleet. This was not enough and Ireland jeopardised construction and modernisation grant aid for the duration of MAGP 3.[19]

BIM recognised the tightening of the noose as the end of the 1980s approached by observing that the European Commission's attitude to fleet enlargement had become "more restrictive". However, the agency pursued, with assistance from Brussels, a planned programme of fleet modernisation, with a view to making the fleet "more efficient" and to up-grading safety standards. The principal concern of the European Commission was for the fate of quota species whose stocks, by the 1990s, were clearly under considerable pressure. BIM's *raison d'être* was to dispense finance. What it needed was another pretext to grant-aid the fleet. Two materialised.

The first sought extra fishing power to exploit non-quota species in which the European Community had no stated interest. Quota species currently amount to some 60% of total landings and, as had been the case in earlier decades, one never knew what might be available until one went to look for it, or such was the justification for unlimited enlargement of fishing effort, despite the *precautionary principle* which came into existence in 1982.[20] The precautionary principle makes the common sense assertion that, if one does not *know* that a biological resource is in robust condition, that assumption should not be made. The principle has been effectively ignored throughout the subsequent history of the CFP.

The second pretext was the perceived deficit between Ireland's actual landings of quota species, compared with what was theoretically available and cynically described as "paper fish".

...the most urgent need is that modern all weather whitefish vessels are brought into the fleet which would catch our underutilised whitefish quota off the west coast and also fish not non-quota species in the deeper waters of the Atlantic.[21]

Paper fish might have been a figment of the December council of ministers' meeting in Brussels which regularly set TACs above scientifically recommended limits. Or they might have been no more than an illusory quantum which never existed.

In 1990 the Commission approved grants for five new vessels to fish edible crab, a non-quota species. The decision sowed the seeds of a tragedy which deserves its own account.[22] That same year, a total of 22 applications for new vessels was granted, some of them for curraghs and 43 for second hand boats. The credit crisis had eased and the collection of boat loan repayments amounted to 94% of those due in the year. The level of arrears had been "substantially reduced", an improvement which may have been assisted by the policy of writing off bad debts in the year in which the relevant loan agreement was terminated.

MAGP 3 (1991-1996) sought even greater fleet reductions. The Commission's first demand in the negotiation was for a 30% reduction in capacity which all member states refused to agree. Eventually a 10% reduction was accepted by everyone, except Ireland, which resurrected the injustice of having cuts foisted upon a fleet which was comparatively underdeveloped in the 1970s. When, eleven months after the deadline, MAGP 3 targets were eventually agreed; its objectives were more detailed and specific: the Irish pelagic fleet was regarded as better adjusted to its fisheries than the demersal one. These were important points because, in future, there was a threat that failure to comply with targets would result in structural funding being withheld from the fleet. Despite all of these precautions, MAGP 3 was only partially successful. Seven of twelve states failed to comply with GT restrictions. There was a more serious difficulty with power (kW) targets because a vessel might have as many as three engines (two of them auxiliary) but was required to declare only one and this fact had not been taken into consideration at the outset.[23]

In 1991 BIM tentatively approved the purchase of fifteen vessels provided European Community funding was forthcoming. It was not and the transactions, described as necessary to land the full demersal quota which was not being caught at the time, did not go ahead. BIM's 1992 annual report announced:

Restructuring of the fleet by taking out older vessels and bringing in newer large vessels continues to be a prime objective of development strategy for the industry.

A fourfold strategy for development was announced: the fleet would be restructured and renewed within the context of evolving European Community policy; finance would be provided for fleet investment; alternative fishing opportunities would be developed and expanded and

alternative appropriate development opportunities would (be devised) to promote a successful review of the Common Fisheries Policy

which was due for revision in 2002.

In December 1992 the council of ministers in Brussels decided to allow Ireland a 1% increase in fleet size over the following four years while other countries would have to accept large reductions. Six new vessels, all less than 20 m oal, were approved, five to be constructed in Ireland. These, the reader was assured, would provide 40 full-time jobs and others on-shore processing their catches, the kind of justification which was new to these reports. However, without any apparent sense of irony, the annual account went on to report that 90% of all loan accounts were up to date, a creditable performance

...given bad weather, lack of fish in inshore waters and poor prices for certain fish varieties...The main bulk of the fleet is quite profitable but certain segments, such as the old inshore vessels, are finding conditions very difficult...

No doubt this tug at the heart strings was intended to garner sympathy for *fleet renewal*; it should have posed sharp questions concerning further increasing dependence on a diminishing resource in the future.

In 1993 loans for the purchase of new vessels were phased out. However, the addition of crab fishing vessels, 18.5 m oal, constructed in the Netherlands and similar in design to those introduced in 1992, took place. In 1993, repayments of boat loans fell to 89% of the amounts due. This was the first year since 1988 in which the collection rate of boat loans had fallen below 90%.

In 1994 a new "operational programme" was announced under which the fleet would be restructured and modernised, jointly financed by the state, the European Union and the private sector. Its cost would be IR£36 million. Restructuring would involve decommissioning (removing) existing vessels from the fleet and reducing its capacity.

The objective was to reduce catching power in order to comply with targets for the fleet set by the EU. The scheme was expected to start in 1995 and run to the end of 1996; seventy applications, accounting for 3,328 GT, were received, and the following year 59 applications for decommissioning were approved.

An imaginative way of maintaining hire-purchase instalments by a sophisticated form of subsidy had been devised. Money was available from the EU for "hardship" and this was dispensed by BIM to purchasers who fell into arrears. They, in turn, passed the money back to BIM as repayments. In 1994 repayments on boat loans rose to 108% of annual requirements.

In 1995 more than IR£1 million was invested in 31 modernisation projects (17 "large" and 14 "small") representing the first round of new approvals, and a decommissioning scheme was introduced. Demand for modernisation of fishing vessels and loan transfers was increasing and 128 applications, an increase over the year before, were received. BIM grant aided construction of two "inshore" vessels of 10 and 13 m respectively. A total of IR£9 million was invested in the fleet, twice the amount in 1994.

Recurring problems with loan repayments had become onerous; a backlog had accumulated once more, and a decision was taken to phase it out. A countdown began. Loans to fishermen decreased from 181 in 1984 to 164 in 1995 and indebtedness fell from IR£13.5 to IR£12.1 million. Outstanding loans further declined 17 in 2001 with the outstanding residue standing at €1.7 million.

By the time MAGP 4 (1997-2001) came up for consideration, the exercise had gained some respect. European Community fleets had increased their engine power more than fourfold between 1970 and 1996. Landings of fish by Community fleets had also risen by a more modest 1.5 times to 1987 and a decade later they had begun to decline. More ominously, the necessary fishing power units to capture them in 1997 exceeded the tonnage of landings.[24] MAGP 4 was therefore too late to avoid the considerable damage inflicted on fish stocks, but it anticipated the necessity to remove large numbers of vessels from the Community register. An assessment of quota fish stocks, for which it was tailored, revealed a deteriorating situation. Yet, in spite of the inevitability of more serious consequences, member states cited socio-economic reasons to maintain their fishing effort. Arguments that cutting landings meant cutting jobs were persuasive but, in the slightly longer term, as we discover year after year, depleting biological resources do not sustain jobs either and the socio-economic rationale simply pushes unavoidable and worse consequences a short distance down the road.

From the mid-1990s vessel modernisation was the main activity of Irish boatyards. Exchequer and EU funding amounted to IR£10 million in 1996 when

Almost all of the modernisation projects aided involved some element of vessel and crew safety and surveys of grant-aided vessels by BIM ensured that statutory lifesaving and fire fighting equipment was on board...

But BIM had other ambitions too:

As (1997) drew to a close (BIM) in collaboration with the Minister...had made considerable progress in devising an incentive package for the much needed renewal of the whitefish fleet. Progress on these measures will be vital if we are to approach with confidence the new millennium and the many challenges in its wake...

The supporting rationale ran as follows:

The objective of the fisheries development programme is to promote the renewal, modernisation and restructuring of the fishing fleet, the development and pursuit of new fishing opportunities and techniques, the improvement of conservation measures and the enhancement of on-board handling practices.[25]

The first programme for the renewal of the whitefish fleet (operational programme 1994-1999) introduced 29 grant aided vessels through investment of €9.3 million and upgraded 97 boats to comply with statutory safety requirements at a cost of €1.4 million. Fifteen existing vessels in the fleet were modernised at a cost of €2.3 million. Just as this scheme was drawing to a close the new one was about to begin.

The second "fleet development measure" was to be financed by the national development programme 2000-2006. A sum of €25 million in grant aid would generate total investment of €95 million in new vessels, modern second hand boats and the modernisation of the existing fleet plus the installation of statutory safety equipment.

A noteworthy exception to the otherwise invariable rule of grant assistance for new vessels occurred in 2003 when five new pelagic boats were introduced independently of any subsidised assistance. The pelagic sector of the fleet had relatively few vessels and, until then, had been able to operate profitably.[26] Apparently, heavy state and EU assistance was available for vessels which could not fish at a profit and the subsidies had the effect of pushing those parts of the fleet which had become dependent on them further into the red.

The arrival of the five new pelagic vessels

...(completed) a significant renewal and modernisation of the Irish fishing fleet and positions the catching sector well for the future...

The revised CFP in 2002 announced the phasing out of any further national and EU grant approvals for new vessels or the modernisation of existing ones.

As set out in the 2004 annual report:

The final three vessels in the national fleet development measure will be added in 2005 and will complete the introduction of 62 vessels and many more existing vessels modernised and made safe with a total investment of €119 million over the past six years...

That would complete two successive fleet renewal programmes.

Thereafter, readers were assured, there would be a new emphasis on conservation, practised by a safer, refurbished fleet.

One characteristic of BIM's *modus operandi* was the careful groundwork leading up to the introduction of a particular subsidy scheme. Rarely did anything appear as a surprise, out of the blue. The second whitefish renewal scheme, for instance, had been under discussion for some years before its launch. Abruptly, in 2005, a decommissioning scheme, worth €45 million, intended to withdraw capacity from the scallop and whitefish fleets was announced.

This scheme which is being implemented by BIM, delivers on the central objectives and recommendations of the White report. By December 2005 the first phase of the new scheme was well underway with 18 of the 22 vessels appointed for decommissioning having exited the fleet.

The facts were that the recently completed modernisation and vessel purchase schemes having been completed, there was a sudden realisation, which should have dawned years before, that fish stocks simply could not sustain the pressure exerted on them. The realisation also coincided with the discovery of widespread heavy landings of "black" fish, whose capture had been illegal. The necessity to curb these practices or face heavy fines imposed by the European Union effectively closed fishing activity down. The fleet was bankrupt. The house of cards had imploded. BIM, which had always reported its activities in detail in its annual reports, had made no mention of a White report in gestation during 2004.

The White report, bearing the imprimatur of another celebrity chairman recommended

...a national decommissioning programme is the best way of permanently removing sufficient capacity to realign the economic demands of the whitefish and scallop fleets with the fishing opportunities available to them...

Unfortunately, the worsening imbalance between resource and fleet was appreciated late in the day. True to form, the annual report for 2007 described the earlier purchases and modernisation as

... transforming a significant proportion of this sector into a modern, safe and efficient fleet, comparable with the best in Europe...

The delivery of the last vessels purchased under the fleet development measure 2000-2006 in August 2007 overlapped with the emergency decommissioning scheme to remove surplus capacity.

This account of the evolution of the Irish fishing fleet is based on the reports issued by BIM describing its own activities but the machinations surrounding the whitefish renewal plan and its aftermath are worthy of a chapter all to themselves.[27]

MAGP 4 was succeeded by FIFG (the Financial Instrument for Fisheries Guidance) to cover the period 2000-2006.[28] It too would seek a balance between fishery resources and catching capacity. It would also seek greater competitiveness among those engaged in a range of fisheries related activities, including aquaculture. Its brief was far wider than those of the MAGP plans. You could say it embraced as extensive a range of activities as BIM.

plans. You could say it embraced as extensive a range of
activities as BIM.

Chapter 13

THE SCRAMBLE FOR FISH

More than anything else, the expanding fleet needed fish to maintain it. A rapidly evolving capture technology operated by appropriately trained staff was indispensable to an explosively expanding fishing effort. By its second decade, BIM was committed to a variety of activities aimed at sourcing and exploiting greater quantities and varieties of sellable marine produce.

There comes a time in the unrestrained exploitation of finite marine resources – it happened in Ireland in the 1970s - when reports of record breaking landings level off, become scarcer and then decline. For as long as they lasted, these superlatives played an important role in the publicity efforts sustaining BIM.

Increasing marine landings required the development of more effective catching techniques, the preparation of personnel through training and instruction courses, the trialling and provision of appropriate gear and the search for exploitable resources, all of which were organised by the agency. The suggestion by Holt that the industry might be left to do its own exploring would have set too slow a pace to maintain the requisite volume of fish.

The tendency towards fitting more powerful engines became established within BIM's first decade. It was symptomatic of what lay ahead. Throughout the 1960s the Advisory services division introduced improved methods of fishing to various ports. Staff of the White Fish Authority in Hull co-operated in the development of gear technology. Training in new skills, particularly mid-water trawling, became a priority. Advances in trawling technology over rough ground also occurred and had the effect of enlarging fishable areas of sea bed, with the result that refuges in which fish had to that time been untouchable and secure, were opened to exploitation by mobile gears. Attempts made to introduce purse seining were less fruitful. Instruction in hook and line fishing was added to the range of choices.

When they visited in the early 1960s, the American team may also have perceived signs of over-fishing because they proposed a search for under-fished inshore whitefish stocks.[1] Investigations on shrimp, scallop and lobster were undertaken with a view to introducing more effective gear. Experiments commenced on the performance of a new crustacean trap, the American parlour pot, and the installation of hydraulic hauling equipment became a standard fixture in shell-fishing boats which greatly boosted the catching power of all operators of static gears. Edible crab, a species for which there was little demand during the 1960s, was sought off southern and western coasts. BIM also assisted companies engaged in mussel fisheries and carried out surveys and some transplantations of seed mussel for on-growing in Cromane, Co Kerry, Carlingford, Co Louth, Wexford Harbour and the Boyne estuary, Co Meath. Oyster development projects were assisted at Clarinbridge, Co Galway.

At the conclusion of the 1960s a number of shellfish surveys, in particular on crab, mussel, oyster and lobster, had been completed. The emphasis in these was to locate exploitable stocks rather than to assess their capacity for sustainable production.

Extending the herring fishing season was a long-held objective and work was conducted on the feasibility of setting up a fishmeal factory, supposedly to be supplied with sprat from the Irish Sea. The factory would prove an embarrassment when it was commissioned.[2] Fish stocks, described as "industrial resources" (meaning they were intended for reduction to fishmeal) were also sought off the south west coast. There was no consideration of the inter-dependence of organisms in the marine environment, the consequences for the remaining species of substantially removing one, nor for disrupting the food web. This was the mindset which permitted the fishmeal plant at Mornington on the River Boyne to be fed with juvenile herrings, thus depleting the adult biomass of that species with consequences for human food fisheries.[3]

In the early 1970s another administrative reorganisation took place at BIM after which the agency functioned as four divisions: Market development, Fisheries development, Investment development and Boatbuilding (which became a stand-alone division). The Market development division was sub-divided in two, one for the home market (containing ten sections), another for export (containing seven). Fisheries development contained eight sections and boatbuilding four. Investment development had eleven sections. The agency was gaining confidence. For the first time, divisional managers were identified in the annual reports by name. Hitherto, only members of the board and the chairman were given that exposure. Otherwise, and with rare exceptions, usually visiting consultants, BIM had been protective about disclosing the identity or numbers of staff it employed.

What had begun as a casual development, sourcing fish for an ever more ravenous fleet, had become an essential tool in the agency's repertoire by the 1970s. It was now incumbent to capture the fish as soon as they were discovered.

The work of the resource development section of the fisheries development division continued to be orientated towards immediate commercial objectives, consequently rapid assessments have to be made of the commercial potential of particular resources. The growth of markets and the resultant demand from processors and fishermen led to an increased need for resource information.[4]

The administrative section in question produced a series of papers described as "Investigating fisheries resource development potential" which, in the annual report for 1972, appeared as a list of papers and leaflets published between 1967 and 1971: 26 resource record papers (on oysters, scallops, mussels, edible crab, industrial fish stocks and sand eel), five resource development notes, nine advisory leaflets and eighteen interim reports. Some were produced in association with staff from the inspectorate of the fisheries department. The list would become a feature of the annual reports for a number of years. It would be displaced by other topics, notably reports on marketing fish products when BIM's fleet and resource programmes began to show diminishing returns and progressively more negative results.

The repertoire of fishing techniques increased during the 1970s as fin and shellfish stocks were assailed from ever more angles. Pair seining was adopted by smaller vessels as were more automated long-lining techniques for whitefish. Tangle netting was brought into widespread use for plaice and other flat fishes off the south and south west coasts and was deemed beneficial to small boat owners. A new Scandinavian combination trawl design yielded increased catches of quality whitefish and industrial species on the east coast. In 1974 trawling by small vessels of 12 m was introduced to the coasts of Cos Clare and Mayo. Pair demersal trawling was developed for whitefish. Extension of the herring season was still an objective and the interminable search for supplies of seed mussels to be transplanted for on-growing continued unabated. Exploratory cruises off the south west and south east coasts continued to seek species like sand eel which might be harvested for fishmeal. These exploratory cruises sought fish very close inshore as well as in deeper water.[5] Experiments were also taking place on shrimp fishing off the Galway coast. The expressed purpose of developing a shrimp fishery was to extend the fishing season for inshore boats. In instances where there was a possibility of starting a new fishery, free gear might be distributed by BIM to a local community to get the venture underway.

The extension of fishery limits was a spur to discover more but the imminence of the CFP intensified the urgency to uncover whatever there was to find. Nothing was apparently too insignificant to warrant attention. Fishing trials with crab pots were undertaken off the south coast while spurdog, of hitherto no commercial value, were pursued using long-lines in the Irish Sea.

Automated jigging machines for the capture of finfish also represented technological advance. The mechanically operated lures for species like mackerel which fed on smaller fish or the juveniles of other species substituted effort which, at an earlier time, had been undertaken by manual labour.

From its earliest years, BIM's technical personnel had worked closely with the industry, tracking the distribution of herring shoals during fishing campaigns. Surveys of sprat and mackerel were later added to the list of practical tasks undertaken on behalf of the fleet. Another service was the preparation of charts of trawling grounds (areas of seabed over which a trawl might be drawn without encountering obstacles) which the agency made and passed on to the industry.

By the late 1970s, Irish territorial waters were intensively fished with mobile gears. In 1978 it was reported that up to 1,000 nm of trawling grounds had been charted by BIM in order to "reduce pressure on stocks on established grounds". This kind of explanation suggests there was some logic underlying a development strategy. On the contrary, opening new opportunities simply enticed more vessels into a fishery which was already over-capitalised and heading towards collapse. The result was that individual boats, in due course, would specialise in fewer capture techniques than in earlier days when small inshore vessels, operating at lower densities, targeted a succession of species using a variety of fishing methods as the year progressed. An administrative process to assist management, introduced as the CFP evolved, corralled vessels into fleet "segments" whose activities were confined to certain fishing techniques or species. Segments targeting quota species were capped so that their GT and kW were prevented from expanding further.

As this is written, in March 2012, the Irish fishing fleet has *pelagic, beam trawl* (beamer), and *polyvalent general* segments among which quota species are shared; their GT and kW are capped. The polyvalent general segment, accounting for 66% of the fleet by number, 56% of engine power and 50% GT, contains a wide variety of gears any of which can be used by vessels which are included within it. However, the *aquaculture* segment, which contains mussel dredgers, is not capped because its vessels do not pursue quota species - nor do vessels in a *specific* segment which are engaged in dredging bivalve species like razor clams. Both dredge segments cause major disturbance of the seabed and so are environmentally destructive. The introduction of a *polyvalent potting* segment is relatively new and will be described in greater detail later.[6]

Smaller fishing vessels within the polyvalent general fleet, would have targeted different species as the seasons progressed but their range of options declined along with the availability of fish. In a review of inshore fishing in 1999 BIM presented an idealised selection of fishing opportunities for the inshore fleet.[7] Pooled information from the entire coastline suggested a variety of possibilities but local realities were, in some areas, grim. The decline and closure of the salmon drift net fishery (for which BIM dispensed compensation to the redundant fishermen) was a major loss but there were other, less celebrated ones. A survey of fishing activity by small vessels in three ports in 2006 indicated they were far from fully occupied and there was a tendency for boats in a particular harbour to target a single species, such as edible crab, where and when it was available in sufficient quantity.[8] Abundance and diversity had both declined beyond the point where a number of alternative species were adequate to support commercial activity.

Throughout the 1970s landings records were broken year after year. In the mid-1970s some 42 "varieties"- mainly species, but including some groups of several species (rays for example) - of fin and shellfish were listed in the official statistics.[9] These were presented as demersal (14), pelagic (4) and shellfish (24). Invariably, a year revealed a spectacular landing figure for a group or one of its constituent varieties. And, if the volumes did not make praiseworthy reading, then their values might. Values were of course current, unadjusted for inflation. During the 1970s and into the early 1980s, inflation rates were enormous, occasionally exceeding 20% annually.[10] The highest landings volumes recorded between 1987 and 2010 were: demersal in 1987, pelagic in 1995 and shellfish in 2001. All combined achieved their highest values, adjusted for inflation, in 2001 also. Since 1995 the total weight of annual sea fish landings recorded has trended down, as have pelagic species, their largest contributor. Shellfish landings have been declining annually since 2001 and demersal landings have had a more or less level trend since 1981.[11] This is according to the official statistics which, as we will see, may have been wide of the mark. It also takes account of the brief contribution of deep water species which we have yet to encounter.

The inability of the rise of landings to match the rate of increase in catching effort and, especially, the disimproving capture rate in herring fisheries led to the statement in 1976 that there was a need for "greater exploration of the exclusive economic zone (EEZ)".[12] The year after, intensive training courses on hydraulics and acoustics were designed with the specific purpose of locating new resources which would be exploited at greater depth with more advanced fishing techniques. Appropriate courses on gear technology were organised in association with Hull Nautical College.

The use of bobbin trawls opened new ground by making hard rocky substratum accessible to mobile gears off the north west of the country, and paired bobbin trawls were experimented with. New seining grounds were opened off the coast of Co Mayo. Experiments were also conducted with Danish gill netting techniques. Efforts were made to encourage small boats to diversify from salmon and lobster fishing. Shrimp pots were distributed to fishing communities in east Cork. Similar efforts were made to encourage pot fishing for whelk in the south west Irish Sea. New scallop beds were sought on the west coast.

The procedure of dredging up mussel reefs and translocating them to sheltered embayments for on-growing was an established routine by the 1980s. Later in the decade it was euphemistically referred to as "enhancement". A decade earlier there were reports that the beds in Carlingford Lough had been fished out, but that did not discourage the practice, and the search for undiscovered reefs continued; a large one was located off the Wexford coast.

In the early 1970s, the depletion of inshore waters directed thoughts further offshore. In the course of its explorations off the south west coast, BIM's efforts happened upon initially small quantities of blue whiting in 1973.[13] Blue whiting, a small member of the cod family, was easily recognised because the species turned from white to blue colour after being brought to the surface. Although known from inshore waters – it moved landwards to spawn - it was more abundant at depths of 500 - 600 m. The species was known to occur in large biomass at the continental edge. In addition to its value as an ingredient of fish meal, blue whiting could be processed into convenience foods, like fish fingers, for human consumption. In January 1974 Norway expressed an interest in harvesting blue whiting. Until then there had been little interest in doing so.[14] Later that year a number of Irish steel vessels began to fish for it, their landings fetching stg£30 per t. At that stage the availability of larger quantities had been confirmed and a vessel was capable of landing 100 t in a cruise of two days' duration.[15] In association with other member states of the EU, a number of cruises were undertaken in 1975 to further assess the newly discovered resource.

Agitation from within the industry for a new fishery harbour situated between Killybegs and Castletownbere coincided with fishing trials on the new high volume and low value species. Rossaveal was earmarked as the location, and predictions were made of blue whiting landings upwards of 100,000 t per year, creating employment for 2,000 persons.[16] Such optimism overlooked some now well established and frequently encountered obstacles. Before fish are harvested, current practice in western Europe requires them to be pre-allocated among participating fishing nations. Waters of the western continental edge were shared by EU states with Norway, Iceland and the Faroe Islands. The share-out allocated 31% of the TAC to the EU, among whom Ireland, obtained some 12% of the Union's share. The record of landings to Ireland shows an increase to some 75,000 t in 2004 after which they declined to 8,000 t in 2010.[17]

For a time, the discovery of blue whiting was greeted as a likely substitute for cod, which, with a number of other "traditional" species, was accepted to be in decline.[18] Rather than try to correct the over-exploitation of whitefish species, which had traditionally been part of the national diet, the industry now embarked on the capture of an unproven substitute which was only accessible to the largest and most modern vessels.

Anticipating another bonanza, BIM chartered a Killybegs boat to explore the best trawling possibilities on the Porcupine Bank.[19]The survey duly reported heavy concentrations of blue whiting, some of which notably burst the trawl net.[20] Yet, despite the supposed abundance of the species off the coast of Ireland, occasional landings of more than 100 t from the west of Scotland made headlines,[21]suggesting blue whiting may not have been equally abundant all along the edge of the continental shelf.

Blue whiting stocks currently consist of older fish, and recent recruitments of juveniles have been disappointing. The stock has proved difficult to assess, hence manage. Falling catches over the best part of a decade must, on track record, trigger alarm. Experience instructs us that diminishing stocks rarely become re-established at their former strength because industry insists on maintaining the heavier landings for which it has geared up.

In retrospect, it is unlikely that the new fishery would have been an adequate substitute for those which were in the course of being eliminated. Blue whiting did not live up to expectations as a provider of human nutrition. A brief note in the trade press in 1977 cautioned that perhaps, after all, it might not be the panacea to yawning fish shortages. The realisation was dawning with the discovery that the species was of small size and of slender dimensions and it was available to trawls for only three months of the year.[22]

Another species revealed by BIM's voyages of discovery was scad, also known as horse mackerel.[23] Regarded as an industrial species in Ireland, the fishery would supply juveniles to Japan and larger individuals to Africa. The species was hailed as an alternative to salmon and lobster, both of which were recognised to be over-fished.[24]However, like blue whiting, scad could only be exploited by larger vessels – typically the RSW fleet - so the beneficiaries of fisheries for these species could not be the inshore fleets which traditionally fished salmon and lobster. It was another example of the dispossession of small boat operators.

A trawling survey in the south west, reported in late 1978, found scad to be plentiful.[25] Everything pointed to another bonanza of heavy landings exporting to the vast untapped demand of the African continent.[26] A review of projects to create new wealth and employment opportunities two years later featured scad on its agenda.[27] But exploitation of this species took longer to get underway. Various experiments to process the species, undertaken in the Torry research centre in Aberdeen, uncovered practical difficulties handling the fish. There were also problems with quota – Ireland was initially offered a derisory tonnage but secured almost 24% of the larger of two stock management divisions. The western one of these is the more economically important. Nonetheless, orders were placed for larger trawlers on the prospect that scad and blue whiting would prove profitable.[28] Landings of scad rose to peak in the mid 1990s, then declined, although they recovered again slightly in 2009 and 2010.

Recruitment of juveniles to the stock appears to have been low in recent years. Officially, the western stock is regarded as sustainable but it cannot be said to be as robust as in the first years in which it was exploited.

In the later years of the 1970s funds for exploratory fishing ran short. Another re-shuffle created two new or reorganised sections in the administrative evolution of BIM: one, Investigating new fisheries, the second, New fishing methods and gear. The activities of heavy bobbin trawls in the north west having been seen to be successful, the method was extended to the Celtic Sea. A variation of trawling technique: fished by a single vessel, the pelagic whitefish net – essentially a pelagic trawl fished deep but above the sea bed – was trialled in the Irish Sea.

In the early 1980s, the drive to find new untapped resources moved up a gear. The rising costs of fuel stimulated attempts to find cheaper ways of making landings; passive rather than mobile techniques were identified as the solution. The more accessible fishing grounds had been, not merely cleaned out, but polished and scarred by the constant movement of heavy nets over them. But not all possibilities had yet been exhausted. One curiosity which satisfied requirements was the realisation that shipwrecks hosted large number of gadoid fishes, like saithe. The wrecks were probably the only remaining effective obstacle to mobile gears. Their fish populations could however, be harvested using gill nets. BIM obtained a magnetometer to locate the sunken hulls and set out in search of the last refuge of fish life still untouched by Irish fishing effort.

As the decade progressed the search for "new" or "non-traditional" species was frequently referred to in Ireland's commercial fishing literature. A species like angler might have claimed this label because it was apparently not harvested to any extent before other member states of the EEC demonstrated its value[29] but species like whelk, shrimp, and spider crab were all consumed at earlier times in the history of this island and could hardly be described, as they frequently were, by the adjective.[30]

The fact that much effort was put into developing a fishery for spider crab, which would yield low tonnage and value, was a signal that the more lucrative possibilities of the territorial sea were close to exhaustion. Research by BIM personnel came up with some startling conclusions: "the bulk of the larger (spider crab) females" we were told, "are located offshore and remain under-exploited". On the contrary, the females are fairly sedentary, in very shallow water up against the shore.[31] Another group of species, brown shrimp, albacore tuna and the two we have just seen, scad and blue whiting, not traditionally fished by the Irish fleet, might have been accurately described as "new" or "novel".

Interest grew in the deeper fishing techniques used by the Spanish and French off the west coast and studies of their methods were undertaken. Long-lining for hake and ling was explored off the south west coast at depths of 200 - 600 m. Effort was gradually moving deeper and farther off shore. *Nephrops* was sought on the Porcupine Bank, some 190 nm (350 km) west of Galway. By the mid-1980s exploratory fishing by vessels of 27 m drawing heavy bobbin trawls was taking place west of Rockall, more than 300 nm (570 km) north north west of Killybegs. Intermediate areas, to the north of Malin Head at the Stanton Bank and its vicinity west of Scotland were combed. A section of BIM labelled "Production services" in 1986 (formerly known as "Resource development") had a policy of encouraging the medium sized fleet (24 – 27 m) to target hake, monk, megrim, *Nephrops*, squid and associated species at 200-400 m on the grounds west of Scotland and in the Celtic Sea.

In 1988 another series of records was broken. Deep water trawling targeting blue whiting and argentine made the largest ever landing to an Irish vessel and the largest single landing into a European Community port from Community waters. Fishing took place below 700 m and individual net hauls of 400 t were registered. The following year, in the course of a joint BIM and European Community expedition in search of new pelagic resources, 8,860 t of silver smelt and 4,000 t of scad were brought ashore. In 1984 the IFO asked RTE, the national broadcaster, to extend its weather forecasts to the Porcupine and Rockall areas to keep the growing number of vessels fishing there informed of weather prospects.[32]

But near-shore waters were not abandoned, or rested, and in 1985 five vessels were "given assistance extending the fishing grounds in the Irish Sea", a water body which, 140 years previously, was considered over-fished![33] To induce further effort into the fishery for edible crab, 1,500 crustacean pots were distributed to fishermen at reduced cost. The newly acquired purpose-built *vivier* boats (constructed around tanks containing aerated sea water) were dispatched to survey offshore edible crab on migration, in the early 1990s.

Studies were undertaken over a number of years "to identify commercial fisheries... and to assist fishermen and processors engaged in these fisheries" for velvet and green crab and bivalve "clams" - the last mentioned would, in time, include some fairly infrequent species such as the warty venus.

As exploration extended further from shore the rate of innovation heightened speed to keep abreast of developments. Experiments were undertaken to improve otter trawl door design. Twin rigging was developed for prawn trawling and seining. Variations in trawl design, such as the "fork-rig", were introduced. Trials with double and pair rigged trawls for *Nephrops* and whitefish took place on the south coast and the Irish Sea. When a double rig trawl proved successful capturing prawns its use was extended to megrim and angler. At the beginning of the new millennium, a 12,000-hook automated long-line system was installed on a 12 m boat at Castletownbere.

All lines of endeavour, identifying new target species, devising new methods of capturing everything and refining the technology to increase its capital to labour ratio, continued into the 1990s. Vessels of between 22 and 24 m focused their attention on albacore tuna, a species long pursued by Spanish and French fleets on the Great Sole Bank (known by the Spaniards as *Gran Sol*), 100-300 nm off the south coast of

Ireland. In its first two years this fishery claimed a profit of IR£1 million. The value of landings increased to IR3.5 million in 1993 and had the double benefit of "…considerably relieving pressure on hard pressed inshore stocks…" suggesting that the people who organised these fisheries were scrupulous about the stress imposed by the industry on the ecosystem in the course of their endless drive to harvest more. The albacore tuna fishery was initially prosecuted using drift nets to catch the fish. This was unacceptable to Spanish interests which dominated the fishery, so experimental trials were sponsored by vessels using pair trawls and hook and line fishing with lures, similar to the methods used by the Spaniards.

The initial enthusiasm for deep-water species, optimistically envisaged as a viable and sustainable source of landings, was becoming tempered by reality. One species, argentine, was not living up to expectation. Exploratory fishing by two large pair trawlers, the *Atlantic Challenge* (91 m) and *Western Endeavour* (65 m), examined the potential for this and blue whiting, and the results failed expectations. But that was not the ultimate verdict, and exploratory fishing at the edge of the continental shelf was still expected to offer exciting opportunities for larger vessels. Additional species earmarked for harvest were grenadier, black scabbard and orange roughy. The European commission provided IR£0.7 million for a study of the biology of these species before an onslaught on their populations commenced.

The majority of this exploratory work was undertaken by vessels whose purchase had been grant aided by BIM which then arranged exploratory funding from the Commission in Brussels and the exchequer in Dublin. In 1997 BIM assisted five of the large pelagic RSV vessels to increase their share of the blue whiting fishery. Two years later vessels from Killybegs and Greencastle (Co Donegal) were assisted to examine the potential of fisheries for deepwater species in areas to the west of the Shetlands, St. Kilda and north and west of the Porcupine Bank. From the outset, contracted vessels, some of them newly recruited to the fleet, were depending, to some extent, on funding from BIM, rather than being obliged to earn all of their income from landings, and this fact must have softened the economic realities of declining stocks for their owners and may even have fostered a false sense of security.

Deep water exploration continued into the new millennium. In 2000 five of eight contracted vessels had been newly purchased under the whitefish renewal scheme. They surveyed an area extending from north west of the Shetlands to south west of the Porcupine Bank and east from the Hatton Bank to 200 nm west of Rockall. The species they recorded included Greenland halibut, redfish, siki shark, grenadier, blue ling, black scabbard and moramora.

In 1991, abruptly and uncharacteristically, BIM announced that "many skippers were desperately seeking alternative fishing opportunities". But it was not feasible to apply the brakes to technology creep which would become more deadly as knowledge of the behaviour of target species accumulated and better ways of fishing were devised, and innovations in hydraulics, gear technology and modern navigational electronics made their unstoppable advances.

One announcement, which, in retrospect, appears rather pathetic, hailed the discovery of a new "shiny clam" fishery in south Connemara.

This was the first time this species has ever been commercially exploited in Ireland. It will provide an income for several inshore boats at a time of the year when they would normally be tied up.[34]

I have never heard of "shiny clams" and have been unable to find any reference to them in the scientific literature. The species in question was most likely a surf clam (*Spisula* sp) and, contrary to what was claimed, it had probably always been part of the national diet. Whatever it was, "clams" appeared in the national statistics in 1990 when landings of 6.3 t, with a first sale value of IR£22,000, were registered. The following year 56.4 t, valued at IR£136,000, were landed. In 1992, eleven t, worth IR£30,000, were recorded and in 1994, less than one t, fetching IR£560, appeared on the books.[35] Shiny clams were not going to put bread on many tables. And this was typical of what was happening. By the 1990s there was considerable surplus capacity in the fleet. Any new discovery was rounded on and promptly fished out. That was the mentality of the industry, acquiesced in by the establishment that should have regulated it.

Policy concerning wild capture fisheries within BIM boiled down to three often stated objectives: to expand the catches of non-quota species – there was little choice in that, quota species were monitored (not very accurately as things turned out) and there was no room to legally expand their landings; to develop new fishing opportunities – not very different from the first but, charitably, it might have meant devising novel ways of harvesting either group (quota or non-quota); and to improve stock conservation. We have had a demonstration of the establishment's view of conservation as manifested in the activities of its development agency. But that agency was not alone in the policy it pursued. If the attitudes of other bodies, particularly the government department ultimately responsible for the rational exploitation of the resource, were not complicit, they were certainly indifferent. Some conservation projects were undertaken: on gear selectivity and the interactions of seals with fisheries for example, but they amounted to no more than window-dressing, incapable of achieving any significant turn-around of rapidly deteriorating fish biomass. What was urgently required was a concerned effort to halt fishery expansion but that was alien to the concept of a thriving, hence growing, industry to which the establishment had become moronically attached and on which BIM had become dependent.

In 1996 a mission statement of sorts by the Fisheries development division of BIM announced its purpose as:

...to promote the sustainable development of the sea fisheries sector through pursuit of new fishing opportunities, conservation measures and the renewal, modernisation and restructuring of the fishing fleet...

The alteration in policy was highly significant. As fish stocks were thinned out and fished down, more powerful vessels and gear were required to, quite literally, sieve greater volumes of water in order to make profitable landings. Construction of a modern fleet had not taken place in isolation from other policy developments. The infrastructure, by placing resources in a few large fishery harbours had concentrated or "centralised" services to the industry and the establishment, increasingly dominated by POs, was moving in the same direction. Fewer, larger, more powerful fishing vessels were just another element of the consensus.

The late 1990s witnessed a number of anomalous and bizarre attempts to generate landings. Another tuna species, bluefin, the most valuable finfish in the world, was experimentally fished for a period. The bluefin trials were made with assistance from Japan, and the landings were sent to that country which pays the highest prices for the species. However, Ireland did not have a track record exploiting bluefin which would have entitled it to quota share and the fishery had to be abandoned.

Squat lobsters, members of the Crustacean family Galatheidae, were considered candidates for exploitation. Some six species occur in the waters around Britain and Ireland. They are said to be delicious to eat and in 1995 one species, *Munida rugosa*, provided landings of 10 t in the Clyde, Scotland.[36] Attempting to generate wealth from such a miniscule rarity was scraping the bottom of the barrel.

Capturing *Nephrops* in creels operated by small inshore vessels was trialled in Inver Bay (Co Donegal), Dunmore East (Co Waterford) and Rossaveal (Co Galway). A creel fishery for prawns had been in existence in the Irish Sea at an earlier time but trawling, encouraged by BIM, had displaced it. A trap fishery is less environmentally destructive, the animals it captures are in better condition and hence, fetch a higher price at market, and small individuals can be discarded on capture with good prospects of survival. However, sponsoring creel fishing for *Nephrops*, after promoting the established trawl fishery for it seems more akin to creating investigatory scientific work for the sake of doing so. There was never any question of a creel fishery displacing one using mobile gears.

By 2000 the paucity of fish and rising fuel costs necessitated the introduction of more fuel-efficient plant and machinery aboard fishing vessels. A note in the annual report revealed how threadbare the fisheries development programme had become:

The impact on the catching sector (whitefish fleet) was particularly severe due to a significant decline in key fish stocks, a substantial rise in operational costs because of the global increase in fuel prices, crew shortages and the loss of a large number of days as a result of prolonged periods of bad weather...

The dream of discovering untapped fisheries resources faded as the twentieth century drew to a close but BIM would devise other ways of occupying its personnel and exercising influence in marine business. Aquaculture and marketing would become its main preoccupations in the new millennium but the agency was resourceful and had not abandoned its traditional interests or established clients for which it would create a service – not always necessary or desirable - whenever a suitable pretext became faintly discernible.

Retrospectively, the frantic rush to construct a grossly disproportionate fleet to devour everything that swam, slithered or crawled through the water column can only be described as suicidal. But it was acceptable; perhaps there was a deep belief that the worst would never happen. Or, more likely, it would have been too troublesome to correct the headlong rush by the industry. Capable biologists worked on the staff of the agency whose ethos displayed no sympathy for the principles of prudent and sustainable management, both critical to the medium and long-term future of any renewable enterprise. Some scientists participated with enthusiasm; others, generally outside the agency, were dismayed. Tourism and recreational angling interests, for whom healthy fish stocks were essential, were angry but what of the commercial fishing industry itself? Ken Whelan tells the story of a skipper whom he reports saying

...to make money you need to plan two species ahead of BIM...[37]

In other words, make your money wiping them out before it does!

If subsidy had become an essential source of income, everyone needed a share of it. In those circumstances one did not question the wisdom of what was happening, one tried to live with it. A cliché describes it aptly: "situation hopeless...but not serious". There was after all, within what had become a robust establishment, an acquiescence to what was taking place. There was little point in an individual, particularly a small-scale fisherman, trying to oppose it.

Chapter 14

THE OFFICIAL FATE OF SOME FISH STOCKS

The expansion of fishing effort arising from, successively, a multiplying then a modernising – technologically improving - fishing fleet, and the frenetic search for species to satisfy its growing appetite exacted a toll on marine life. Exactly how much we will never know but the official record is a good place to begin looking for answers.

The exploration programme embarked on a major new expedition, investigating and harvesting fish from the deep ocean in 2001. It was to be the last great exploratory initiative. Twelve vessels, ten of them recently purchased under the whitefish fleet renewal scheme, would probe the last uncrossed frontier. The often-repeated justification for the investigation was by that time familiar:

...all of these fisheries could provide potential alternative fishing opportunities for a number of larger whitefish vessels and help diversify effort away from traditional whitefish stocks...[1]

Exhausting "the mine under water"

From the coastline, the continental shelf descends gradually to approximately 200 m below sea level; the sea bed to this depth is known as the continental shelf and it the area which as concerned us thus far. Thereafter the gradient steepens, descending along the slope to the abyssal plain, whose depth averages more than 4 km beneath the surface.

South of Ireland, the slope is a hard, rock surface but west of the British Isles it is covered with sediment. Two offshore banks, Rockall and Hatton, are separated from the continental slope by the Rockall Trough. The northern boundary is marked by the Wyville Thomson and Iceland-Faroe Ridges and the south by the Azores. To the west lies the mid-Atlantic Ridge running south from Iceland (Fig 1).[2]

Throughout the basin, isolated seamounts break the monotony of the otherwise featureless plain. The cold water coral *Lophelia* grows on the continental slope, on offshore banks and seamounts in association with rich and diverse faunal communities including aggregations of fish which congregate around these features where they are readily scooped up in trawls.

The uppermost 200 m of the eastern ocean is warmed by the north Atlantic drift moving north from the Gulf of Mexico. Below it, cold currents move south from the Labrador and Irminger Seas. Beneath 700 m, the temperature averages 7 - 8° C and there is little seasonal variation. Photosynthesis is confined to the uppermost 200 m and it may be limited there by a shortage of nutrients. Upwelling around the pinnacle-like seamounts can enrich their vicinity. For much of the abyssal plain the only nutrient input is introduced through carcasses and other decaying matter descending from the upper layers through the water column.

This is a still, stable, dark world of slow natural change, exposed to little environmental variability. Life spans can be long, maturation late and replication slow; *K*-strategy life histories are to be expected.[3] Information no more specific than these environmental conditions should alert fishermen and investors that cropping yields, if they are sustainable, must be very low.

If ever the analogy between fishing and mining, referred to by Sir William Temple in his letter to Lord Essex in 1673, were relevant, then it had to apply to the deep ocean. The regeneration of harvested species is, relative to the human life span, so gradual that their replacement will never be seen by the fisherman who captured them.

Exploitation of deeper water by Irish vessels began during the early years of EEC membership as Spanish and French fishing techniques became more familiar. The first species to be harvested in quantity by Irish vessels was argentine, a herring-shaped fish with characteristically large eyes, frequenting the edge of the continental shelf, down to 550 m. The greater argentine has a maximum length of some 56 cm and no culinary value. It is harvested for fishmeal. From the late 1980s, Ireland shared the bulk of the catch in the north east Atlantic with Norway, the Netherlands, Germany and Scotland. The total landings, initially high, rapidly stabilised at a much lower level in the 1990s. Ireland harvested 6,000 t in 1989 and 6 t in 1995 but its low commercial value may have discouraged efforts to catch it. Notwithstanding, the scientific prognosis for this fishery should have counselled caution; the species grew slowly and the population had a low rate of renewal.[4]

Another species at the upper end of the deep water range (100 - 600 m) was ling which occurred in the Irish Sea as well as the north east Atlantic. Attaining two metres in length and 35 kg in weight, it is placed among the largest species of the cod family. Unlike argentine, ling was valued for human consumption. It was taken on hook and line and by trawl. Some 54,000 t of this species were captured in European waters in 1998 when little was known of its biology.

Other deep water species were landed irregularly by Ireland during the 1990s. A maximum of 73 t of blue ling, smaller than ling but more esteemed for its culinary qualities,[5] were taken in 1994. It ranges between depths of 400 and 1,000 m. Roundnose grenadier, growing up to 1 m in length but having very little commercial worth, has a depth distribution similar to that of blue ling; before 1999, the maximum annual landing recorded by Irish boats was 59 t. Ireland also made small, irregular, though increasing from the mid-1990s, landings of torsk. Growing in excess of 1 m and to 12 kg in weight, it was utilised mainly as stockfish. The greater forkbeard frequents depths of between 100 and 350 m; it was captured in increasing quantities by Ireland from the mid-1990s. The species grows to 75 cm in length and, though palatable, is not in great demand for human consumption. Small landings of black scabbard fish, which occurs from the shelf edge down to 650 m, completed the list of species most commonly landed by Ireland before 2000. The list is not exhaustive and various deep-water sharks, along with a number of teleosts, were also exploited.

The patchy performance of Irish deep-water fisheries was attributed to the lack of vessels of greater than 24 m and 735 kW in the Irish fleet.[6] That shortcoming would be made good by the whitefish renewal scheme at the end of the 1990s.

The charismatic orange roughy

Many deep water species have extensive distributions, moving through abyssal homogeneity from one ocean to another. One of the best known is the brightly coloured orange roughy which would be emblematic of what occurred next. Usually the species is associated with the continental slope from 200 to 1,800 m but is more concentrated between 700 and 1,400 m. It forms feeding and spawning aggregations around seamounts, at certain times of the year.

Commercial exploitation of this species commenced in 1978 in New Zealand. The following year, fisheries for it developed off south east Australia and the year after that in the Rockall Trough. In 1995 a fishery started off the coast of Namibia and another, three years later, off the Chilean coast in the Pacific; in 1999 another got underway in the south west Indian Ocean.

The species is a popular food fish. Almost half of all landings were imported by the United States of America in 2001. Orange roughy epitomises *K*-strategy life strategies and, because of its economic importance and vulnerability, has been investigated to a greater extent than other deep water species. It has a slow growth rate, matures at between 25 and 30 years old, has a low fecundity and lives for more than a century. It is well adapted to the environment it frequents and has a low natural mortality. Such well-adjusted populations are incapable of dealing with the higher rates of mortality imposed by fishing. Most stocks were fished down to 20% of their original biomass within five to ten years of exploitation commencing. Uncontrolled fisheries are known to have been depleted to commercial extinction.

Orange roughy stocks are more abundant in the southern hemisphere and New Zealand's fisheries have yielded most, her ships searching seamounts for exploitable populations. In 1995 New Zealand put 80% of the world's orange roughy landings onto the market. Total world landings peaked at 91,500 t in 1990 but declined sharply afterwards.

France was the principal fisher of orange roughy in the north east Atlantic up to 1999 when Ireland was recording landings of approximately 65 t. France's almost 19,000 t to that date accounted for more than 99% of all landings off the Irish coasts. They had peaked in 1991, two years after Ireland registered very small landings and they rapidly declined to 1995 when they stabilised at one quarter their peak volume.[7]

Despite having form as a rapidly and irreversibly depleted species, orange roughy was identified as the most valuable deep water target for some of the largest vessels in the Irish fleet.[8] Once again, entering the last decade of the twentieth century, the level of orange roughy landings began to rise. The north east Atlantic closest to the Irish west coast yielded 1,300 t in 1998 increasing to almost 6,000 t in 2001, exceeding the highest previous landings of less than 5,000 t in 1990. The new explosion in catches overloaded the market which could not absorb them. In 2001 the first sale price of orange roughy fetched €3.13 per kg but by 2003 the price had collapsed to €1.29 and this extraordinary and attractive exemplar of the deep seas was being disposed of as fishmeal. In 2001 orange roughy was the tenth most valuable species landed by the Irish fleet[9] but that rating was not repeated, nor ever will be.

Alarmed by depletion rates in its populations, the EU designated orange roughy a quota species in 2003. EU vessels were constrained to taking less than 1,400 t annually of which Ireland's share was 310 t. The quotas were not well received by the Irish industry which felt it had been duped into betting on a horse that did not run.

The orange roughy initiative was the last deep water adventure by the Irish fleet. In 2001, 6,000 t of deep water species raised €12.4 million at first sale. The following year, 8,000 t were sold for almost €18 million. Thereafter, the entire deep water fishery followed the familiar path of individual species. In 2003, 3,000 t were sold for €4.6 million, in 2005 2,000 t sold for €2.4 million and so on to 2010 when 455 t raised less than €1,000 each. Current commercial prospects are summarised as follows:

While deep water species still present an opportunity for the Irish fleet, with increased fuel prices, declining stocks, low TAC share and sustainability issues, they are now of minor importance.[10]

Some whitefish staples of the industry

And what of the whitefish from which pressure would be relieved by the diversion of new vessels to plumbing towards the abyssal depths? The official record of landings of a selection of these is set out in Appendix 6 but they are far from being the complete story. Before embarking on an account of what happened, it is appropriate to delve back into the origins of the industry.

One interpretation of history might date the origin of fisheries management in Ireland to 1842[11] when legislation provided for the appointment of fisheries inspectors and stipulated that all fishing craft should be registered. Technology has made matters infinitely more complicated in the meantime but the essentials remain unchanged: to manage a fishery, data are required to measure the amount of fishing effort (numbers, size and power of vessels engaged), their yield (the landings) and biological information on the exploited fish stock. Things were a lot less complicated at the beginning of the last century.

The recorded history of fish landings began in 1899, under British rule. The routine for collecting statistics on the industry, through the coast guard, was meticulous and standardised. During the early years, few varieties of landings were monitored – 23 to be exact – of which seven were whitefish. It should not be supposed that the species which were recorded were the only species utilised but it has been observed elsewhere and it is probably true in this case too, that as fisheries intensified throughout the twentieth century, and particularly in its latter half, the number of categories of landings increased considerably. An explanation for this was the necessity to squeeze more commercial value out of the catch by developing markets for a greater percentage of it.

Landings of two of the species we are going to examine here – cod and haddock – were reported since statistics were first collected. Plaice joined the list in 1903 and, at this stage, we will look at the official account of how the three fared. In the early years the "prime" species among the demersals were brill and sole; cod and haddock were referred to as "coarse" or "common" species. Seventy years later prime species would be angler, megrim and hake.

The landing records for the Gadoid species, cod and haddock, are intriguing. Haddock has for many years been referred to as a "spike" species; it has large recruitments to the fishable stock which mushroom into a burst of abundance, manifest in heavier landings.[12] More recent observations suggest the two species assume reversible predator-prey roles: when cod are abundant they consume the earliest life stages and juveniles of haddock and *vice versa*.[13] If this theory is correct, it might explain the distinctive landings patterns. Until the 1950s, it is suggested, cod were sufficiently abundant to suppress population spikes in haddock. Thereafter, as cod was fished down, haddock landings undertook a number of brief population explosions which played a major part in their fluctuating abundance since.

The three species, cod, haddock and plaice, are common and were, at one time or another, abundant over the territorial sea. All have been managed by TAC and quota since the CFP got underway: their populations are scientifically assessed and an annual allowable harvest calculated. This is divided into shares or quotas on the basis of pre-determined percentage allocations to EU member states; it sounds an easy and straightforward procedure but it is not. The landings of cod and plaice observe similar trends, rising to a maximum in the 1980s and then falling back. Because the power of the fleet continued to increase after the peak in landings, it can be assumed that catching effort was rewarded with progressively lower returns. From 1950 to the late 1980s the landings of cod and plaice might be regarded as a proxy for the growth in fishing effort. In general terms then, it can be assumed that the abundance of cod and plaice is in decline.

The relationship between landings and fishing effort is less straightforward for haddock.

The fact that all three are regulated by TAC has been cited as a reason to doubt the landings figures. Many in the industry maintain that scientific assessment under-estimates the population size on which TACs are based. If that were the case, imposing a TAC, would be a way of artificially understating the biomass of a species and landings would not necessarily bear any relation to its abundance. Others point out that setting limits on what can be landed leads to misreporting and that is verifiably correct. The data set out in Appendix 6 refer to landings, country-wide. However, scrutiny of what occurred in specific areas ("stock units"), for example, the Irish Sea, confirms the worst fears.

For a number of years TACs for the three species have been set far above and out of reach of landings. This being the case, there is was no reason to suppose that TACs were pessimistic estimates of stock size; nor was there any motive for under or mis-reporting what was caught.

Concern expressed by scientists and the general public prompted the collection of detailed information on fishing operations and, as a result, the amount of capture effort over many areas of the north east Atlantic can be expressed in precise terms. But such data are not new.

The earliest accounts of fishing effort in these islands made a distinction between trawling and other gears. Trawling was more profitable, required greater input of power to operate and was often accounted for independently in official annual reports of the fleet. The logical consequence of this kind of thinking was to relegate other fishing methods to a lesser status or to discount them altogether. From the late nineteenth century, trawling was reported separately and additionally to what the rest of the fishing fleet did. In the early years of the twentieth century, trawlers accounted for 3 - 6% of the fleet by number but 50 - 70% by GT.

Ireland's trawling fleet at the beginning of the twentieth century consisted of some 400 sail-propelled boats whose numbers steadily declined to the mid-1930s although a handful survived into the following decade when motorised vessels replaced them. Few steam-driven trawlers were based in Ireland but British-registered boats of this type fished Irish waters. A census in 1907 reported more than 2,200 steam trawlers in Europe; almost 60% of them were registered in England and Wales and a mere six in Ireland.

The Irish fleet was operated by the Dublin Trawling, Ice and Cold Storage Company Ltd[14] at Ringsend, Dublin. It supplied the Dublin fish market which was the only organised one in the country and, when demand was slack, the boats landed into Milford Haven (Wales) instead. We do not know how many days in the year they operated or their precise characteristics but we can make a stab at guessing those. After 1914 their individual tonnages were reported as above or below 100 GT. If we attribute 80 kW engine power to each, that would probably be on the generous side. In 1923 nine of these, in today's terms, tiny trawlers, landed between them 560 t of cod, 79 t of haddock (more than one third of all landings of this species sold on the Dublin market in that year) and 57 t of plaice. We do not know how many days they needed to spend at sea to capture these fish but their diminutive fishing effort, 87 years ago, harvested almost as much cod as the considerably greater combined fishing effort of EU fleets (comprising vessels from Ireland, the UK, France, Belgium and the Netherlands) in the Irish Sea in 2010. That combined fleet effort also harvested only seven times more plaice and eleven times more haddock than did the Dublin steam trawling fleet in 1923.[15] It is an indicator, albeit not a very precise one, of the devastation that occurred in the meantime.

The individual history of every fin and shellfish species illuminates the industry from its own perspective. The examples of whitefish we examine tell us something about species whose regulation by TAC and quota was not sufficient to rescue them from near elimination. Another, whose story tells us much about the industrial environment of its time is a little shark species, the spurdog.

Rock salmon – abundance to near extinction in less than forty years

The origin of fish dipped in batter and deep-fried as a fast food is credited to Sephardic Jews who immigrated to England from Spain and Portugal in the seventeenth century. Charles Dickens referred to a fish shop as a "fried fish warehouse" in his novel

Oliver Twist published in 1830; at that time the cooked fish was accompanied with bread.[16] Later, combined with chips, it made a nutritious comfort food for the poor and became a British institution. George Orwell, a century later, rated its products highly, saying they relieved stress to the extent of averting revolution among the impoverished working class.[17] In Britain, members of the National Federation of Fish Friers absorbed a significant proportion of whitefish landings to process as fish and chips. The Federation was formally consulted and involved in plans for the whitefish industry.

The pre-eminence of fish and chips has declined over the past fifteen years, displaced by fast foods originating in the United States of America, Italy, India and China. A factor contributing to the demise of the traditional fast food is scarcity of some whitefish because these species are the most suitable for deep fat frying. As early as 1977, the paucity of appropriate species for the fish and chip trade was used to support the argument for Britain obtaining a fifty nm exclusive fishery limit. A spokesman claimed that at least one fish and chip shop was closing every day.[18] Cod and haddock were traditional favourites; efforts to substitute them with oily fish like mackerel are still at an early experimental stage, and there does not appear to be a lot of enthusiasm for the change.

While the burgeoning of the fish and chip trade is associated with Britain, Italians took it further and became its ambassadors in Ireland. A number of families, some of whom are still in the business, settled here following the First World War. Antoni Matassa had the Roman Café in Marlborough Street in Dublin; it passed to the Nardone family in 1922. Peter Ferrari immigrated from Parma to Ringsend in 1915 and passed the business there on to his son on his death. Dominic Macari moved to Dublin from Naples and founded a business which still bears the name. Mario Cafolla established another familiar retail outlet. Brief occasional descriptions of these operations were published in the early issues of the *Irish Fishing and Fish Trades Gazette* but, while their names associate the fish and chip trade in Ireland with Italian operators, they suggest the provision of this fast food was less plentiful than in Britain, probably another reflection of the Irish lack of appetite for fish and an indicator of the absence of a similar manufacturing industrial background here.

A layer of deep fried batter can conceal species identity and a number of whitefish species have probably been proffered for sale under its camouflage. One non-teleost species which proved popular was spurdog, a small shark species, referred to by BIM as a whitefish species. Being an elasmobranch, its fillets did not contain bones but cartilage from which the flesh readily peeled away. Spurdog was popular in the British fish and chip trade where it was sold as "rock salmon". In addition to being providers of fish suppers, the Italians were also known as promoters of elasmobranch flesh. Another of their products was "long ray" (ray wing sliced along rather than across the supporting cartilage).

In their recommendations for a re-organised fishing industry the Commission of Inquiry into the resources of Ireland, in 1921, considered the use of spurdog as an ingredient in fish fertilizer because the flesh contained as much as 10% nitrogen and 10-12% phosphoric acid.

There is, moreover, a further advantage in the development of such an industry. For dogfish are a perpetual menace to fishermen's nets; they devour great numbers of valuable fish; often their numbers are so great as to impede the continuance of fishing; and in several countries the organised destruction of dogfish is the subject of constant investigation...We conceive that an enterprise of this kind would offer a favourable opening and would certainly prove worth special inquiry, since every opportunity should be given to fishermen to increase the prosperity of their craft.[19]

Shortly after the commission was established, Irish fishermen were confronted with an opportunity which they failed to grasp. In its report for 1923-1925, the Department of Fisheries observed:

The prevalence of dogfish (spurdog) *among herring shoals all around our coast led the Department of Fisheries to experiment with various forms of fishing gear suitable to the capture of this pest which is so destructive to the drift nets. As this fish is not sought by our fishermen it has multiplied until it has now become so numerous that the herring fishermen are now frequently deterred from fishing through fear of damage to their nets. Unsuccessful efforts were made in 1924 to induce the Galway fishermen to capture and market this fish which has a good sale in London.*

The report went on to show comparable prices at Billingsgate Market where spurdog obtained 80% of the value of hake and 93% of the value of haddock. The fact that spurdog had been regarded as a worthwhile target species in the previous century[20] is further indication of the alienation of Irish people from the sea.

All western European nations have landed spurdog but the principal catchers in the early twentieth century were Norway and the United Kingdom. Norway had an inshore fishery for spurdog as early as 1930 but after the Second World War her effort expanded into an offshore long-line fishery ranging to the Orkney and Shetland Islands. Between 1950 and 1970 Norway accounted for 70% of all spurdog landings in the north east Atlantic. Landings by the UK were made mainly by demersal trawl, a minority by long-lining, and they reached their maximum at 7,000 - 8,000 t in the late 1930s. Total landings of spurdog peaked in the mid-1960s at 60,000 t a year, followed by a downward trend over the next twenty years, at which time the Irish industry had woken up to the lucrative possibilities of the species.

The first mention of spurdog as a commercial prospect in Ireland appeared in the trade press in 1970. During what was described as "the worst lobster season in memory", fishermen in west Cork began long-lining for porbeagle shark which they exported to France *en route* to processors in Italy. Tope and spurdog were also acceptable to this market. The head and fins were trimmed off the carcass which would fetch stg£0.1 s/6 d per kg. The fishery spread to the Irish Sea where BIM seized the opportunity to promote it.

The momentum for this shark fishery was gear overspill from another, a further demonstration of the inter-linkage of various types of fisheries. Drift netting for Atlantic salmon at sea commenced in the nineteenth century. Up to 1968, some 20% of the landings were harvested in this way but after that the figure rose to 80%, coinciding with the introduction of monofilament or multifilament plastic nets. Reports indicated that the length of drifting gill net in use were enormous; anecdotally, in excess of 20 nm (37 km) of net panels attached end-to-end were reported to be stretched across the migratory path of the returning fish by several vessels combining their efforts in Co Donegal. In the late 1980s the problem of salmon drift net fishing ran out of control. Individual vessels, many of them unlicensed, fished up to 10 nm of net, 30 - 60 meshes deep. In 1987 a review group recommended that gill netting should be either strictly controlled or abolished altogether although it would be another twenty years before that happened.[21]

Plastic drift nets could be fished for up to three years but after that they were less effective on the surface and had to be replaced with new gear. An alternative use for older nets was to sink them onto the sea bed and operate them as gill nets for other species. Whitefish were harvested in this way and so were spurdog. The discovery of large quantities of the latter off the coast of Co Donegal coincided with its depletion in the North Sea. A market opportunity had arisen, and the Irish industry threw its energy into satisfying it. The Irish fishery commenced in the late 1970s and peaked in 1985 when 8,800 t were landed. Most of these were sourced in the vicinity of Greencastle and Burtonport in Co Donegal. As the Donegal spurdog population eroded, fishers directed their attention southwards and the fishery focussed on Co Kerry where spurdog was landed into Carrigaholt, Dingle and Fenit.[22]

Spurdog is a long lived and slow-maturing species, another *K*-strategy exemplar; some of those captured off the west coast of

Ireland were more than 35 years old. The species is a live bearer with the longest gestation in the animal kingdom, almost two years. The number of whelps borne by a female depends on her length. Older, larger mothers may give birth to as many as fifteen but the youngest and smallest mature females have as few as two or three. The size and life prospects of a new-born are also dependent on its mother's dimensions, an important management consideration: younger females produce smaller and less fit juveniles. Selective gill netting removes the larger females and the progeny of the smaller survivors are less successful, as well as less plentiful, an additional complication when a stock is exploited.

Biological investigations in Ireland commenced as the fishery got underway. Landings were of two kinds: trawl catches from deeper waters – which were landed from the time the market for them opened – consisted of mixed sex shoals of younger, immature individuals of fairly similar size. Gill-net caught spurdog were predominantly older and invariably pregnant females and they were targeted later, as the market developed.

Although its importance as a whelping (reproducing) area of the Celtic Sea off south west Ireland was not immediately appreciated, the vulnerability of spurdog, based on its biological and reproductive characteristics, was. Attempts were made to alert fishermen to the implications of harvesting pregnant females and one responded by letter to the trade press, addressing the subject from a typical perspective: welcoming research into the impact of gill nets "as long as it is not carried out with the negative attitude often encountered nowadays...". Money was being made and interference was not welcomed. The letter did not suggest any form of regulation or gear limitation by the industry and instead concluded with some diversionary musings:

An interesting subject for research would be the effect on other stocks of decimating the huge predatory shoals of dogfish. It may be reasonable to regard them as a kind of vermin or weed that are better cleared, just as a farmer clears scrub to let the grass grow. But then there are some bright sparks around who would much rather see acres of scrub and rushes than well tilled fields.

They might even style themselves friends of the Earth or some such: it's doubtful if they could be described as friends of the human race.[23]

It was some time before the effects of the removal of spurdog from the vicinity of Burtonport (Co Donegal), where the first heavy landings of the species were brought ashore, were fully realised. A note in the trade press in May 2002 provided the essential statistics. During the 1970s and 1980s, Burtonport had been a leading fishing centre to which whitefish and Atlantic salmon were landed in large quantities. In the early 1970s more than 100 boats were engaged in the pursuit of salmon and there were 26 trawlers based at the port. A decade later, another 55 boats were engaged in the gillnet fishery for spurdog. One by one these fisheries collapsed to be followed by an equally brief brown crab fishery.[24] The industry, in the form of either individual or collective operators, had no word of regret on hindsight to contribute on the removal of the stocks on which it depended and there was no hint of remorse for the way it had greedily damaged its own prospects. Instead, the solution was to seek funding to develop in another direction. The note concluded with the statement that finance was required so that the harbour might diversify into marine tourism.

Because spurdog has economic value, various attempts had been undertaken, worldwide, to understand and describe its biology and behaviour. No other shark species has been so intensively harvested or so vigorously studied. However, investigations were invariably associated with the commercial fishery developed by a single nation, hence they tended to be undertaken where its vessels were operating. Not unreasonably, successive investigators described a number of separate stock units. In 1966 an imaginative marking experiment was set up and operated for twenty years. Spurdog were tagged off the south coast of Ireland and their subsequent progress tracked through the locations at which tags were recovered. The fish were found to pursue a slow migration course (which was sufficiently gradual to be more accurately described as a "diffusion") north, around the west coast of Ireland, past the west coast of Scotland and into the North Sea, which was accomplished in some seven years. In the course of its lifetime, a spurdog may have completed the circuit several times. Whelping (giving birth) had a southern emphasis, being concentrated in the Celtic Seas where the gravid females moved inshore, possibly to make distance between their new-born and older, deeper swimming, males. It was in the course of these whelping migrations that gill nets intercepted and harvested large numbers of pregnant females.

The consequences for the Irish fishery could be confirmed from several of the monitoring schemes in operation. A discard recording system observed the smallest spurdog in the Celtic Seas. Large angler-caught spurdog, recorded by the Irish Specimen Fish Committee, all but disappeared from the sea area north of south west Ireland after 1983, possibly indicating their slow migration northwards had been stymied by the removal of larger individuals from the population further south.

If the total landings of the species in the north east Atlantic had been declining until the mid 1980s, that decline then intensified. In the early years of the new millennium less than 1,500 t were landed into Ireland annually, approximately half of them taken in gill nets. In 2005, 8,400 t were harvested from the entire north east Atlantic and thereafter restrictions on catches were introduced. However, as late as 2008, the industry in Ireland was still seeking access to spurdog which were in a very depleted state. A discard rule stating that spurdog must not exceed 5% of landings by a vessel required animals above that percentage to be returned to the water – discarded. Elasmobranchs are believed to survive discarding. This time a plea to retain all spurdog emanated from MEP Sean O'Neachtain (Fianna Fail, Ireland north west) who pleaded that losing this species so recently after the salmon fishery had been closed would impose considerable hardship on the industry.[25] EU regulations temporarily imposed a zero TAC on spurdog; landing them for a time was prohibited. Discussion of the fate of spurdog has been forced onto another agenda, an unforgiveable trait of an industry which depletes fish stocks and then abandons their rescue to others, in this case an international conservation initiative.

CITES (the Convention on International Trade in Endangered Species) was formed in 1963 and in 2009 it had 175 nation members, referred to as "parties". It lists approximately 5,000 endangered animal species in its three Appendices. Until the 1990s marine species were not seriously considered and it was not until 1994 that the vulnerable characteristics of sharks and allied species were debated. Five years later member states of the Food and Agriculture Organisation of the United Nations (FAO) adopted an international plan of action on sharks aimed at securing their long-term sustainability. Species which are protected by CITES are allocated to one of three Appendices: 1, lists species threatened by extinction which may be affected by trade; 2, species which, while not currently under threat, might be if trade were not controlled and 3, species which parties manage within their own jurisdictions but for which assistance from other parties is required.

Spurdog qualifies for membership of Appendix 2 by virtue of its biological characteristics and fishery history. When intensive commercial exploitation commenced in the last century, the species proved susceptible to population collapse from which stock recovery was problematical.

Vulnerable to commercial fishing, spurdog is, in other respects, a highly successful species. The group to which it belongs evolved in the post-Jurassic period, approximately 145 million years ago; the basic design of the group of sharks to which it belongs is believed to have remained substantially unchanged throughout geological history. The majority of species within it occupy deeper waters, but spurdog frequents the shallower continental shelves of temperate seas.

Chapter 15

TWO LAST CASTS OF THE DICE

During the 1980s and 1990s, as the fleet honed its catching ability and fish stocks declined, it became more obvious, and was eventually admitted, although perhaps not in so many words, that the resource was simply not capable of supporting the industry. Smaller inshore operators, accounting for an estimated 50% of those employed in the catching sector, were ignored, written out of the equation, but the remainder had to be accommodated or, rather, the survival of the most successful among them would be fostered according to otherwise Darwinian principles. As the established lines of demand for its services became less reliable, BIM, which justified its existence by facilitating the business of fishing, needed to invent new means of making itself relevant. An event which may have alerted the agency of the necessity for a policy swerve was a report published in 1997 by Price Waterhouse which recommended a major reorganisation of the state's marine services. Among many suggestions, it proposed establishing a new commercial sea fisheries and aquaculture body in place of BIM.[1]

If the Irish authorities were in ignorance or denial of the gathering storm, the Commission in Brussels foresaw difficulties ahead. The Commission's fourth multi-annual guidance programme (MAGP4) was attempting to reduce fleet size, and had effectively capped it, and Brussels had become reluctant to contribute funds towards vessel construction and purchase. But this was an industry with a number of drivers towards modernisation and change. In the normal course of day-to-day activities, fishermen improve their catching abilities by improvising new techniques and introducing new technology to their operations. And there were precedents for state-sponsored fleet make-overs. France, for one, had implemented "fleet renewal" in anticipation of Spain joining the Community.[2] Fleet renewal meant essentially replacing older vessels or modernising existing ones within the capacity limits of the existing fleet (its tonnage and kiloWattage), with assistance from the taxpayer, of course. The issue simmered for a number of years, occasional public mention a reminder it had not slipped off the agenda. When, in 1996, Martin Howley of the KFO raised it again, the minister, Sean Barrett (Fine Gael, Dun Laoghaire), expressed concern that more powerful larger vessels would wipe out the livelihood of smaller operators.[3] The following year the way ahead for a fleet renewal scheme was cleared in Europe[4] and later that year, the minister, Michael Woods (Fianna Fail, Dublin north east), indicated he was in favour of it. Martin Howley registered his satisfaction with the way the matter was being handled by the Department and BIM and Joey Murrin confirmed the KFO had strongly backed setting up the scheme. Details of the programme were announced by BIM in July 1998.

Four aspects of fleet renewal would be supported with grants and tax concessions: installing safety equipment, constructing new vessels, purchasing second hand boats and modernising existing vessels.[5] The scheme, it was anticipated, would renew 5% of the fleet. Some 600 vessels targeted whitefish and, assuming that each had a working life of twenty years, 5% was the appropriate annual level of replacement.[6]

Ireland would have two fleet renewal schemes. The first, in 1998, focused on larger vessels exploiting deep water and non-quota species. The second was geared towards smaller ones: however, only 22 of the 38 to be grant aided were under 15 m while the balance were between 16 and 46 m oal.[7] In other words, owners of larger vessels were disproportionately assisted.

The value of grants towards the purchase of new vessels was to be 40% and there were separate schemes for second-hand purchases, modernisation and acquisition of safety equipment. The 1998 finance act, referring to the first scheme, entitled participants to write off their investment in certain types of fishing vessel against other income at the top rate of tax in the first year, half their investment in the second.

Prelude to the perfect storm

On 13 May, 1999, Michael Finucane (Fine Gael, Limerick), shadow marine spokesman, asked the minister, Dr Michael Woods, if he would submit BIM's Seafood Industry Agenda 2000-2000...

...to be considered for funding under the next EU operational programme and if he would include its proposals for funding in the national development plan...[8]

The minister replied he had a high regard for the document, which highlighted economic opportunities and set out an ambitious agenda backed by the necessary supporting analysis. BIM's proposals envisaged a total investment in the industry of IR£364 million, 54% of which would be supplied by the taxpayer. The whitefish fleet would be "renewed" − more powerful vessels added to it - there would be increased investment in fisheries infrastructure and a new impetus for growth of aquaculture and fish processing. The minister added that there had been other suggestions too and he intended to have all considered together and reconciled so that a comprehensive package might be put forward.

The sum of money involved was considerable for the time and it might have been thought prudent to request a second opinion on how it might be spent. Instead, the question was referred to another committee, the National Strategy Review Group (NSRG), not far removed from the first, which was deliberating various issues relevant to the reform of the CFP due to take place in 2002. The review group was convened under the chairmanship of Padraic A. White and its membership included, among others, representatives of government departments, POs, the IFO and the Marine Institute. BIM had two representatives on it and the agency supplied the secretariat. The two largest political parties in the country were in favour of the outcome, the operational programme currently in progress was deemed a success and BIM was in a position of influence, essentially dealing with packaging a programme it had already formulated.

The industry was considered to consist of four elements: landings, exports, processing and aquaculture. In the early days of BIM the fourth would have been the fleet but that was no longer in rude good health and better left unmentioned. All of the four indicators had expanded in the years 1993-1997 and this was selected to be the reference period for prognosis. It is difficult enough to predict future events on the basis of economic statistics but looking nine years into the future using biologically related criteria is hopeless. It is relevant to revisit chapter 9 and refer to the fatal attraction of predicting from statistics which was probably less reliable than fortune-telling for BIM; but that was long before, in the first twenty years of its existence, and the experience and any lessons arising from it had clearly been erased from memory.

The method of prediction on which the committee embarked appears to have been very simple-minded. I was reminded of the technique outlined in Michael Lewis's book, *Liar's Poker*, on which I believe it was modelled:

...When it came to speculation European investors shared a weakness for get-rich-quick schemes. And, like a crap shooter who has a pretty lady to blow on his dice, the speculators had an amazing array of irrational systems to help them win money. They had to guess which direction the American bond market would take: up or down. The systems usually involved staring for hours at charts showing the history of bond prices...The chartist, for that is what such a person called himself, would then use his

ruler and pencils to draw the future of bond prices based on the assumption that the historical pattern would be projected forward...

Lest this view be considered gratuitous, it is worth examining the projections that emerged from the NSRG.[9] The changes in value between1993 and 1997 established the trend of future increase. From 1993 to 1997, an average increase of 11% annually was recorded for aquaculture. Joining the dots for the two years and projecting them on to 2006, the pencil-and-ruler rise would have been 99%; the value that emerged from the committee was 108%, fairly close. In the case of processing, the projected increase would have been 76.5%, the committee's estimate 72%. For exports the projected increase would have been 52%, the committee's figure 62%.

The only consistency in this exercise was the similarity of its outcome to previous attempts by BIM to forecast what was likely to happen to the industry over which it had so much control and influence but with which it apparently had no empathy. The outcomes, adjusted for inflation, were as wide of the mark as any the agency had attempted in earlier years. Inflation of 32% occurred between 1997 and 2006 and this has to be taken into account when the disparities are quantified. Aquaculture rose in value by 30%, not the predicted 108%; the value of processing fell by 79% rather than rose by 72%; exports declined by 6% in value instead of growing by 62%.

Central to the conclusions of the committee was the value of landings, which had risen by 33% between 1993 and 1997. Their projected rise was a modest 28%; they declined by 4%. Landings are a significant driver of exports and it is relevant to ask whether the projected 28% increase, had it materialised, would have raised export values by the projected 62%.

Most of today's secondary school pupils would be able to do a pencil-and-ruler projection but more would have been expected of a committee which included appropriate technical expertise. Administrative civil servants tend to defer to their scientific colleagues in these matters; representatives of the industry would have done likewise. Little was to be expected of BIM which had been conducting exercises of this kind for almost forty years, making extravagant forecasts and then changing the subject around the delivery date. But what of the Marine Institute? Three named members of its staff sat on the NSRG; one of its divisions was routinely publishing an annual review of fish stocks, dealing mainly with quota species. There was no sign of dissent or disagreement from that quarter.

Instead, another state agency, the Economic and Social Research Institute (ESRI) had published a paper reviewing national investment priorities for the period in question.[10] I wonder whether this paper was what minister Woods referred to when he spoke of having various views reconciled? Nobody from the ESRI was a member of the NSRG – omitting dissenters is an guaranteed way to ensuring unanimity – but its report was referred to in the group and comprehensively rejected. It is relevant to revisit some of the concerns expressed by the ESRI which, if taken as seriously as they deserved to be, would have scuppered BIM's intentions to facilitate further expansion of the catching sector.

The ESRI pointed to the limitations of quota allocations which were declining as the fish stocks declined. This drew the pathetic riposte that while many of the most valuable species were regulated in this way, Ireland, with 13% of EU waters on its doorstep, was sure to find something else to exploit. In technical terms, this phenomenon is known as counting your chickens before they hatch and, as was found fairly early into the new investment spree, there was no unrealised harvest awaiting discovery.

The NSRG observed that much of the national fleet worked close inshore – "largely outside the scope of TAC and quota-managed species" were the words used and, it could be supposed, outside the reach of the Commission's writ. An interpretation of this is that the industry could land what it wished without hindrance. In view of what materialised later this is not an unreasonable assumption.

The ESRI's final objection was something that should have been obvious for a long time: the limited returns generated by fisheries in relation to the amount of public moneys spent. That was summarily trumped by the NSRG prematurely trumpeting its projected increase of 63% in the value of the industry (obtained by combining the expected increases of its four elements); in real terms the combined values fell by 27%.

The deliberations of the NSRG were published in an attractive cover. That probably helped to sell the idea; the authoritative committee line up and a chairman of national renown, would not have hindered the message either. In 2001 BIM was overjoyed when an investor of great wealth expressed an interest. Denis O'Brien, the media and communications tycoon, had received IR£250 million for his share of the company Esat Telecom Group plc, purchased by BT Group plc. The whitefish renewal scheme allowed private investors to offset their investment against their top rate of income tax and this was believed to be a strong enticement to become involved. Otherwise O'Brien's decision to invest was seen as something of a leap of faith because he had no known involvement in the fishing industry.[11] John McManus, who broke the news of BIM's delight in the *Irish Times*, summarised it thus:

The involvement of high profile businessmen was the icing on the cake. It was a powerful sign that fishing was in fact quite sexy...

Exactly what is had to do with BIM's mission was another matter. A justification for their existence read:

Developing the sea food industry, sustaining coastal communities

Because the size of the fleet was capped at this time, the introduction of a new vessel meant that one or more of equivalent tonnage and power (GT and kW) had to be bought out and removed. The whitefish renewal scheme was seen as highly successful but new introductions to the fleet, being more modern, would have put slightly older vessels in the fleet at a considerable disadvantage. The cost of the renewal scheme had been IR£80 million of which IR£17 million was contributed by the Irish government.

Five years into the new development programme, coinciding with the discovery that the industry was surviving and utterly dependent on illegal catches, it was suddenly realised that the fleet was far too big. There was no alternative: an emergency decommissioning scheme had to be hastily cobbled together. BIM wheeled out its current celebrity chairmen to break the news.

The trade press received it positively:

Highly respected throughout the various fishing sectors, Padraic White's report[12] *is recognised as not just a* quick fix *to reduce capacity by removing 25% of whitefish and scallop fleets...*[13]

The fact that the chairman of the NSRG, which had advocated the increase in fishing capacity which precipitated the crisis, was now announcing the emergency solution did not elicit any published surprise. Between 1998 and 2006, 62 new vessels were delivered to the Irish fleet along with 17 modern second hand ones; a further 77 underwent modernization and 650 were modified for safety reasons, within the terms of the whitefish renewal scheme. Oddly, the main plank in the argument for reducing the whitefish fleet by almost 11,000 t (one third of the 33,000 GRT whitefish fleet), now proposed, was the deteriorating status of quota species "... quotas have fallen by almost 50% since 1990." Yet that was the argument against expansion proposed, in the first place, by the ESRI and summarily dismissed by NSRG. The activation of the decommissioning programme, the second in the history of the industry, and its policy implications are discussed in chapter 18.

The sudden acute dearth of fish presented enormous difficulties for the industry and its administration. Decommissioning was a minor element of the climb back to solvency. Another review was sought, this time from Dr Noel Cawley who would look at underlying structural problems. After all, according to the chief executive of BIM, Jason Whooley:

(there is) no point in overcoming one crisis if fundamental weaknesses remain...With world demand for seafood increasing there is still a silver lining for a restructured sector...[14]

To deal with the latest crisis effectively, some €600 million would be needed. This was possibly as much again as BIM had channelled into developing and building the fleet in the course of its history.[15] Measures envisaged included a tie-up scheme for boats whose owners would be compensated for doing so, and some kind of fuel subsidy to enable vessels to capture ever scarcer fish. BIM also envisaged grant aid for more effective marketing. Ironically, increased funding for measures to counteract illegal and unregulated fishing by nations outside the EU who, by their efforts were supplying the Union with cheap fish and undercutting fish prices, was also considered relevant.[16]

The Whitefish Renewal Scheme was the last great fleet initiative involving BIM. The agency continued with other aspects of fleet management, notably the mopping-up operations of the decommissioning scheme but, for the moment at least, the fleet was complete. Boat building, for which it had engaged the largest number of employees in its history was its primary objective when the agency was established in 1952. However self delusionary the administration in Ireland, the Commission in Brussels had grasped the message: the tap supplying funds for vessel construction from that quarter was switched off. In the course of its 58 years, up to 2010, approximately one third of all monies invested in BIM had gone to fleet development. Less than 20% had been allocated to, respectively, administration and aquaculture; in further descending order, marketing and exploratory fishing completed the list of topics on which monies had been invested. The promotion of aquaculture would likely become BIM's most significant occupation in the future but there was one further fisheries project in which the agency had got involved.

The Shellfish Framework

Throughout its history BIM had resourcefully become involved with anything and everything concerning marine, and even some freshwater environmental and fisheries issues.

While the official departmental view, *de facto* if not *de jure*, was that small-scale fisheries operations could look after themselves, some effort had been made by BIM to facilitate developments within that category of fishermen. As the national source of grants, applications were made to BIM by local groups seeking assistance and thus drawing the agency into local projects. BIM became involved in lobster management in the early 1990s, and could take credit for its part in the administration of the lobster V-notching scheme, successfully promoting conservation of this species. BIM assisted with the formation of lobster fishermen's co-operatives. Local applications for funding for aquaculture, for which it was distributing grants, introduced BIM to membership of a number of local fisheries committees under the umbrella of the LEADER, rural development programmes; WORD (the Wexford Organisation for Rural Development), which hosted a local lobster fishermen's co-operative, was one of them. There were similar groups in north west Connemara and south Kerry. European Community monies channelled through PESCA were used to promote lobster fishing; mussel harvesting and scallop fishing were added to the list.

In 1996 some IR£53,000, for small scale coastal fisheries, appeared in BIM's accounts and the following year reference was made to the work of its aquaculture promotion officers working with local groups when it was announced that BIM was continuing its involvement in measures to improve coastal zone management. BIM looked forward to the formulation of a coastal zone management policy in 1998.[17]

In 1998 small-scale fisheries were more firmly ensconced in BIM's work programme. Assistance was proffered to local communities who wished to embark on mackerel fishing with automatic fishing (jigging) machines. An eighteen-month programme, aimed at family members in twelve coastal communities, was launched that year. The prospectus offered training in tourism, business and safety awareness and women were encouraged to participate.

BIM's 1999 annual report announced a discovery. The chairman's statement put it like this:

For some time the Board has felt that the inshore fisheries sector has not been receiving sufficient development focus and following a request from the Minister of the Marine and Natural Resources to examine the sector in terms of structural and employment issues, conservation and management of inshore stocks and the identification of existing and new development opportunities for the resource, BIM launched a Report on the Inshore Fisheries Sector in May. The Board attached considerable importance to the Report's recommendations, especially those pertaining to a national Inshore Fisheries Advisory Committee and a number of pilot Inshore Fisheries Development Committees and major priority was given to getting these underway.

There could be no doubt that small-scale fisheries operating within the exclusive six nm limit required management. Much of the fleet of smaller vessels was not even registered; with the exception of regulations for the conservation of lobster, the laws, including those transposed from EU directives, were not enforced, and there was no protection for the smallest vessels from the activities of the largest in the fleet which were permitted to fish anywhere they wanted. Local fishermen occasionally combined to make representations about these issues to the Department, the ultimate port of call, but they were ignored. There was a general despondency about what could be achieved and little interest in fishing rationally, with due consideration for conservation issues.

Twenty years before, small scale operators moved from one species to another as the year progressed: lobster, crab, salmon, herring... By the late 1990s a fisherman was lucky to have one species from which he might scrape a subsistence. Thus, boats in Malin Head (Co Donegal) were dependent on edible crab, those in Magharees (Co Kerry) targeted spider crab, pot fishermen in the south west Irish Sea exploited whelk and all were virtually single species fisheries. Salmon was still exploited but it was becoming more obvious that this fishery would soon be curtailed. Herring stocks had been depleted and were being exploited by larger boats and mackerel was almost exclusively the quarry of the Killybegs RSW fleet.

The first fruits of BIM's newly discovered interest in small scale fisheries were revealed in its 1999 annual report. Exploratory fishing sought new grounds for whelk; fishing trials attempted to land more crab and scallop. It was business as usual: catch as much as possible, no question of regulating anything. BIM simply did not know how to behave otherwise.

Reappraisal of commercial inshore possibilities was under full sail. The trade press reported a seminar jointly organised by BIM and the Central Fisheries Board in November 1999 entitled "Sea angling inshore fisheries, the future of shared resource" containing "recommendations for the future development and enhancement of both..."[18] In the same year BIM began to grant aid sea angling vessel safety measures.

The exclusion of BIM from any fisheries committee would have been an extraordinary oversight as the end of the twentieth century neared. CLAMS (Co-ordinated Local Aquaculture Management Systems) committees sought to locally reconcile aquaculture with other marine enterprises and interests. As part of its contribution to their work, the agency devised and erected interpretative signs demonstrating the benefits of aquaculture. BIM also worked with Meitheal Forbartha na Gaeltachta and Udaras na Gaeltachta in Gaeltacht areas. It even participated in establishing a bereavement support service for the families of those lost at sea.

In 2000 BIM recruited three "facilitators" to pilot inshore development committees in south Wexford, north Kerry and west Galway, areas coinciding with or adjacent to its LEADER projects. A strategy was unfolding:

Following the launch in mid-1999 of a BIM report on the inshore fisheries sector, the Board oversaw the appointment of facilitators for the establishment of pilot inshore fisheries development committees...These initiatives, of which more were being planned, were showing signs of promise, but it must be stressed that to achieve a reasonable degree of success, all the public and private stakeholders involved must take ownership and responsibility for the resources involved. The process may involve some short-term unpalatable measures (my emphasis) if these resources are to be utilised wisely for the benefit of peripheral coastal communities which must be sustained and assisted to retain their marine skills and traditions.[19]

The first task for the facilitators was to assemble committees of "all relevant interest groups" in their areas – these would later be formally labelled Local Advisory Committees (LACs). It was probably an easy enough task. The difficulty of extracting information from government departments, and particularly those dealing with fisheries matters, sharpened appetites for any crumb of intelligence to fall or, better, be authoritatively delivered from those in the know. A facilitator would obviously be such a person. Curiosity alone would fill a local meeting on the half promise that those in attendance might ascertain what future was being mapped out for them. The topic was explored by Michael D. Higgins in his essay "The limits of clientelism" which used as its illustrative matter, appropriately, the shenanigans accompanying the sale of an oyster fishery in Co Galway.[20]

The following year four more facilitators were appointed. A fisheries inshore diversification and safety programme was launched in Carna, Co Galway with fisheries minister Frank Fahey (Fianna Fail, Galway west) in attendance. To give the man some credit he appeared to be genuinely interested in the proceedings and he stayed to the end of the day; not many ministers do that. A galaxy of speakers had been assembled from as far afield as the Antipodes, North America and continental Europe to address inshore fishermen from all over the country. Every speaker had experience in managing near-shore fisheries and each cautioned restraint in the way they were exploited. Stock assessment should precede harvest and management regulations should be enforced. They might as well have stayed at home. Minister Fahey blew the gaff when he generously dispensed automatic jigging fishing machines to catch pollock at the end of the conference. As the only remaining finfish species that was caught in any numbers in Connemara, pollock was indispensable to the Irish angling experience offered by local tourism operators who wanted their clients to capture the fish by manually operated hook and line. As further inducement to small scale fisheries operators, €9.2 million worth of new and second-hand sea angling and marine tourism vessels and safety equipment was proffered by BIM. That could only have further blighted the plight of fish populations in inshore waters. To make matters worse, a proportion of the allocated sum was spent on purchasing new rather than second-hand boats, introducing additional fishing effort to the equation.

One of the major scandals to be tackled was the number of unregistered vessels fishing with impunity in plain sight. When the CFP was introduced, anyone then fishing was automatically assigned, without charge, tonnage and kiloWattage for the vessels in their possession. Some inshore fishermen did not claim theirs. There were various explanations; a fisherman may not have wished to be "officially" fishing because he had other interests or commitments and thought it better to remain invisible to the authorities. Whatever the explanation, twenty years had gone by and many fishermen had become compliant by registering, purchasing GT and kW from the brokerage market in the meantime. In that period also, two attempts had been made by the Commission to regularise the stragglers by dispensing free GT and kW but a large number were still operating illegally. The straightforward way of dealing with this would have been to tell them to stop and register but that was politically unacceptable, so instead, the Department applied to Brussels to sanction another tranche of free handouts.

In December 2002 the Department concluded negotiations with the Commission on the introduction of "a limited scheme to licence and register inshore fishing vessels." Those boats which had been operating illegally, were now awarded, free of charge, a "polyvalent P(pot) licence", entitling them to fish crustaceans and molluscs using traps. There was no obligation to purchase GT or kW but the beneficiaries were precluded from using other types of gear. It was an eloquent illustration of how out of touch officials in the Department and BIM had become. Crustaceans and gastropod molluscs were the only species left for the majority of legitimately licensed small boat operators who were now confronted with an influx of legitimised fishermen who could target the same animals at considerably lower operating cost. In defence of what had taken place, BIM would argue that the earlier purchase of entitlement had enabled its owner to fish a wider range of species but in many areas there was no choice because there was nothing else to catch. A two-tier system of licensing had been put in place.

The number of boats entitled to use pots rose by one third (from 1,300 to 2,000). The customary subterranean grumbles greeted the scheme. One which broke the surface described the prospect as "disastrous".[21]

When it emerged from its shell as a new inshore administration, the Framework could trace its roots back for the best part of a decade. Its launch at a trade exhibition in Galway in February 2005 by the minister of the marine, Pat the Cope Gallagher, otherwise champion of the industry in Killybegs, took place in appropriately bizarre circumstances. The exhibition hall was besieged by scallop fishermen who congregated there from various parts of Ireland, brandishing placards and protesting about the shortage of scallop fishing opportunities. Inside, BIM presented a number of lectures showcasing its inshore competencies. One of these demonstrated an improved technique of fishing greater quantities scallops at lower cost. In view of the fact that the species was in short supply, it could only accentuate the difficulties of the demonstrators outside.

The structure of the framework was set out in a glossy leaflet. It was complex to Byzantine heights and should have aroused the suspicions of those who would become enmeshed in it. The local advisory committees (LACs) assembled participating fishermen who were "facilitated" – organised – by BIM but its deliberations were passed onto another committee, a species advisory group (SAG) populated with as many officials as fishermen. The purpose of the SAG was to refine the raw contributions from the industry and package them in the appropriate scientific and institutional context. Parcelled up in this process, matters were referred on to yet another in the chain, the inshore fisheries review group (IFRG) on which representation of the fishing community had dwindled to the secretaries of the SAGs. Even that however was not the end of the line. The IFRG referred the ultimate distillate to the minister, in other words the Department. In one sense it created jobs – on committees; in another it was comical. Riding instructions for committee members dictated they should observe confidentiality, participate impartially and act corporately. The approach would be "inclusive, consultative and transparent." It would not be.

The Framework plan, as outlined on the leaflet, was to manage Ireland's inshore fisheries, generally understood to be located within six and definitely within twelve nm of baselines. Because it dealt with shellfisheries, the Framework extended its remit to manage those further afield also. The most significant of these was the fishery for edible crab, the majority of which was concentrated in the Atlantic, north west of Co Donegal.

Male edible crab belonging to the north west stock frequent coastal waters but the females undertake long migrations to the shelf edge, against the prevailing current, carrying and incubating their eggs; when these hatch, the larvae are carried by the current back to make a landfall along the Co Donegal coast. In order to mate once more, the females must then return to coastal waters in which the males remained. Brown crab constituted a typical inshore fishery, fished in most months of the year but yielding its main bounty in the late summer when mature - "ripe" – females formed the basis of a processing industry. In the 1980s, as a new departure, "super-crabbers", craft containing large sea water tanks, whose construction was grant aided by the Commission in the drive to exploit non-quota species, began to target crab outside the territorial sea, moving shelf-wards and back above the migrating females. The financial benefits were immediately obvious and the fishery gradually switched from an inshore to an offshore model exploited by, as might be expected, vessels belonging to a PO, the KFO.[22]

It is one among many ironies that the offshore element of the north west crab stock was possibly the best documented shellfishery in Ireland. Its first assessment was to address the lack of information on what was regarded as a very valuable fishery: in the mid-1990s, 75% of all edible crab were landed to Ireland from the Co Donegal coast. Exploitation by the offshore fleet generated an exponential decline in landings per unit of fishing effort – per pot fished over a standard unit of time - in the period 1982-1997; by one third in the six years following 1990.

A steep decline of this order should have triggered some kind of precautionary reaction or at least an expression of concern. Instead, official reports alluded in reassuring terms to short periods of apparent stability in the record.

In 2003 the EU introduced a management regime for vessels above 15 m in length fishing crab in marine areas west of Scotland, and the following year specified the maximum number of kW-days which could be fished annually. At that date, the offshore fishery had lost almost half the value it had in 1990. However, there was still money to be made and, in November 2005, Ireland, represented by BIM, applied to the Scientific Technical and Economic Committee for Fisheries (STECF), a halfway house between the Commission and the scientific assessment body ICES, to place additional fishing power in the north west brown crab fishery. The exponentially declining data set depicting fishery performance was presented in support of the case, presumably prompted by the fantasy that a nation which documents the destruction of a resource cares too much to allow its ultimate disappearance. The STECF saw the data series for what it represented: "a decline in abundance". The STECF went on to recommend removing fishing effort rather than adding more. Needless to say, nothing was done.

Meanwhile, what had become of the traditional inshore Donegal fishing community dependent on edible crab? Its members expressed concern as early as 2001 when a large-scale tagging programme was undertaken to inquire whether "inshore" crab were affected by what happened offshore. Both components belong to the same stock. There was little point awaiting a seasonal inshore influx of crab when they were being targeted further out to sea by vessels working all the year round. If you could not beat them, there was no alternative to joining in (another rule of sea fisheries practice). And so the inshore fleet was converted to "fast-worker" vessels less than 15 m in length - so not restricted by the kW-days limit - which could fish alongside the super-crabber fleet. Incidentally, this also contravened the advice from the STECF to remove fishing capacity from all fleet sectors.

What should have been harvested in perpetuity as a summer-autumn inshore yield for the benefit of coastal communities was allowed – no, encouraged – to evolve into an unstable, costly, get-rich-quick, offshore fishery for female brown crab which were intercepted on migration and captured before they reached coastal waters to moult and mate, an essential phase in the reproductive cycle. On occasion, landings overloaded the market and collapsed prices. It was only a matter of time before the line of declining crab abundance cut across the line of rising costs as skippers invested in more gear and more costly vessels and travelled further to intercept the diminishing migrations, and the fishery became uneconomic.

Even in its depleted state in 2008, edible crab was our fifth most valuable marine species. In November 2010, it was officially acknowledged that the north west edible crab fishery had collapsed. Had the Framework, which shouldered the management of the north west fishery, made any contribution towards mitigating the disaster? The answer is no and the fact that it occurred posed serious questions about how competent the Framework was to fulfil its objective, stated in its explanatory leaflet to

...(provide) a more secure investment and planning environment for fishermen, thus ensuring a more viable industry.

The commentary which accompanied the disclosure that the crab fishery was no longer viable read

...Although the position of the fishermen now appears to be very difficult at least 80% of them invested in the fishery when it was clearly profitable. They had few other employment opportunities, there was a strong tradition of fishing in their families and their investment in the fishery was both logical and sound at the time.[23]

An obvious statement of what happened. Although it was in a position of influence and had the information and even prompting from scientists outside the country to do otherwise, BIM had behaved true to form; the agency allowed policy to be dictated by money; the most fortunate acquired wealth at the expense of traditional inshore fishermen and in the end, everybody who was involved, including those who did best, lost something; traditional inshore fishermen lost most. The political fisheries establishment, of which BIM was an intrinsic element, proved once more that it was as averse to formulating durable policy as it was to implementing regulations.

Chapter 16

A ROT ON THE COAST

One of the most extraordinary features of this story is not the amount of damage that was inflicted on Ireland's fisheries but how they remained as productive as they did for so long. In the autumn of 1958, a meeting of the Permanent Commission of the International Fisheries Convention took place in Dublin. This intergovernmental organisation had been founded in 1946 and later became the Northeast Atlantic Fisheries Commission. The consensus of the gathering was that neither the regulations sponsored by the convention nor the laws or bye-laws formulated by the Irish government for the conservation of fish stocks were being enforced. The trade press observed:

...To make laws regarding the size of net meshes or the sizes of fish which may not be landed, while at the same time allowing the machinery for enforcing such regulations to fall into disuse, is not only futile but dangerous. It is dangerous because it creates the impression that the authorities are not seriously concerned about the conservation of fish stocks. Such an attitude very rapidly leads to a rot on the coast. Fishermen cannot be expected to make themselves amenable to regulations unless those regulations are constantly and impartially enforced...it is simply that a law which is not enforced is no law.[1]

It is as true today as it was then.

It did not require an international organisation to tell the catching sector what was going on. In 1958 the industry was conscious that regulations were not being observed. The trade press reported at the height of the controversy on fisheries limits that "…some Irish fishermen are doing more harm inside the existing limits than any foreigners (outside)…"[2]

The disadvantages of expanding fishing effort became obvious as early as 1959:

…In relation to our coastal fleet, there is…cause for concern. The significance of the fact that the increase in production of whitefish is slight when related to the increase in catching power and fishing effort should be fully assessed…Compared with other European fish industries we are fortunate in that we can introduce controls as the industry develops. Unfortunately there seems little awareness on the part of the authorities of the necessity for such controls…To stimulate production haphazardly in the belief that everything will work out for the best is to invite trouble…[3]

Later that year there were complaints about landings of undersize plaice. In 1961, the trade press referred to a "hidden crisis in the whitefish section of the industry…" and inquired "are we suffering the effects of overfishing?" The list of affected species was lengthening and problems were already known to exist for plaice, cod, haddock and hake "…the list we know about…"[4] Another report in 1961 referred to a lack of fish "…due to overfishing of the whitefish stocks in inshore waters…"[5] while, later that year,

…(it is) hard to escape the conclusion that too much capital and too much effort (has been) put into catching too little fish…[6]

In 1961 the GT of the national fishing fleet was approximately 15% of the size to which it would grow in 2006. A lot more damage lay ahead.

In mitigation, it might be argued that the expansion of Ireland's fisheries during the latter half of the last century was not rooted in a maritime tradition, because sea fisheries developed later here than in other nations. But Ireland's population was largely agriculture-based and that might have been expected to imbue some cautionary sense of the relationship between the size of a parent stock and its progeny. Some individuals within the industry and scientists associated with it were worried about the consequences of the removal of large quantities of adult fish biomass combined with landing undersized individuals of various species but the prevailing ethos smothered any concerns.

John Molloy was involved with the management of herring throughout his career in the Department's inspectorate and, latterly, the Marine Institute. The Celtic Sea herring stock, the second largest in the country, presented a variety of administrative, international, political and biological challenges, associated with an increase in landings after the 1960s, their decline in the following decade and recovery to a lower level of yield until the mid-1990s when they fell further than before. With regard to a welter of fishery management committee arrangements in the 1980s, supposedly informed by scientific advice based on the state of the stock, Molloy observed in his memoir:

...It soon became clear, however, that the biology and the state of the stock were not high priorities for many of the (fishermen) and that the fishery was simply a means of creating as much wealth as possible for different sectors in the shortest possible time...[7]

Intensive fishing enhances competition among fishermen who must generate sufficient profit to pay their outgoings. Problems of this kind are not confined to Ireland but there may be some exacerbating factors in our case. There is, for example, the atavistic explanation: that conservation was, if not irrelevant, then beneath the dignity of a people who associated it with the oppressors of their earlier history from whom they had gained their independence.

The mindset was well illustrated by an article written for the *Irish Times* by John Horgan in August 1992. Horgan had a most distinguished career in journalism and Labour Party politics and became Ireland's first press ombudsman in 2007. His emotional broadside at attempts to regulate the salmon drift net fishery offered no solution to the problem; rather he seemed to suggest that, if extinction of Atlantic salmon were to be avoided at the cost of penalising fishermen, then perhaps extinction was preferable.

...the core of the current argument over salmon (is that) the drift net fishermen have been singled out by powerful forces in our society as the scapegoats for an economic and environmental problem which has been caused, not just by over fishing but by gross mismanagement of a precious natural resource...

Horgan conceded that not all fishermen were angels and then dismissed misbehaviour as a contributor to the crisis. He also wrote as though drift netting were an ancient tradition although the practice had recently greatly expanded and new materials had made it far more effective and deadly than in earlier decades; the dilemma of balancing traditional harvesting rights against a more effective killing method was not confined to the fishery in question and always posed difficult choices wherever it manifested itself. Reading through his article again, and appreciating that such views were not confined to Horgan, one realises why Ireland was the last European country to ban drift netting and why that ban was forced upon us by the EU. It is instructive to revisit some of his outburst:

...seeing the LE Deirdre *(naval vessel) steam off in hot pursuit of a couple of half-deckers crewed by fathers and their sons is redolent of nothing more than an imperial tradition against which we once understandably rebelled: the thesis that a whiff of grapeshot will keep the natives in their place...Did the youth of our cities join the naval service in order to wage war on the last, and increasingly desperate, survivors of a difficult and dangerous way of life – and their own countrymen to*

boot?...Meanwhile, the zealous servants of the boards of conservators roam the rivers and estuaries, armed with the technical and legal firepower of a Wyatt Earp, but without a vestige of his moral authority...Nor is it strange that some of their colleagues are more hated than Cromwell...A law which has such inbuilt reservoirs of arbitrariness is bad law and it has...turned almost every fisherman into a poacher...If this is a triumph of the legislative process and of democracy, despotism might be preferable...[8]

And there is more along these lines to this rant.

Horgan was obviously and sincerely sympathetic to the plight of inshore fishermen but every fishery has to be controlled if it is to remain viable; killing off commercial species is even more impoverishing for those who would make a living from catching them and there are numerous examples of that having happened. Another aspect not given appropriate weight was the fact that Atlantic salmon was, at the time, the foundation for several kinds of commercial endeavour which generated value other than by its value as dead flesh, some arguably more beneficial to the nation.

Horgan's sentiments had wide currency and some who professed them may have had more self-serving reasons for doing so than his. On one detail we may be assured: no politician wanted to be identified with any of Horgan's pantheon of roguish stereotypes. Becoming involved with the industry would have meant plunging into a maelstrom of anguish, very possibly being scarred with a nasty label and achieving relatively little for one's pains. This was, after all, an industry which contributed relatively little value to national wealth. BIM's published estimated investment in the fleet, the combination of shore-based processing plus the values of landings and exports, suggested their contribution to gross national product (GNP) had risen from 0.28% to 0.58% (slightly more than half of one percent) between 1967 and 1974. And yet, however miniscule its share of the economy, the industry was disproportionately vociferous and troublesome to deal with.

For all that, fisheries problems could not be ignored. There was still a need to do something; individuals within the industry and others among the general public demanded it. In due course the European Community would impose some discipline but nationally, particularly in near-shore waters, controls were lamentable. One solution for a cute politician and the Department was to devise a regulation and then not enforce it. The official response would then run along the lines: legislation has been enacted to rectify the problem, neutralising concerns expressed by conservationists, while the minister in charge would not be harassed with petitions from individuals, representative organisations and the press to show leniency, thus satisfying the industry. This simple device made life within the ministry uncomplicated. Another, infinitely more subtle ploy, was to devise a defective regulation which could never be enforced.

An illusion of protection

An excellent example of the latter was the legislative provision to protect the purple sea urchin, a small inshore echinoderm species occurring along western shorelines of the eastern Atlantic and in the Mediterranean. Its flattened spherical shell, clothed in long spines, attains a diameter of 6 - 7 cm. One location in which it is clearly visible from the shore is where the karst limestone of the Burren in Co Clare enters the sea. There, countless generations of these tiny creatures have eroded cup-like hollows in the rock in which they sit, secured against the surge of waves. The species is easily accessible here and down to a depth of 30 m offshore by divers; in fact the species can only be harvested by manual collection. Slow moving and feeding on algae, it is not vulnerable to trapping nor does its habit of wedging in rock interstices expose it to capture by net.

Purple sea urchin is a great delicacy sought by continental holiday makers who ravaged its populations in the 1960s. Some action needed to be taken. The solution was a bye-law which forbade the gathering of shell fish by scuba divers. [9] While the law afforded protection to other species of shellfish, notably crawfish and lobsters, the purpose of its enactment was to conserve sea urchins.[10]

Bye-laws are minor regulations, brought into force by ministers and not subject to debate or vote in parliament. They are devised within the legal constraints of parliamentary acts. The regulation in question was enacted within the terms of reference of the 1959 Fisheries Consolidation Act and that defined shellfish as crustaceans and molluscs; echinoderms were not included in the definition. Hence, the supposed protection afforded to the purple sea urchin was a sham.[11]

Eight years after its enactment, some doubts were expressed in the Dail about the efficacy of the bye-law. The minister, Michael Pat Murphy (Labour) explained:

The taking of sea urchins by skin divers is prohibited by bye-law. I do not consider any other conservation measures are necessary at present...

His answer to a supplementary question revealed that protection might be less than perfect:

...While scientifically there seems to be a doubt as to whether sea urchins are shellfish. The official attitude is that sea urchins are covered by the bye-law...

So there! Inquire a little further: was there ever a successful prosecution for the capture of sea urchins by divers?

The parliamentary question to the minister came from Flor Crowley (Fianna Fail, Cork south west) who was prompted to ask it because foreign tourists were depleting populations of the animal in the vicinity of Schull (Co Cork) thus depriving his constituents of their livelihood, which suggests that those same constituents were up to similar tricks themselves. Further inquiries to the minister drew a dismissal of the topic:

...Landings of sea urchins are deemed to be of little significance and are not recorded separately in the fisheries statistics...

But of course, how could they be if the only method of catching them were illegal?

The minister was incorrect. In 1972 only 41 varieties of marine landings were officially recorded, nine of them shellfish, and sea urchin was among those. In 1972, 150 t were brought ashore; landings peaked in 1975 at 353 t and they declined steadily each year after that. In 1981 a mere 100 t were registered and five years later half that amount. In 1991 there was a resurgence and 105 t were landed. But six years after that the total was six tonnes and since 2003 the species has not been listed. If purple sea urchin had not been cleared out completely it was severely depleted.

The hypocrisy of a sham regulation masquerading as protection was exposed once more in April 2008 when celebrity chef Gordon Ramsay went scuba diving in Co Clare and removed what he described as a "hapless bunch of sea urchins" which he cooked and distributed *gratis* on the street in Galway city. Placed on the spot, the explanation of the responsible department (Environment in this case) was the customary hand wringing. It was regrettable that Ramsay had not consulted the department before his dive. He would have been discouraged. Ramsay was not prosecuted and one can only wonder how he would have been advised had he asked.

Purple sea urchin is not among those species listed as protected in the fifth schedule of the Wildlife Act, 1976 so it is available for exploitation. The only defence offered to it, contained in the 1966 bye-law, was the false security of an empty gesture. As I write this, it is 46 years since purple sea urchin was supposedly granted the protection of the law. It is hardly surprising that so many sites which once had thriving populations of the species have been stripped of it.

If national conservation principles are not enforced by the authorities, the responsibility for doing so devolves onto individual fishermen. But conservation is rarely free of financial implications. Avoiding a species or not fishing in a particular area or returning undersized shell fish to the sea all impose costs on an operator: additional labour, revenue foregone. And the fisherman will see little point in courting disadvantage unless his fellows are required to do so too.

In 1997 two scientists from Baylor University in Texas elicited responses to a number of value statements and questions on the industry from fishing communities in five fishing ports: Helvic Head, Dunmore East, Youghal, Clogherhead and Burtonport.[12] The ports were classified as "stressed" where much of the fishing fleet had disappeared in the recent past and "viable" where vessel numbers were stable. Additional queries regarding whether respondents needed additional work to supplement their fishing activities and whether their incomes were increasing or declining reinforced the categorisation of the locations.

The experiment consisted of gathering responses to 85 statements and questions such as: what are the characteristics of a "good" – meaning moral or ethical – fisherman? The responses from viable and stressed communities differed significantly. Almost half of the respondents in a viable fishery regarded a "good" fisherman as one who had a strong work ethic (went to work on time and was dependable) and one quarter characterised "good" as inculcating respect for one another among crew members. In stressed communities "good" had difference connotations; 40% interpreted it to mean having a large boat and one tenth believed "good" fishermen were "greedy".

Members of viable fishing communities regarded overfishing and disobeying conservation regulations as wrong while members of stressed communities did not refer to them at all in their responses.

The organisers of the survey found a realistic appreciation of what was going on in the industry at the time. The widespread perception in 1997 was of decline, for which two thirds blamed the indigenous industry and one third other European Community states. Over-capitalisation was the culprit and trawling was identified as a cause of much damage. When it came to proposing solutions, the two community types chose different mechanisms. Operators in viable communities opted for wider mesh sizes and reducing fishing effort but those from stressed communities favoured widening fisheries limits, enlarging fishing territory.

A strong impression emerging from the exercise was that stressed fishing communities become more focused on survival and, in dire circumstances, even greed can become a virtue. Viable communities hold conservation regulations in higher esteem but, as the shoals are thinned out, such ideals are less tenaciously cherished. Eventually they are discarded altogether. It is easier to ignore regulations, provided you can get away with it and get on with grabbing whatever you can harvest regardless, blaming foreign fleets for stock depletion. According to the findings, stressed communities tended to be less dependent on one another within their immediate circle; instead they formed links outside it and proceeded along more political lines in the way they did business.

Bratton and Hinz's study was persuasive. It modelled in a few pages what the history of industrialisation demonstrated to us. Solutions to the gathering crisis were not proposed. Perhaps none existed.

Instead life went on, increasingly a matter of survival rather than prosperity for a large proportion of those who made a living from the sea. The phenomenon of increasing fishing power exacerbating the decline of fish stocks was adversely affecting more of the fishing community. There was little point in complaining, you made whatever money you could, however you were able. These are the circumstances in which a black economy thrives.

A case of alleged overfishing came to light early in 1997. The EU received complaints that Ireland was landing more than twice her mackerel quota. The complaints, which originated in the Netherlands and Norway, accused Ireland of landing 200,000 t of the species in 1995 and mis-identifying it in the records as scad, also known as "horse mackerel". Mackerel straddles the divide between EU member states and Norway which together exploited it. At the time, the Norwegians were proposing their quota should increase by 30%. The KFO, the PO most associated with pelagic species, denied the claims and the Department stated it was not carrying out any investigations into them. Norway referred to the Eurostat statistical compendium prepared by the Commission which reported that Ireland had, in 1995, exported 199,544 t of mackerel whereas its quota had been 79,030 t. [13]

Mackerel became a contentious issue again in the early years of the next century, and the episode will be alluded to again. Police raids in Scottish processing factories in late 2005 revealed illegal landings of more than 40,000 t of mackerel, worth at least €48 million. The Scottish authorities also reported to the Department of the Marine that they believed Irish boats had landed up to 35,000 t in 2003 and 2004. [14] According to the Department, some 75% were attributed to Irish RSW boats. The KFO made much of the fact that nobody had been prosecuted. The issue was a matter of intense debate between Ireland and the Commission in search of a way to proceed and, while that debate was in progress, licences to fish in 2006 were witheld. In January of the following year the KFO offered to take a cut of 30% in their quota in order to get some fishing in before the end of the season. [15]

In 2002 the abject status of fish stocks was openly debated. In the run-up to the December council meeting of ministers in Brussels, the prognosis was dire. Cuts of 80% in the landings of cod were proposed along with similarly drastic reductions for other whitefish species. In order to allow the stocks to survive and recover, the fleet would have to be slashed by more than 30%. In advance of the meeting, the usual complaints were rehearsed: Ireland's large proportion of Community waters contrasted with her small quota allocations. Once again, the solution identified by the industry was an increase in Ireland's share.[16]

A pervasive inkling as to what was going on persuaded the authorities to take a closer look. In 2003 Revenue launched a major fact finding enquiry into the fishing industry in south east Ireland which "yielded millions in unpaid tax."[17] A deepening interest was developing in the way the industry operated and the industry did not like it. Early in 2003 the navy was accused of "over-policing" the Irish fleet. Industry spokespersons observed there appeared to be a bias against Irish vessels which were being disproportionately targeted and arrested.[18]

In October 2004, the dam of pent-up frustration burst. In July 2004, a Donegal fisherman had approached the minister and Department with information concerning irregularities in the way fisheries were being conducted in the north west. Several meetings and exchanges of correspondence followed but the complaints were not satisfactorily dealt with. In October a letter issued from Pat Cannon, a disaffected founder member of the KFO, to the fisheries authorities in twelve EU countries which began as follows:

Dear Minister,

As a concerned fisherman with forty-four years experience in the fishing industry I am contacting you to call for a full enquiry by the EU Fishery Commission into blatant irregularities regarding catches and quotas and the falsification of documents regarding catches which, if allowed to go unchecked or penalised, will cast a huge cloud on the future of the industry in this country.

Cannon went on to write of disregard for quota regulations; he claimed to have information concerning daily landings exceeding permitted amounts, falsified receipts and evidence of misreported fishing locations. The chief executive of the KFO, to whom the complaints were presented, advised Cannon to take them to the Department. Cannon concluded his letter by stating that he had been received with "respect and sincerity" when he brought his claims there. And that might have been the end of it.

In the course of his representations Cannon approached two ministers with his story. Pat the Cope Gallagher (Fianna Fail), champion of the Irish industry, a west Donegal deputy at the geographical heart of the problem had, in the same month as Cannon opened the can of worms, been appointed junior minister responsible for fisheries at the Department of Communications, Marine and Natural Resources. On being told of the disclosures, Gallagher was quoted as asking "what do you want me to do about it?"[19] His senior ministerial colleague, the Meath west deputy, Noel Dempsey, was more fortunately placed: nowhere near Killybegs or Co Donegal, the eastern boundary of his Co Meath constituency was some 50 km inland from the Irish Sea. Dempsey was an effective minister because he was prepared to take decisions on fisheries matters in the interest of the country and independently of the PO-driven establishment. He referred the matter to the Commissioner of the Garda and a fraud enquiry got underway.

The chief executive of the KFO responded that his organisation had no knowledge of the supposed illegal activities in Killybegs and he found them incredible because of the level of controls existing in the port. He acknowledged "...an allegation that Departmental officials colluded in such activities"[20] which was repeated in several contemporary newspaper accounts.

Despite protestations of ignorance there were questions to be answered. Much of the evidence proffered by Cannon could hardly be described as inaccessible. He referred to the activities of fishing vessels recorded by the "eye in the sky", satellite system. Satellite surveillance and tracking systems had been developed in the mid 1990s, the "starfish" model being one of the prototypes. Using an array of satellites and a computer system known as Vega, the position of any trawler carrying a tracking device was accessible at any time.[21] The version adopted in Ireland was INMARSAT which became compulsory for vessels greater than 24 m from 1998[22] and was later extended to vessels of 15 m and over. Where they had been fishing could be readily compared with the information reported on their log sheets when any discrepancies would have been obvious.

Cannon told the press that what he had done was for the protection of "ordinary" – meaning small-scale - fishermen; he claimed to have had support throughout the country from people who had similar tales to tell. He also said he had received "a lot of abuse" when he socialised in Killybegs, and it would have been strange if he had not.[23] Whatever his motives for opening this Pandora's box – several were suggested – Cannon was a courageous man.

Cannon's claims caused consternation at government level because earlier the same year the Advocate General of the European Court of Justice recommended that France be fined €115.5 million for failing to police its fisheries. Departmental sources feared that the regulation of the Netherlands and Ireland's industries were next in line for scrutiny.[24] There was certainly a precedent for looking into the matter. In 2004, only 26 successful prosecutions of Irish boats had taken place, against 3,153 in Spain.[25]

In November 2004, approximately one hundred gardai from the National Bureau of Criminal Investigation and the Fraud squad descended on co-operative and PO premises, fishing vessels, fishing agents, fish processors and the private homes of boat owners in Killybegs and Greencastle (Co Donegal), Rossaveal (Co Galway), Castletownbere (Co Cork) and Dingle (Co Kerry). Computers and documents were seized and an incident room was set up in Donegal town.[26]

Half of the investigating personnel concentrated on Killybegs. One of the trade papers based there, the *Marine Times,* reported the port "in turmoil". Some factories closed; the town was reduced to "ghost" status, a possible indication of the extent to which it was dependent on illegal landings. KFO senior personnel described Killybegs as the most controlled port in

Europe but assurances were given that the PO would fully co-operate with any inquiry.[27]

The garda investigation uncovered overfishing in every big fishing harbour along the west coast:

We can't bring charges against every captain and fisherman but in a lot of cases there'd be people who would own both the boats and the factories and they'll certainly be looked at...[28]

The extent of the fraud was massive; up to three times the quota was being landed but not reported.[29] A second source corroborating the amount of the deception was the files of the Marine Institute which, as part of its monitoring duties, placed observers on fishing vessels. Their routine involved checking how the landings captured in the course of a trip were subsequently recorded. Mandatory log sheets showed that quota species were being under-reported four to six fold while non-quota species were being over-recorded by two to four times.[30] The reason for over-recording non-quota species might have had two applications: one to explain the number of boxes of fish landed on the quayside to officials checking them, the second to create "track record" of "entitlement" for species which had not yet, but would in time as their populations were depleted, become subject to regulation by TAC and quota.

The Marine Institute was extremely unhappy at being involved in the investigations. Its work of assessing fishing opportunities could only be undertaken with the co-operation of the industry and the institute was apprehensive that, should this be withheld, it would not be able to function.[31]There was no confidence that either politicians or the Department would go so far as to instruct the industry to facilitate scientific enquiry. In fact, to the best of my knowledge, the institute never revealed details of individual vessels which might have assisted prosecution of their owners or crew; rather, it provided bulked data supporting the existence of massive deception within the industry without implicating any individual.

In July 2005, the garda announced that files from Killybegs, Dingle, Rossaveal and Castletownbere were being prepared for the director of public prosecutions.[32] In October, files on alleged fraudulent landings and alleged "collusion" with Departmental officials were actually sent.[33] For the moment, the game was up and the realisation that the industry drew significant quantities of revenue from illegal activity – metaphorically, the industry was trawling over barren bedrock – prompted emergency measures. A decommissioning scheme worth €45 million was launched, to take boats out of service. However, the full extent of illegal activity had not yet been uncovered.

In November, Stephen Collins publicised an elaborate scam, described earlier, involving a number wealthy trawler owners. Huge catches of mackerel were being pumped into underground storage silos at a fish processing plant in Peterhead, Scotland. This time the Criminal Assets Bureau joined the National Bureau of Criminal Investigation. Almost 40,000 t of mackerel had been secretly put ashore over a four year period and the activity had continued after the fraud investigations commenced the year before.[34] The mackerel scam had netted an estimated €40 million for a handful of fishermen.

The Sea Fisheries and Maritime Jurisdiction Act, 2006

Two supreme court rulings, in 2003 and 2005, had "significantly undermined" existing fisheries protection laws and hampered the abilities of the state to impose fines for illegal fishing.[35] That was a stimulus for new legislation which would carry heavy penalties, appropriate because much fisheries law-breaking netted huge rewards. Punishment had to be substantial if it were to be effective and the level of penalties had to be sufficiently high to deter, and that meant that the offences had to be "criminal" rather than merely "administrative" because administrative offences carry lower penalties in the Irish judicial system. Conviction for a fisheries offence in future would be a serious stigma.

By October 2005, the damage inflicted on fish stocks by the home fleet was no longer a secret. The imminence of retribution from Brussels increased with every day that passed. Early in the month new legislation was drafted and published. The night before it was due to be debated, the bill was withdrawn. Stephen Collins described what happened:

...the government abandoned its plans to protect the national interest and instead capitulated to a small vested interest group of wealthy trawler owners, backed by a handful of Fianna Fail TDs...(it) is a salutary tale about the way the clientelism in Irish politics works against the common good...[36]

The Sea Fisheries and Maritime Jurisdiction Bill, 2005 was long overdue; it set out a tough regime of fines and seizures for misconduct. However, the only deputies to welcome it belonged to the Green Party. That party's Eamon Ryan described the Bill as hugely important

...it represents an attempt to deal with one of the biggest environmental scandals facing us – the illegal destruction of our fish stocks...

The government party responsible for enacting the legislation was openly hostile and Pat the Cope Gallagher, the minister expected to shepherd it into law, was identified as the deputy who withdrew it from consideration. The secretary of the Department wrote a letter to each member of the joint oireachtas (parliamentary) committee on marine and natural resources warning them of the financial penalties which were likely to be imposed if the legislation did not go through.

Stephen Collins concluded his article with an apt commentary on the state of the industry:

Incredibly we are now in the position that the government has decided to expose the Irish taxpayer to massive fines so as to appease a small number of wealthy trawler owners who are not prepared to obey the law. That a few (deputies) who represent

fishing constituencies are able to dictate government policy reflects all that is worst about Irish politics...It is even more astonishing that the main opposition parties appear to be colluding with the government against the interests of the taxpayer, not to mention the environmental destruction involved...The Greens are the only party to emerge with honour from this sorry saga.

Dempsey, however, stuck tenaciously to his responsibilities and continued to crowbar the new legislation into existence. Fishermen organised a substantial campaign to oppose it. In February 2006 debate on the bill was postponed for the third time. Dempsey insisted that a small number of fishermen were involved in "systematic criminal activity".[37] Irish members of the European Parliament of all parties united to oppose the bill; fishermen organised demonstrations against it. A stream of delegations from a variety of fisheries-associated organisations negotiated to have details of the legislation trimmed in an effort make it more palatable. But the strain was too much for junior minister, Pat the Cope Gallagher, who moved to another department rather than ferry his own party's bill into law.

In mid-2005, while the protracted squabble on the legislation was in full swing, another scam came to light. Even while the industry was protesting its innocence and minister Dempsey was trying to get his bill onto the statute books, two vessels were discovered unloading mackerel onto a fleet of fifteen lorries which, employing a succession of drivers so as not to leave a coherent trail of evidence, brought the fish to a processor's premises. The VMS blue box tracking devices had been tampered with so their true position was not being accurately reported by satellite.[38]

Early in February 2006, several hundred fishing vessels planned to sail into four ports, Dublin, Cork, Galway and Waterford, on the morning tide in protest against the impending legislation. The organisers were careful to describe their action as "not a blockade". The main opposition party's John Perry (Fine Gael, Sligo north Leitrim) expressed his support for the protest in opposition to the "harsh and destructive" legislation. Eventually there were more than 200 amendments to the Bill, 100 of them from the government side of the house.

In the same month, the Sea Fisheries and Maritime Jurisdiction Bill failed on its third attempt to get a reading in the Dail, opposed by Fianna Fail combined with the opposition parties. Only Eamon Ryan (Green Party) and one Progressive Democrats deputy backed the minster. Stephen Collins, perceptively, identified competing elements in the argument; those opposed to the legislation portrayed themselves as small-scale, hard-working fishermen wrestling a bureaucratic state apparatus. In fact, the struggle was between a small group of very rich trawler owners and the authorities, in place of which he might have substituted "the minister". He observed that the conventional image of the Irish fleet composed of little trawlers battling with big Spanish boats was misleading. But there is no doubt it was cultivated in collusion with the political-fisheries establishment which characterised the model of Ireland's fleet as "inshore".

The latest technological devices, expensive legal advice and raw political power have been deployed successfully by the fishing industry to defeat the efforts of the state to enforce the law. [39]

The argument continued when the Federation of Irish Fishermen appealed over Dempsey's head to Bertie Ahern, taoiseach. A letter from the POs stated they were

...appalled that some influential elements here in Ireland seek to represent our industry in the most negative possible light, without placing the problems we face in proper context... [40]

What could that context have been? That a too powerful industry simply did not have enough fish to support its infrastructure?

The chief executives of three POs under the FIF umbrella combined to write an article in the *Irish Times* in an attempt to scupper the new legislation. Admitting that some fishermen had benefited from substantial ill-gotten gains, the article argued that "the vast majority had not made €500,000 profit in the past ten years combined". Specific cases of wrong-doing could not be extrapolated to the entire fishing community. Relatively minor offences could not be seen as criminal in intent. Ireland must not be regarded as different from the EU and responsibility for the "present mess" must be shared with the Commission in Brussels. Confiscation of catch and gear was not "an appropriate automatic consequence of conviction in a Circuit court". A plea of injured innocence concluded: "why is this vital debate being shrouded in obfuscation?". [41] As though the circumstances were not crystal clear.

Dempsey's campaign to pass the bill involved compromise. Gradually, the maximum value of a fine was halved and the threat of confiscation of gear and vessel removed.[42] On the evening of 22 February 2006, the Sea Fisheries and Maritime Jurisdiction Bill passed its final stages in the Dail. Last demands that taoiseach Bertie Ahern introduce fixed penalties were adroitly side-stepped. A final swipe by the leader of another PO questioned minister Dempsey's fitness for office.[43]

The Bill became an Act in April 2006 (number 8 of that year's enactments). The fisheries protection staff of the Department was transferred to a new agency, the Sea Fisheries Protection Authority (SFPA), seen as a move to "de-politicise" its functions, a development welcomed by the industry.[44] The authority, overseen by three independent commissioners, would be answerable to a joint Oireachtas committee. A mechanism for dialogue between agency and industry, through a fourteen person statutory consultative committee, was put in place; in 2007 five of them came from POs.

The SFPA was launched in 2006 and began operating in the following new year. Its annual report for 2007, its first year of operation, gave a detailed account of the number of inspections of fish landings by its staff, and of naval inspections at sea, by registered nationality of vessel. Inspections of landings were undertaken for 7% of Irish boats, 32% French and 41% Spanish vessels. The majority of naval inspections at sea were directed at

Irish boats (53%), Spanish next (21%) and British after that (15%).[45] Irish-registered boats were scrutinised more than others but then, they were more numerous in the seas around Ireland.

The SFPA presented a conciliatory exterior. Its customer charter addressed everyone as just that – individuals engaged in fishing were "external customers" while staff were "internal customers" – in much the same way as everyone in Russia was addressed as comrade after the revolution.

This conciliatory approach did not endear it to the industry. In 2008 the SFPA embarked on a five-week charm offensive, explaining its role to the public over the radio. The agency stated that the advertisements conveyed "a simple message that a compliant fishing sector is in the interests of all stakeholders". The FIF was infuriated by the "...unprecedented public spin-doctoring against them in this manner" and the Federation reminded the taoiseach that he had given a public commitment to examine the feasibility of amending the legislation.[46]

The UK administration adopted a more business-like attitude to law breaking. Early in 2007, the skippers of twelve vessels fishing out of Kilkeel misrepresented their landings of quota as non-quota species. By the standards of some of the cases publicised in the Republic their misdemeanours were puny. Liverpool Crown Court heard pleas that the fishermen had been motivated not by greed but commercial necessity because of lower fish quotas.[47] They were fined a total of €391,000, more than all penalties (fines and forfeiture of gear combined) imposed for illegal fishing in Irish waters in 2007. In the case of the Republic, in that year, only 13% of the €388,000 levied on illegal fishing was paid on behalf of boats registered in Ireland.[48]

The SFPA employed the same people as the Department but the industry was highly suspicious of their new brief, and the agency was consistently characterised as malevolent and ill-disposed towards Irish fishermen. Although statistics gathered by the naval service indicated Irish boats complied with regulations, the

SFPA was accused of believing the worst of its "external customers".[49]

Revelations arising from the garda enquiry led to a revision of some practices. There was a suspicion within the Commission and the Department that the weigh-in of landings on the quay at Killybegs might be less than precise. The industry agreed to the appointment of temporary staff to check the accuracy of measurements. When the officers arrived for work, they were forbidden entry to twelve factories and, following further negotiations, a high court injunction was sought to ensure they were able to perform their duties.[50] That was in March 2006. When, a year later, another Bill, this time dealing with criminal justice issues, was introduced in April 2007, it contained powers enabling fisheries protection officers to search premises. The Criminal Justice Bill was introduced to the Dail by the Minister for Justice, Michael McDowell (Progressive Democrats, Dublin south east), whose party was in coalition with Fianna Fail; the principal target of the bill was gangland crime. The outraged tones of contributors to the ensuing debate deplored insults to the sensibilities of those in the fishing industry being confused with common criminals. Its sponsor pointed out that the relevant provision was merely a mechanism to secure a particular end and, presumably, avoid the stand-off that occurred the year before in Killybegs. Dinny McGinley (Fine Gael, Donegal south west) spoke of the loss of boats from Bunbeg in the course of his lifetime; the local fleet had declined from 30 to 40 half deckers down to two.[51] It had not occurred to him to delve beneath the figures and inquire why.

Noel Dempsey moved to another ministry, Transport, in 2007. He had displayed greater tenacity than any other minister for fisheries in the history of the state and left a tougher, if largely watered-down, administrative act after him.

Elections to the 30th Dail took place on 24 May 2007. Fianna Fail leader Bertie Ahern was again returned as taoiseach in a coalition, this time with the Green Party, a development which did not find favour with the KFO. Its chairman, had considerable confidence in the KFO's powers of persuasion with the dominant government party. He told the local press:

We spoke to Fianna Fail this week-end and told them in the strongest possible terms that we do not want them to be involved in a Green Party coalition. We actually got assurances that is would not happen...Explaining his fears, Mr Howley said Green Party policy documents virtually suggest getting rid of the modern fleet and going back to environmentally friendly inshore type fishing...They seem to favour small vessels and stricter quotas for fish ...[52]

In the event, the Greens and Fianna Fail went into coalition together but the plight of small inshore operators did not benefit from the arrangement.

In the course of investigations into his finances by the Mahon Tribunal, Ahern's position as taoiseach became untenable and he was replaced by Brian Cowen almost a year later. Cowen appointed a deputy from Donegal south west, Mary Coughlan, as his second in command (tanaiste). Coughlan had been minister for Agriculture and Food in the previous administration and in 2007 she assumed responsibilities for Fisheries. The industry in Killybegs must have welcomed their new minister.

Coughlan, in common with all politicians, looked forward to as uncontroversial a tenure as might be arranged. Before the most recent election her new department had dealt with agriculture and food, but not fisheries and she was uneasy about some of the issues she would have to confront. Before leaving office Dempsey had approved a more formal role for the Garda in fisheries control. The ban on drift netting for salmon was also likely to place its ministerial enforcer in an unpopular light and this she wanted transferred to another department. Coughlan was not prepared to talk to the *Irish Times* but she had earlier given an interview to a trade paper indicating the benefits of linking marine with her department.

Referring to criminal investigations in several west coast ports over the last three years, the minister indicated she intended to take a different approach, based on "working with the sector".

People want to see an end to these investigations, Ms Coughlan said. Regardless of the outcome, closure of this matter will help lift the air of doom and gloom and let those in the industry get on with their jobs.[53]

It was correct to say that the disclosure of wrongdoing had had a massive adverse effect on Killybegs which was reported, in its aftermath, to be

like a ghost town, as landings have dried up since the regulation of the port was tightened up...[54]

The latest in a long series of reports on the sea fishing industry, known as the Cawley report after its author,[55] published in December 2006, recommended that the regulation of the industry, in keeping with the views of Coughlan, should involve

The adoption of a more responsive, customer-based approach (which) will pay dividends in terms of building bridges with the industry in due course...

The new beginning failed to explain why the supposedly more rigid approach before the discovery of so many offences had failed to spot the wholesale lawbreaking. The passage from Cawley was cited in the trade press, which reported on the contrary,

...ever increasing levels of ultra rigid enforcement of regulations and a general feeling of being picked on...Fishermen...are claiming that there may be a hidden agenda behind the increased levels of monitoring and inspection in recent months...[56]

A cavalcade of prosecutions arising from investigations in 2003 and 2004 had begun their tedious caravan through the courts.

Four trawler owners fishing out of Killybegs had 66 charges levelled against them for under-reporting landings and making false logbook entries. All demanded the right to have their cases heard in Irish.[57] The industry was already complaining, through the trade and national press, that more effort was being invested in pursuing fishermen than drug importers.[58]

Incidents that occurred in 2004 and afterwards demonstrated the lawlessness and brazen disregard for regulation of an industry that was accustomed to getting its own way and was out of control. Much of what happened could not have taken place had there not been substantial sympathy for it within the political and fisheries establishment. Perhaps there existed, somewhere in the system, a residual tinge of guilt for the reckless manner in which the industry had matured as an insatiable, ravenous monster. The chief executive of the new protection authority declared as much:

(He) is aware that many were encouraged by the State to take financial risks in recent years, only to find that new restrictions on particular species made their vessels less than viable...[59]

That was probably as close as anyone would ever get to an official admission of culpability.

As the December council of ministers' meeting approached in 2006, business appeared to be back to normal. The *Irish Times*, canvassing opinion in the ports, quoted the KFO chairman:

...we've seen continuous reductions in quotas every year more or less...(despite BIM's fleet renewal scheme) new vessels are still finding it difficult to survive...[60]

The possible role of illegal landings in reducing quota size did not rate an allusion: as we will see, the industry was anxious to distance itself from the scandal.

The straitened circumstances of a more closely scrutinised fleet caused unbearable strains. It was not so much a problem of forcing a quart into a pint pot as trying to squeeze the same pot to yield a quart. The fish were not there and the fleet was literally starving. If ever circumstances agitated for light touch regulation this was the occasion. The SFPA found itself in an impossible situation, facing a blizzard of complaints. It is easy to criticise any enforcement agency for lack of courtesy in stressful circumstances when tempers run high, such as during inspections of vessels, gear, documentation or landings, and complaints of the kind were hurled at the SFPA. The trade press recommended that everyone involved, "must operate in concert and in a mutually respectful manner... "[61]

One of the trade papers, *The Irish Skipper*, gave prominence to complaints throughout 2008. However, in a grovelling U turn, the paper carried a front page retraction in January 2009, acknowledging that articles in recent issues had been:

*...entirely untrue and misleading...*The Irish Skipper *wishes to make a full retraction of these articles...Allegations that sea fishery officers abused their powers and abused members of the public are acknowledged...to have been totally without foundation...*

As late as 2008, politicians were promising to review the controversial legislation and introduce lower fines.[62] The SFPA was still treading on egg shells. Instead of coldly going about its business where a prosecution for breaking regulations occurred, the agency expressed its "disappointment" that a small number of Irish fishermen were depressing market prices by making "illegal" landings.[63]

In April 2009 an employee opinion survey on conditions within the SFPA was leaked to the *Irish Times*. Questionnaires had been circulated to staff, and the return of completed ones was regarded as exceptionally high and representative. The survey, published a month later, revealed low morale among "internal customers". "Rumour and poor communications structures, perceived lack of transparency have a very negative impact".[64]

The industry, meanwhile, had added to its shopping list of concessions, which included higher quotas, demands for decriminalisation of offences and lower fines. Political parties in opposition were always ready to oblige, a typical example being the move by Jim O'Keeffe (Fine Gael, Cork south west) who suggested introducing fixed penalties for infringements with a maximum fine of €1,000. He prefaced his comments with the remark that neither he, nor his party, had any wish to devise a "pirates' charter" but he obviously had no idea of how profitable illegal fishing can be.[65]

Chapter 17

THE TREACHEROUS REASSURANCE OF FISHERIES SCIENCE

In terms of current values, grant assistance to developing the sea fishing industry, channelled through its development agency, cost the nation relatively little: expressing all investment from national and European sources into €-equivalents, a crude conversion, which does not take inflation into consideration and is not the full extent of state assistance – infrastructure and the provision of services from central government are not included – may not have exceeded €1 billion.[1] Another way of looking at what happened is how much environmental deterioration the €1 billion worth of subsidies purchased.

It is not feasible to provide a comprehensive evaluation of the reasons for and consequences of biological change over the period of approximately sixty years on which this account is focused; many factors are involved and recent years have witnessed dramatic variations in significant causal factors, atmospheric and marine temperature among the most significant. However some devastating consequences can be confidently directly attributed to human endeavour, specifically over-fishing.

Nephrops *replaces cod in the Irish Sea*

Nephrops, prawn, is widespread in the north east Atlantic, its range extending south from Norway into the Mediterranean Sea. Usually referred to by its generic Latin name, the species has a number of vernacular labels, "Norway lobster" and "Dublin Bay prawn" being two of the common ones. Dublin Bay prawn supposedly originated from the practice of trawler-men, returning to Dublin port, shooting their gear in the Lambay Deep outside the Bay to secure a haul of the crustaceans for their personal enjoyment on arrival home. There is no evidence that sizeable quantities of the species occur within Dublin Bay itself.[2]

In the first half of the last century there appears to have been little appetite for *Nephrops* in other areas of the British Isles.[3] Between 1939 and 1950 the landings to Ireland averaged 23.5 t annually but after 1951, demand for the species increased considerably. New processing methods facilitated its more cost-effective presentation to the market. In 1940 the value of the catch was stg£360 and in 1959 it had risen to stg£48,425.[4]

Nephrops could be captured by trap – and a fishery of this kind existed up to the 1970s in the Irish Sea – or by trawl. The trawl in question would have a small mesh and smaller meshes offer greater resistance to water and require greater expenditure of energy to tow them. The early fishery for *Nephrops* was reckoned to be economical if a vessel could harvest 200 kg per day; otherwise its time was more profitably devoted to the capture of whitefish. In the mixed fishery of various species in the first half of the last century, *Nephrops* would have been a relatively small proportion of the total available biomass.

There is a strong feeding association between whitefish and *Nephrops*. In the early 1980s, some 20 - 40% of stomach contents of cod in the north west Irish Sea consisted of prawns. The cod stock in this region accounted for an estimated annual consumption of between 1,700 and 5,400 t of the prey species.[5] In 1991, the year in which that study was presented, *Nephrops* fetched a first sale price of IR£2,505 per t, twice the value of cod, IR£1,212.[6]

The removal of whitefish, particularly cod, a predator of *Nephrops*, had several consequences: fish biomass declined and, thus relieved of a predator, populations of fodder organisms were able to multiply without constraint.

Nephrops was first entered in the national statistics in 1955 when 209 t valued at stg£13,815 were landed. In the same year the combined tonnage of cod, haddock and whiting, predatory species which feed on the crustaceans, was 5,625 t with a combined value of stg£264,154. Over half a century later, in 2008, landings of *Nephrops* had risen to 9,544 t and those of the three whitefish species to 8,285 t, with respective values of €40 and €14 million. Their percentage contributions to the values of the national landings had vastly altered: *Nephrops* increased from 1.6 to 16.9% while the whitefish species fell from 30 to 6%.

There was at one time some soul searching within the fisheries and scientific communities as to whether the Irish Sea might be more profitably devoted to the production of the crustacean rather than the fish. However, to attribute what subsequently occurred to a dastardly experiment in eco-manipulation would be crediting those in a position of influence with more organisational ability than was their due. Instead, over-fishing was allowed to wreak its damage and the fish were progressively cleared from the highly productive Irish Sea which provided prawn populations with the opportunity to burgeon.

Cod was a staple food for countries in the northern hemisphere with a considerable global historical importance and role, particularly as sustenance for explorers on voyages of discovery, and it was the species for which Iceland extended her fishing limits, provoking a series of conflicts with Britain.[7] Cod fisheries of considerable size, notably in Newfoundland, had failed as a result of industrial fishing. Exploitation there had been rising since the fifteenth century but in the mid-1950s landings increased dramatically, before collapsing in 1992. There were efforts to implicate climate change as a contributory factor and that has always remained a possibility, but the main cause of the problem was excessive harvesting.

Cod is known for its high fecundity, which suggests a life history undergoing high mortality rates during its planktonic and juvenile phases. The presence of large individuals in the populations in Newfoundland and Iceland was believed to have been an important element of their reproductive potential. When, in 1974, Iceland decided to extend her fishing limits to 200 nm, a cited reason for doing so was the decline in the numbers of older fish in the population: previously cod of eighteen had been common but by the mid-1970s it was rare to encounter fish of twelve years of age. In the landings from the Irish Sea in 2006, cod had a maximum age of six years and 90% were younger than four. Declining adult stock levels, age and egg numbers were all associated with the fall in recruitment to that stock.

In 1981 the Irish Sea yielded a total of 14,000 t of cod. The year before 6,500 t had been landed from the area to Ireland; a decade later Irish landings had declined by one third. From 1980, scientists recommended, through the European advisory agency ICES, that cod fishing mortality be reduced. Despite the gradual decline in landings, TAC levels were, as a result of the annual bargaining charade in Brussels, set out of reach of landings and that, in turn, precipitated further deterioration in stock levels.

The cod stock of the Irish Sea, had been declining for twenty years before fire-brigade action was taken to arrest its slide. In 1982, the size of the reproductive proportion of the stock was estimated at 20,000 t and four years later the number of young cod entering the fishery was the highest in more than the 30 years prior to 2000. For a large proportion of that period the landings approximated 75% of the tonnage of mature cod in the population.

In November 1999 ICES recognized that the cod stock in the Irish Sea was at serious risk of collapse and proposed a recovery plan to deal with the crisis. The immediate objective was to allow as many cod as possible to spawn between mid-February and the end of April of the following year, for which purpose a spring closure of the fishery was mooted. To that end, the use of a range of gears with the capacity to capture cod (the list ranged from towed, through static nets to hook and line) was prohibited. However, the industry which was the cause of the problem, had to be accommodated and it was decided that fisheries for *Nephrops* and species other than cod should not be disrupted. The end result was a complex of sophisticated regulations which, as a general rule, is a hopeless way of implementing fisheries controls. Thus, the conservation initiative was diluted and its effectiveness compromised from the outset.

The proposals were not favoured by the industry. Despite investment by the state in setting up controls and administering the management of fisheries in the Irish Sea, mis-reporting of catches was widespread and one estimate suggested that only one third of cod actually landed were reported.

In 2001, a cod recovery programme was again scheduled but this time, to appease the industry, the area closed to fishing was reduced. Northern Ireland fishermen, who claimed they had suffered real loss as a result of the closure in 2000, demanded compensation even though the responsible ministry there reported they had filled their quota allocations.

The spring spawning closure was supplemented with technical conservation measures designed to relieve cod of fishing pressure. Measures of this kind included variations in gear design, such as mesh size, to preferentially exclude cod. It was however known that all gears, particularly mobile nets used in the Irish Sea, retained the species. It was even suggested by BIM that it would be feasible to undertake a fishery for haddock which avoided cod, which was nonsense. The new technical emphasis compromised the spawning closure and, paradoxically, in more recent years, most of the landings of cod have been made during the spring months.

Pitted against the vested interests of the commercial fishing industry, the efforts of scientists to save this valuable fish species were simply ineffective. Their main significance was to provide reassurance to the public that "something was being done"; that efforts to restore the natural resource to good health were in competent hands; meanwhile the passage of time eased the panic, lower landings became acceptable and the stock declined further.

The obstacle to cod recovery was an industry which was not prepared to forego fishing opportunities in order to restore a fish stock endangered by its activities. Scientists are familiar with the phenomenon of rapidly depleting fish numbers which requires years of painstaking effort to repair. At a meeting on the Irish Sea in 2002 a speaker from a PO stated that fishermen

...continue to feel betrayed by a process that unjustifiably singled them out to bear the brunt of the Irish Sea cod recovery programme...

It was a perfect illustration of the way the industry regarded its entitlements.[8]

In 2003, another approach to solving the problem, which showed no signs of resolution, was launched. Fishing effort would be rationed on a days-at-sea basis. This was greeted by the CEO of the KFO as "illogical and in breach of the EU's stated policy of fairness and transparency" while the leader of the IFPO described it as "totally daft". The Department volunteered the suggestion that the best way forward was to reinforce technical adjustments which had not secured the desired objectives, rather than keep boats in port. Franz Fischler, EU commissioner, reminded the industry it could claim compensation for hardship but he was adamant that cod should be saved.[9] A review of the recovery programme issued by the Commission in June 2003 was pessimistic about its prospects. And, by this time, other north east Atlantic cod stocks, notably off the west coast of Scotland, were perceived to be in trouble.

Even in the most favourable circumstances, stock re-building after heavy depletion may take many years. The scientific community had made a stab at rectifying the problem; the effort was of brief duration but the industry was impatient for success. The next move was to long-finger the issue. Regulations issued in 2004 set out parameters for a number of cod stocks which had

...been subjected to levels of mortality by fishing which have eroded the quantities of mature fish in the sea to the point where the stocks may not be able to replenish themselves...

and, as a result, were threatened with collapse. The best way forward would be multi-annual plans for the recovery of stocks which, it was envisaged, would require between five and ten years to achieve a positive outcome. Success would be reached when the biomass of mature fish had rebuilt to an historical level (10,000 t) and remained there for two consecutive years. The mechanism would be by TAC and that would be set to achieve a 30% annual increase.

When the latest proposal was introduced, the TAC mechanism was so historically compromised that there must have been serious doubts about its prospects of success. An accompanying technical modification was to re-direct trawling away from cod towards *Nephrops*, using net meshes of 70 - 90 mm. Unhappily, large quantities of juvenile cod were captured in the process and this, together with misreporting, further jeopardised the objectives of the recovery plan.

As the decade following 1999 advanced, every measure adopted to rescue the sinking cod stock proved inadequate: piecemeal and minimal adjustments, each too little, too late. As its biomass declined, the eroding stock was confronted with an ever steeper climb to recovery. At the heart of the problem was the array of fishing gears used in the Irish Sea. At the northern end trawl nets, further south beam trawls and gill nets on the south coast. All captured cod. To be effective in re-habilitating the species, all should be closed down and that was unacceptable to the industry in the Irish Sea.

By 2006 the industry had lost patience with cod recovery. A prominent member of the industry made it clear that there were other species to catch, notably plaice and *Nephrops*, and restrictions on cod were getting in the way of fisheries for them.[10]

As this is written, cod are still being caught in the Irish Sea. Progressively fewer are landed in Howth, the most central port on the east coast of Ireland, but more are being put ashore in Dunmore East. These have been logged as coming from the Irish Sea although it is likely they are being sourced from the neighbouring Celtic Sea stock. And a cod TAC is still provided for the Irish Sea. Although much reduced from its historical high points, the industry is unable to harvest it. The fish are no longer there in the numbers that made this area of sea a fertile producer of finfish.

The complete removal of a fish species from a sea area is a profoundly unsettling event but it is the logical consequence of giving free reign to an industry that is determined to go on fishing as long as there is anything to catch and whose worsening economic situation reduces its options to do anything else. In theory, fisheries are regulated on the basis of scientific advice which is acquired at considerable cost; in 2008 for example, the cost of running the stock assessment division of the Marine Institute was €5.4 m or 2.3% of the first sale value of all landings to Ireland. It is just one item in the enormously long list of subsidies to the industry, courtesy of European and Irish taxpayers. As an almost invariable rule, the industry contributes nothing to the exercise. On the contrary, it may benefit by having some of its facilities and vessels commissioned to undertake surveys to obtain the information on which stock assessments are based. The industry can also benefit directly from access to survey data. There is a long established practice of making information on the distribution of pelagic fish shoals available to commercial interests in a timely manner thereby saving fleets considerable expense steaming about in an attempt to locate them when fishing gets underway.

Other than these perfunctory civilities, the industry is not generally enamoured of scientific endeavour which it regards as an unwarranted intrusion on its ability to make money. Predicting the yield of any stock is not easy because so many natural factors contribute to it and the industry has never been slow to criticise scientific advice which, in the view of its principals, sold it short. But the science of fisheries management is indissolubly wedded to the industry which harvests such a high proportion of fish biomass that scientists cannot reach conclusions on their own without its co-operation. Industry provides biological samples to scientists and, more significantly, it provides data. When, before 2004, it is known that many multiples of permitted landings were being illegally harvested, the official, under-recorded data. were fed into the computer models estimating future yield and this must have distorted subsequent forecasts. Criticism of the industry by scientists is likely to provoke a sulky bout of non-co-operation – as happened when widespread illegal practices were revealed – and that results in no data, good bad or indifferent, being made available. So the scientific community seeks to cosy up to and win the confidence of the catching sector, a process whose end result is a kind of Stockholm syndrome in which scientists become apologists for the industry.

Fisheries science is a technical process but it need not necessarily be as abstruse as it has become in order to demonstrate what is happening to our major food fish species. Probably the majority of fishermen, with rudimentary scientific knowledge, are able to evaluate the relative abundance of the stocks they exploit from year to year; and yet, despite the effort and investment in fisheries science, such is the requirement to assemble and process vast quantities of data to robustly carry out the exercise, that only a proportion of fish stocks in EU waters are adequately investigated. Fisheries scientists, in order to maintain their relevance and their friendships with the industry, devise ever more complex methods of modelling and evaluating events, none of which has so far been able to halt the decline. Rather than restoring biomass – literally replacing the tonnage of fish lost from over-fished stocks – softer, vaguer and, to the industry, kinder targets for reparation have lately been devised. In common with most ambitions in fisheries, their fulfilment dates will retreat into the future.

Cod may be close to commercial extinction but it has not completely disappeared from the Irish Sea. The amount of attention it receives in the future will depend on public concern, and that is a fickle commodity; when it fades, the file will be closed. In the meantime it is reassuring to know that scientists are monitoring developments even if they are unable to coax the industry to forego immediate gain in favour of a more reliable source of income over the longer term.

Summarising, without corrective action from government which, so far, has not been forthcoming, the fishing industry is a law unto itself. Public opinion, which has been critical, has not raised sufficient momentum to achieve useful change. But public disapproval is not to be under-estimated and, in at least one instance, has raised a substantial obstacle to change demanded by the industry.

Seals in the Atlantic

At one time, perhaps decades and certainly hundreds of years ago, the seas around Ireland supported large populations of fish and co-existing seals which consumed them. A healthy ecosystem can support large biomasses of both but, once the fish element is eroded, seal predation can obstruct its recovery. Those circumstances were encountered during the 1970s.

This story begins with a letter to the *Irish Times*[11] from the wildlife photographer Gerrit van Geldern who recounted details he had learned of the slaughter of 68 grey seals, adults and pups, on the Inishkea Islands off the Co Mayo coast in 1979. Two years later 170 animals, 150 of them pups, met a similar fate at the same location. Van Geldern commented on that incident also. He had witnessed seal culls in Greenland and Canada and they had, in comparison, been humane. On the Inishkeas, adults had been shot in the belly and left to a lingering death while their pups starved.

The seal kill had arisen directly out of competition between fishermen and seals for salmon. One catcher recounted recovering three entire fish out of a net in which 70 mutilated ones remained after seals snapped chunks out of them. These circumstances were first described as unusual. A fisherman from the area recalled that until the mid-1970s, there had not been a problem. Others reckoned it had arisen within the previous five years. At the time, drift netting, legal and clandestine, was approaching its height and the national salmon catch had halved in less than a decade.

Reading back over the newspaper coverage, I got the impression that the people of the Erris area of Co Mayo, were taken aback by the public outrage at what occurred. The perpetrators of the kills remained silent and anonymous. There was a rumour of eight armed men having embarked from a neighbouring port and gardai were reported to have examined firearms but nobody was prosecuted for having killed a species that was protected by law.[12] A scientific investigation to quantify the seal problem had been initiated in 1979 and then abandoned the following year when a local fishermen's representative group withdrew co-operation.

Seal numbers were unquantifiable but potentially large off the Co Mayo coast. Uncorroborated estimates placed them at between 2,000 and 4,000. The largest neighbouring concentrations of the animals were in Scotland where seal counts totalled 73,000. Since the 1960s their rookeries had been expanding by some 7% annually. The grey seal populations of the Hebrides alone numbered 35,000 and produced 9,000 pups per year. Results from tagging experiments suggested that a third of Scottish seals visited Irish waters in the course of foraging. Any action to control the species would need to be considerable if it were to be effective.

In the earlier years of the last century seals were hunted and a bounty was paid on them by Boards of Conservators concerned with the protection of salmon in estuaries, but times and public attitudes were changing. It was still feasible to apply for a licence to cull seals where a local problem was established. In the UK, a proposal to reduce seal populations to their 1960s levels had been greeted with public consternation and abandoned.

As far as the fishermen of Erris were concerned, the kills of 1979 and 1981 had not solved their problems and they approached Tom Fitzpatrick (Fine Gael, Cavan), minister for Fisheries and Forestry (1981 - 1982) in an attempt to authorise a cull. The application was refused. In 1982 a Fianna Fail minister, Brendan Daly, succeeded and was issued with an ultimatum, delivered by the Co Mayo deputy Enda Kenny (Fine Gael) who later became taoiseach, threatening that if a licence to cull were not issued – something his own party had declined to do when in office – the fishing community would take the law into its own hands. Daly, equally reluctant to face an angry public reaction, refused to oblige on the grounds that a 50% reduction in seal numbers (a figure disputed by the fishing community) had occurred, making such a move unnecessary. Daly then pulled the appropriate card from the deck: he ordered a study of the population dynamics of seals off the Atlantic coast. The Irish Wildlife Federation called for a two to three year moratorium on a cull while the facts were established.

Meanwhile tension built. A direct action group, the Sea Shepherd Conservation Society, occupied the Inishkea islands to protect the animals during the pupping season of 1982 and during the two subsequent years.[13]

In 1982 the Department of Fisheries and Forestry had morphed into the Department of Tourism Fisheries and Forestry under minister Paddy O'Toole (Fine Gael, Mayo east) who proved just as reluctant to authorise a cull as had been his predecessors. The following year there was an aerial survey of seal haul-out sites along the west coast from which it was concluded that a cull was not justified. That failed to convince the fishing lobby and the controversy simmered on, the occasional seal corpse bearing gun shot wounds washed up on the beach, a reminder that the issue was still live. Finally, in 1987, the Department of Tourism Fisheries and Forestry was replaced by the Department of the Marine which, together with the National Parks and Wildlife Service and BIM, provided with funds by the Commission in Brussels, commenced an enquiry into the physical interactions between grey seals and fishing gear. Part of the investigation would involve the sacrifice of 45 seals per year to ascertain what they had been feeding on from the contents of their stomachs. That licence was issued in 1997 but, when three seals had been shot, authorisation was abruptly revoked in April 1998 following disagreement about the value of the exercise and reflecting, as one correspondent put it "continued official nervousness over the seal issue".

The investigator to whom the seals and gear contract was issued completed the assignment in 1997 and sent me a copy of the draft final report.[14] The brief had been diligently pursued and an impressive quantity of data assembled. Enquiries had focused on gill nets set in the vicinity of the main island breeding colonies of grey seals off the west coast of Ireland: the Inishkeas, Co Mayo and the Blaskets, Co Kerry. Analyses had been undertaken of fish remains in seal faeces and in the stomachs of 51 grey seals which had mis-adventured into becoming tangled in fishing nets where they drowned. Two of the seals had been tagged in the UK, confirming the migration of those individuals to Irish waters.

The research concluded that 96% of all damaged fish in the gill nets were attributable to marauding grey seals which had snapped chunks out of the enmeshed catches but these were described as "minimum verifiable losses" because fish could also be removed in one piece leaving no recordable trace.

Seventeen tonnes live weight of fish were forensically reconstructed from macerated body parts, eggs and skeletal fragments, in seal droppings. Some twenty varieties of fish (mainly species) were reconstructed, ranging in size from small sand eel to large hake and including molluscs (cephalopods). Salmon were not among them.

It could hardly have been a pleasant task, probing through seal stomachs and faeces and the completion of his contract must have been accompanied by a glow of achievement by the investigator. The outcome of the investigation was hardly surprising. The conclusion that grey seals eat fish was to be expected.

The draft final report was accepted by the European Commission in April 1997. Some weeks later BIM, the agency which had organised the work with the Commission, received instructions from its parent authority – then transformed into the Department of the Marine and Natural Resources – to carry out substantial alterations to the draft. There appeared to be some apprehension that the seal study would reawaken the demand for a cull, thus embarrassing yet another minister. The contractor found the demand unacceptable and resigned.

When, a year later, a second edition of the final report emerged from BIM it was a document of a different complexion altogether. As the contractor described it, the findings of his submission had been "deleted, negated or diluted". The principal factual difference between the two versions was the implication of conger eel as a consumer of fish wedged in gill nets for which there was not a scintilla of evidence.

The contractor complained to the European ombudsman. The substance of his complaint was the fact that the Commission received two substantially different versions of the document without seeking explanations for differences between them. The ombudsman did not have within its remit the power to criticise the Irish national authorities but did find against the Commission for the reasons cited, which were deemed "an instance of maladministration".

The final report on the "Physical interaction between grey seals and fishing gear" is available on the Internet, as is the European ombudsman's report.[15] It still implicates conger eel for the misdemeanours of grey seals. It is published under the logo of BIM but the contractor's name appears nowhere on it.

If the report on the interactions of grey seals and fisheries served any purpose it was to demonstrate the ranking of science in the management of marine resources. When it comes to the demands of industry, greed takes precedence over painstakingly established facts. And when it comes to embarrassing a minister, the facts can be altered to spare his blushes. Either way, we should have no illusions about the precedence of scientific method or procedure in regulating our fisheries.

There is no question that grey seals compete for prey with fishermen. With as many as 200,000 occupying British and Irish waters there is a problem, a problem which even the industry, despite frequent attempts, has been unable to have addressed. In his letter which opened the controversy in November 1979 Gerrit van Geldern acknowledged why the fishermen of Erris had done what they did but he added, presciently,

...(it is) not the seals that will be responsible for the...eventual extinction of the salmon which appears to be only a matter of years...

It did not require much time to confirm the trend to which he alluded.

Jellies everywhere

The depletion of any species has knock-on effects in the ecosystem. Diversity is reduced and the balance among those that remain is disturbed and altered. Populations of fodder organisms thrive when a predator species is depleted and they may provide alternative fisheries, but other invertebrates also avail of opportunities to bloom into monoculture with altogether less desirable effects.

For the last half-century scientists have speculated about the role of jellies whose global abundance appears to have increased, coinciding with the run-down of finfish stocks. Jellies – more familiar to us as jellyfish, a term that is not appropriate because they are coelenterates, not fish – embrace perhaps 2,000 species world-wide; they belong to four taxonomic sub-groups which have life-style characteristics in common. Their adults are free living; the gelatinous, bell-shaped, *medusae* have limited powers of propulsion or they drift passively with the currents.

Typically, jellies bloom in the spring and summer months in the seas around Ireland. Their life cycles are complex; they reproduce both sexually and asexually and studies suggest they are susceptible to and favoured by environmental factors. Light, temperature, salinity and feeding all enhance growth and reproduction and make for greater numbers of them. Local chemical enrichment, eutrophication, can also assist them by creating circumstances which inhibit some species, thus favouring others.

Their ability to rapidly generate enormous numbers in suitable circumstances enables jellies to dominate where they are unrestricted. Two methods of reproduction make for explosive population growth which outpaces finfish, allowing them to overtake and dominate the food chain. It has even been suggested that jellies might be the eventual victors of chronic over-fishing.

The relationships between jellies and finfish are not straightforward and they are much speculated about. Jellies feed on zooplankton, consuming fish eggs, larvae and post-larval fishes. On the other hand, juvenile finfish eat small jellies and the adults of some fish species, like spurdog, consume the medusae. Because of their economic implications these food links have attracted scientific enquiry but quantitative assessment of their significance is not yet feasible.[16] One possibility is that a reversible predator-prey dominance, also alluded to between cod and haddock, may apply to jellies and finfish. When one group is abundant it grazes down the other.

If sufficiently precise methods are not available to predict the outcome of specific encounters between finfish and jellies, there are disturbing, if not entirely conclusive, precedents elsewhere. Sardines off the coast of Namibia had provided heavy landings of up to 0.25 million t annually in the 1950s. Another casualty of industrial fishing, their landings continued to rise into the following decade reaching 1.4 million t in 1968, although unrecorded catches and discarding may have pushed that figure over the two million t level. Three years later they were reduced to 300,000 t and by 1980 a by-catch of only 12,000 t was registered.[17] The collapse of the sardine fishery, one of the most spectacular of its kind the world has seen, was followed by an increase in the abundance of jellies which impacted negatively on the fisheries of the region.[18] Sardines are small pelagic species, planktonic grazers, which would have exacted a high mortality on juvenile jellies. Once the predator had been depleted however, the former prey turned the tables on it. Such is the theory although it must be added that the link with the alternate dominance of the food chain by sardines and jellies which unfolded over four decades has not been conclusively proven.

Ecological change is rarely explained by a single clean event. But, if what occurred in Namibia were not conclusive, so many similar events are taking place around the world that they must be persuasive. In 2010 the Asian countries, Taiwan, Indonesia, Japan, Korea and China were reported to be in consultation about how the effects of jellies might be mitigated. The most visible damage they inflicted consisted of consuming finfish, trailing poisonous tentacles over fishing gear and damaging fish stuck in gill nets, rendering the flesh valueless when brought ashore. Related problems have been confirmed in the Black Sea, the Gulf of Mexico and locally in the North Sea. In Italy a much publicised "jelly watch" surveillance has been set up, and France, Tunisia and Turkey have established similar schemes.

Landfalls by jellies on the UK and Irish coasts have been a feature of our summers at the seaside for many years. The common species striking dread into paddlers and swimmers is the lion's mane jellyfish; the other more newsworthy species, Portuguese man of war, is a rarer phenomenon.

Citizens in many countries have set up jelly watch surveys to monitor the abundance of these organisms. There is also a less formal record which could indicate an increasing abundance or a deepening public awareness of the presence of jellies in the seas around Ireland. The place to search for it is in the electronic archives of the *Irish Times*, regarded as Ireland's newspaper of record.[19] From the mid-1930s the frequency of references to jellies increased gradually but following the 1970s, the rate of increase steepened. Jellies had three times the number of references to them annually at the end of the first decade of the new millennium as forty years before; the period coincides with the expansion of the fleet and deep inroads on stocks of finfish.

There may be a real basis for this increasing frequency of mention. Scientists have observed an increase in the quantity of small stage jelly material in continuous plankton recorders since 1970, and a recent paper published in 2011 discussed its association with declining abundance of finfish.[20]

The scientific process responds to a genuine and urgent need when it supplies data crucial to teasing out and clarifying the intertwined effects of environmental and anthropomorphic factors. In this case, the public is concerned about the increasing frequency or abundance of jellies. Should the likely linkage between the increasing abundance of these organisms and the decline in fish stocks be appreciated by the general public it would constitute a significant advance in awareness of the potential for ecological change. That may happen in due course but how much more damage will have been accomplished when it does?

Chapter 18

ANATOMY OF A TAKE-OVER

In one of its many commentaries on the state of the industry in the mid-1950s, the trade press bemoaned the increasing stranglehold of BIM. The agency had put itself in competition with the catching sector, with wholesalers, boat builders, suppliers of ice and fish processors. The fishing industry epitomised a command economy in which even landings were the property of the state. All together contributed to a coherent policy of centralisation,

...crudely socialist in its inspiration...(opposing) the ideals of freedom in private and commercial life, wasteful of public money...never (to be) successfully implemented...

The words were redolent of threat at a time when soviet communism was topical and feared.

The industry must now choose between crude socialism and free enterprise...[1]

Centralisation involves decision-taking by fewer people and it has intensified in the industry since the 1950s along with an unhealthy and unproductive consensus. Government policies - such as the construction of large harbours dedicated to fisheries -

contributed to the concentration of resources among smaller groups of people. Larger harbours were intended to accommodate larger vessels and larger vessels were inevitably bound to more fuel consumption, greater capital investment displacing labour. Once the power centre – made up of politicians, state agencies, the civil service and more prosperous fishermen – became influenced by private coterie, the levers of control passed to individuals who effectively commandeered a natural resource for private enrichment to the dispossession of those who had laboured in it for generations. Because the resource supported traditional employment, public money could be enticed into it on the pretext of creating local jobs in marginalised areas, whereas the ultimate beneficiaries were a smaller, more wealthy minority. Thus, the result of centralisation has been the liberation of private enterprise to the detriment of the majority of fishermen, the deterioration of the environment and the public's loss.

Four case histories illustrate how the process of centralisation has, incidentally and opportunistically, benefited smaller numbers of more fortunate fishermen or, looking at it in another way, been an inevitable consequence of industrialisation in the absence of corrective policy at the centre. First, we will look at the consequences of capitalisation for employment creation; secondly, how the incidental effects of providing a service – supplying ice to the fishing community and later withdrawing it – further disadvantaged those operating in small craft while larger vessels, as a result of state investment, became independent of the service. Next we examine the results of fleet decommissioning and renewal policies on larger and smaller vessels and, finally, we inquire how Ireland's officially "inshore" fleet became host to the largest fishing boat in the world.

Employment creation as a pretext for investment

A selling point for the Vega satellite tracking system for fishing boats, introduced in the mid 1990s, was the information it could convey to and from vessels. Data identifying the location of plankton blooms and measuring sea temperature would enable a boat to locate concentrations of fish more precisely, and the same medium of communication would inform skippers of where the largest quantities of particular species were being landed, enabling them to market their catches where landings were lower and prices higher.[2] A project promoted by BIM in 2000, developed an onboard automated fish handling and sales programme. It worked like this: the net would be hauled aboard, the fish scanned, sorted and weighed automatically and instantaneously offered for sale over the internet. Bids accepted, the landings would be picked up at the quay on arrival of the vessel.[3] The implications for employment and fish stocks, had such a utopian ideal been realised, would have been considerable. Much of the routine deck work on a fishing boat would have been made unnecessary and more effort could have been re-directed to filling the net. Although its ambitions were not realised, six years later the damage to depleted fish stocks had reduced them to the point where 45% more landings were required to keep the fleet solvent.[4] Individual projects were being pursued independently of any social context. That is one way of characterising the history of fisheries development in Ireland.

The phenomenon of capital replacing labour in many aspects of human endeavour became apparent long before the last century but it intensified after the 1950s. While it was taking place, the most industrialised sections of the fishing industry drew funds into the industry by conjuring dreams of greater employment *via* public investment.

The fishing industry, though occupying progressively fewer people with the passage of time, was always fertile ground for unfulfilled promise and unverified growth projections. Fisheries traditionally occupied the divide straddling social services, into which money is pumped, and wealth generation for its own sake. The first of these two scenarios arises from fishing being the only viable way of life in many coastal areas. Invariably, when investment is contemplated in such circumstances, a case is formulated on the basis of anticipated wealth gained and employment created. Historically, the estimates were enormous, an illusion based on pre-Famine nineteenth century reports of boat numbers, without any consideration for their individual size and fishing power or the available biomass of fish, and later, on-shore processing promises which were never fulfilled. As time went by and catching power and capitalisation expanded and the biological limits of fish stocks became more apparent, the projections have become more modest but they have not disappeared.

Most recently, the Federation of Irish Fishermen, in its submission to the government's "Our ocean wealth" initiative, confided its ability to create, within twenty years, an additional "100,000 sustainable jobs – particularly in peripheral coastal areas" – with a total value of €20 billion (or 5% of gross domestic product, GDP), the fisheries sector contributing €7 billion "provided the necessary supports are put in place".[5] It is as fatal a lure for investment of public funds as an irresistibly baited hook is for a fish and similar proposals have been floated on thousands of occasions. Targets, twenty years hence, around delivery time, will melt into oblivion as they have on numerous previous occasions. It would be a fascinating study for a higher degree student to follow up the actual outcome of the many promising scenarios conjured to date. We cannot attempt anything more here than to recount some of the good intentions paving the road to the projected prosperity of the nation, solving the endemic unemployment problem of the Republic of Ireland.

The collection of statistics on the numbers employed in sea fisheries can be traced back to 1821 when 21,000 men and boys were so occupied. These may well have been part-time and we have no idea of full-time job equivalents. There was a steep increase in fishing employment between 1821 and the onset of the potato Famine in 1847; the preceding year recorded 113,000 men and boys working on 19,883 boats. De Courcy Ireland surmised that these fishermen were inexperienced, equipped with primitive gear and in sharp competition with one another.[6]

The Famine took a heavy toll and there has been speculation about the failure of fishing to rescue the starving population. De

Courcy Ireland cited the Commissioners of Public Works who in 1847 observed that people would not consume fish unless it was accompanied by potatoes whose crop had been extinguished. He also referred to the difficulty of distributing fish inland from the coast, something which was still a problem more than a century later.

In 1800 the Acts of Union united the Kingdoms of Ireland and Great Britain in a single parliament and administration. The Dublin lawyer Isaac Butt was initially conservative but became increasingly liberal and federalist after the Famine. He demanded more autonomy for Dublin although he was not expressly in favour of leaving the Westminster parliament. In 1870 the Home Government Association was formed. A year later Butt delivered a speech on the industry to a meeting of the association in Dublin. The motion to which he spoke was:

That the continual refusal to comply with the strong and repeated recommendations of committees of the House of Commons and Royal Commissioners that aid, by way of a loan, should be given to the men engaged in the Irish deep ("deep" did not have the same meaning then as now) *sea fisheries, supplies a striking illustration of the disadvantage which the Irish industry sustains by the absence of home rule.*[7]

Butt displayed great familiarity with the work of the three inspectors whose report, in 1870, recorded the devastation wrought by the Famine. The high employment numbers and vessels in 1846 were reduced five years later by respectively 43 and 26%. The inspectors attributed the collapse to the Famine, which curtailed the demand for fish so that those who were able found something else to do or simply emigrated. Those who remained became so physically weak as a result of hunger that they were unable to go to sea. Their gear rotted and their boats fell into disrepair.

The sentiments of the inspectors had previously been expressed by a royal commission in 1866 and a select committee of the House of Commons in 1867. The economist John Stuart Mill was in favour of a subsidy for Irish fisheries to repair the damage.

The nub of Butt's argument was that Scotland, which nominally had the same status as Ireland in the Union, was more favourably treated by the British establishment. The Oyster Commissioners in 1870 observed that Scotland had received stg£1.25 million more than Ireland since the Act of Union. Finance provided for Scotland had, among other things, enabled her fishermen to operate a branding system for herrings which gave them a competitive advantage. The Oyster Commissioners observed that if Ireland were a province of France, her fisheries would be subsidised. Butt requested stg£10,000 annually towards rebuilding and administering Ireland's fisheries, a mere tenth of the sum spent on the Royal parks. The case of Ireland's fisheries, he argued cogently, would be more justly dealt with under federal rule.

Year on year the employment and fleet statistics continued their inexorable decline. At the end of the nineteenth century (1898) when more comprehensive recording of sea fisheries statistics got underway, employment and fleet numbers were registered at 26,307 men and boys and 6,690 vessels, reductions of, respectively, 77 and 66% from 1846.

The first comprehensive fisheries report in 1898 and its immediate successors are an eye-opener. More recently we have been informed that a number of new fisheries were developed between 1970 and 1990. In fact there were fisheries for shrimp, a supposedly "new" species, oyster, an export trade in mussels and a number of inshore fish species more than a century before. The demand for what was harvested was subdued and remained substantially unchanged until our accession to the EEC when continental markets sought a wider species range and more powerful technology harvested hitherto inaccessible resources.

Arthur Griffith was described by de Courcy Ireland as the only one of Ireland's founding fathers to actively advocate the development of our sea fisheries. Moving a step further than

Butt, Griffith epitomised the nineteenth century nationalist movement. In 1911 he declared that fishing could be an industry second only to agriculture, employing 100,000 people and providing ancillary employment for four times that number. I have seen similar numbers quoted by various sources and I presume that Griffith is their origin and that he, in turn, had been inspired by the pre-Famine census.

Throughout the history of the Republic, fisheries presented a tantalising possibility for economic growth which, until the biological limits of our resources dawned on some, though possibly not yet all, might be profitably worked (to the boundaries) "only (of) the amount of capital embarked in them"; sentiments expressed in 1883 described "the rich harvests of the sea" as "inexhaustible".

Various initiatives cropped up from time to time. In 1941 the Maritime Institute of Ireland was formed to fight for the implementation of a national marine policy. In 1949 the Institute set up a sub-committee, of which de Courcy Ireland himself was convenor, to furnish data to a commission on population and emigration. Its evidence stated that ultimately Irish fisheries would provide employment, ashore and afloat, for "nearly as many of the 100,000 people visualised by Griffith". The submission, according to de Courcy Ireland, was the first acknowledgement that aquaculture would contribute to these figures. But Griffith had envisaged a much larger employment potential than that in capture fisheries.

Within the industry another trend was taking hold. Technology creep had begun in earnest with the introduction of steam and the construction of larger sailing trawlers during the early years of the last century. At the end of the First World War the number employed in fisheries had declined to 17,000, the number of fishing craft to less than 5,000, twice the fleet size of today but very different from it. In 1935 a new increasingly motorised fishing fleet had reached 8,250 GT; on the formation of BIM in 1952 it had grown to 12,000 GT and by, 1960, to 12,870 GT.

Overkill!

In 1960 the Maritime Institute published a pamphlet *Ireland's Sea Fisheries* which maintained that Ireland should have 10,000 full time fishermen with 40,000 employed ashore in a variety of processing and supply industries, much the same ratio as Griffith had himself proposed. Arthur Reynolds, founder of the *Irish Skipper* in 1964, stated that a modern sea fishing industry should provide six jobs ashore for every one at sea.

As time went by, the boundaries of a resource, once described as unlimited, became more circumscribed. In 1981 de Courcy Ireland declared that, among other factors

...the problems of the industry are not so much the overwhelming penetration of EEC trawlers into our coastal waters...but the diminution of certain species...

It was a realisation with a long gestation.

At the conclusion of his history of Ireland's fisheries, de Courcy Ireland provided estimates of employment within the industry. Prior to our entry to the EEC in 1970, Ireland had a total of 5,860 fishermen 3,900 of them part-time. Onshore processing employed 920 and other full-time employment (excluding distribution), 360. Thus the ratio of sea-going: ancillary employment was 0.2, a reduction far beneath the levels predicted by either Reynolds or Griffith. In 1980 total employment rose to 8,818 at sea and 2,370 ashore (a ratio of sea-going: ancillary employment of 0.3) coinciding with the heavy industrialised exploitation of finfish stocks.

In the early years of the new millennium, Ireland's fishing fleet was approaching 2,500 vessels, historically a very small number, accounting however for probably 80,000 GRT and 200,000 kW, the maximum ever recorded. The Cawley report acknowledged that jobs in the industry were declining but stressed the value to rural communities of those that remained. In 2006, Cawley observed that fishing provided 11,665 jobs (less than 5,000, full and part-time, in fish catching) which would require investment of €597 million over the next six years (2007 - 2013), an average of more than €9,000 annually per person employed, a historically novel but more realistic appraisal.[8] For the catching sector, with which we are primarily concerned, the figure worked out in the vicinity of €5,750 per worker per year.

Cawley did not provide projections for job growth; rather he drew attention to the difficulties of retaining existing ones, a new departure. On the other hand, ever- optimistic BIM in its Strategy document for 2010 – 2012, predicted some growth in numbers would have taken place by 2012. Among key outcomes there would be 600 additional jobs in marine seafood.

What price ice?

Rather than chase after the many and varied strands and disciplines of an expanding industry it would be useful to examine the history of a commodity of central importance to fishing.

Despite its long history in the preservation of food quality[9] the use of ice was no longer common practice in Ireland when independence was wrested from Britain. The 1921 commission of inquiry observed that there was no "ice factory" in any of the fishing ports.[10]Ice could be purchased in Dublin, Belfast or Cork but its transport to the fishing areas was costly and cumbersome. The commission recognized that ice was essential for fresh fish sales and observed that Canada, Belgium, Great Britain and Denmark were all addressing problems associated with providing it.

Fish in Ireland were not iced at sea or in the ports or in retail outlets. Indeed fish was not retailed by fishmongers but,

... sold (where sold at all) by small dealers who keep their supplies in a small barrel or box in the corner. It looks unsightly and unpleasant, becomes sodden and offensive of smell...

The commission cited an employee of the United States Fishery Board who remarked that people in the United States had developed

... a distaste for fish with an insufficiency of ice...this distaste was not present where fishmongers' shops could be found with marble slabs, tiled walls, abundant ice and a pleasing display of fresh fish...

The ISFA had tackled the deficiency in its modest way, providing cold storage facilities in Hanover Quay (Dublin), Galway, Killybegs, Murrisk and Helvic. A quick freeze facility was planned for 1947.

When vessels commenced working industrially, moving progressively longer distances from port to fishing ground, additional approaches to keeping fish fresh were actively considered. A sinister solution was the addition of antibiotics to ice. The expense and effort invested in this line of research are testimony to the urgency of devising a method of keeping fish flesh in pristine condition.

Experiments commenced with the injection of broad spectrum antibiotics into farm animals whose flesh kept in acceptable condition for longer after slaughter than would have been the case had the antibiotics not been used. The Torry Centre, Aberdeen, among a wide geographical spread of interested laboratories, extended these trials to fish. Best results, according to the trade press, were obtained by using a broad spectrum antibiotic, chlortetracycline, sold in the United States of America as Acronise which, it was claimed, was rapidly destroyed by cooking. In medicine it was known as Aureomycin.[11]One authority recommended that while antibiotics could pose a threat to human health, aureomycin should, nonetheless, be introduced without delay. Another dismissed any threat out of hand: it had been proved, the statement went, that one would have to consume 454 kg of food a day before aureomycin-treated fish would have any adverse effect on the human body.[12] Some scientific sources recommended this treatment of fish flesh and for some time, following experiments in Japan, "acronised" fish appeared to be a likely market innovation. An announcement assured "A technological revolution in the world food industry is beginning".[13] Experiments in the USA continued into the next decade but the antibiotic treatment of fish was not adopted.[14] Now that we are aware of the problems of microbial resistance in our own species due to the over-use of antibiotics in farming, it is obvious that the adoption of this method of ensuring good quality would have had profound adverse consequences.

BIM envisaged that fish handling would evolve by a series of developments. The first would be the provision of ice making plant in the ports, a development which had been started by the ISFA. In 1954 it was intended that appropriate infrastructure would be located at Galway, Caherciveen, Schull, Killybegs, Castletownbere, Ballycotton and Limerick and further construction was under consideration for Dingle, Dunmore East and sites on the east coast.

Progress was slow and a shortage of ice became a bone of contention between the infant BIM and its waspish critics in the Association which demanded that government provide grants for ice plant construction rather than entrusting it to BIM which, the Association claimed, was motivated by political considerations. BIM's second annual report presented the information on a map, an innovation – but BIM was always strong on presentation. The map was derided by the Association as

...a resplendent production in glorious technicolor...which puts the best complexion on (BIM's) old tale of woe...

The map demonstrated that, other than Dublin, there were no ice-making facilities between Clogherhead on the east coast and Dingle in the south west, whose plant only worked occasionally whereas Murrisk and Cleggan, whose combined fishing population numbered sixteen, each had one. The Association continued:

*...what (BIM) has done for those inside its "Pale" is very little –
even for Killybegs which, on this map, is tricked out with colours
and badges like a field marshal's chest...*[15]

As early as 1955 the potential of Killybegs, which would later
become Ireland's largest fishing port, was anticipated.

Suspicions of parochial preferment where critical infrastructure
was sited surfaced again the following year under the heading
"Politicians at play" when an "indefensible" decision was
reported to locate an ice plant in Ballycotton. According to the
Association, fish landings to Ballycotton were the second lowest
in Ireland. The capital cost of the plant was stg£5,000 and its
running costs would be stg£1,500 annually. Ice would sell at
stg£8 per t whereas it was already available from commercial
sources in Cork city at 43% of the BIM price.[16]

Three years later ice installations operated at Killybegs, Murrisk,
Cleggan, Galway, Dingle, Schull, Ballycotton and Dunmore
East. With one exception, all made a loss which amounted to
stg£3,095. Their under-performance was explained by the fact
that supplies of fish were insufficient; economies of scale were
required.

In 1959 losses doubled to stg£6,325 and a subsidy was provided
to make ice available at a more affordable price. The following
year the losses climbed to stg£7,410 of which stg£6,000 was
provided by the exchequer.

In 1961 ice was produced at Killybegs, Cleggan, Galway,
Dingle, Castletownbere, Schull and Ballycotton. It was
discontinued at Murrisk and the Dunmore East plant operated
only during the herring season. The production of ice increased
23% over the previous year. It sold at stg£4 per t but that price
was reduced by half if the purchaser took it to sea. Losses on the
year were just short of stg£8,000 of which the subsidy
contributed stg£6,000. A new source of finance, the Fishing
Industry Development Fund provided stg£304 for ice applied to
catches on board, a miniscule amount illustrating the extent of

the practice. The use of ice did however continue to expand and in the year 1962 costs of production stood at stg£7,633 of which sales recouped stg£4,731.

As envisaged by BIM, a good quality fresh product should be handled carefully on shore, processed and/or frozen. The provision of ice manufacturing plant was a first step but added value, secured through the use of deep freeze and processing technology, was the desired objective.

Despite the bickering, BIM persisted and established ice plants at all the main ports by the end of the 1960s while some of the older ones, at Greencastle, Achill, Dunmore East, Kilmore Quay, Killybegs and Schull had already been replaced with more modern units. Work on installing new plant at Castletownbere and Howth was underway in 1967. At the close of the decade and less than twenty years after BIM was set up, a new ice plant was built in Galway. Ice storage facilities were provided at Clogherhead and Arklow. Sales of ice increased to stg£14,610 in 1967 but the financial loss on their production was stg£13,773. In future the cost of ice would be heavily subsidised by the state.

Subsidies to ice production were supportive of BIM's ambitions. The agency was charged with establishing a fishing industry and that industry had to use ice. The subsidised price charged by BIM was stg£4 per t. The Association claimed much cheaper prices were available from other producers. Examples of stg£1.15 and stg£0.17 per t in Denmark and certain English ports respectively reinforced the case for the state bearing much of the cost. The course of development of the industry in Ireland was not exceptional; similar subsidisation was taking place elsewhere in Europe and, indeed the wider world.

Despite subsidy however, not enough ice was being used and the reason was not entirely related to money. A major problem to be overcome was the connection in the public mind between poor quality fish and the preservative. To sprinkle ice crystals was somehow to proclaim that the fish they covered needed treatment to conceal its rottenness. When consultants from the United

States of America reported on the fishing industry in 1964, they recommended that landings should be covered in the proportion of one part ice to three parts fish.[17]

The provision of ice manufacturing machinery and storage was initially intended to generate income for BIM and later to, at least, pay for itself. When, in the early 1960s BIM withdrew from commercial operations, the supply of ice became a service but activities of this kind were not intended to be entirely free. On several occasions the public was reminded:

Some of the activities shown in the development and operations trading account are also covered in the exchequer grant-in-aid. In this connection it should be noted that although boatbuilding and repairs, gear trading and ice production activities are treated as commercial operations for accounting purposes, they are in fact carried on as development operations for the purpose of providing essential services to the industry.[18]

Every annual report of the agency devoted space to ice production which BIM regarded as a core activity. It should therefore be easy to ascertain what happened and how close the industry came to achieving the targets set for it in 1964, but BIM had an aversion to reporting bad news, and the constituent data contributing to the ice accounts are frequently indecipherable, grouped with other items under separate accounting headings. In the 1960s ice production cost approximately twice as much as it earned. In the following decade sales of ice were worth one third of their production costs. One way to reduce balance sheet costs in the 1970s was to eliminate plant depreciation costs from the ice account and report it elsewhere on the balance sheets. It improved the look of things but the operation still made a loss. Other elements making up the ice account included wages and commission (though not necessarily administrative costs), raw materials, operating costs and interest charges. I never encountered mention of an EU grant for ice production in the relevant section of the accounts but I came across it elsewhere.

Ice plants were continually under construction, enlargement and improvement. In 1975 BIM had fourteen, eight directly operated, the balance leased to fishermen's groups. By the end of the decade there were eighteen plants manufacturing 200 t of ice daily, with storage capacity for 900 t. In 1988 there were fifteen large and medium sized plants under BIM's management and thirteen smaller ones leased out. The number continued to grow and in 1991 the number of smaller units had risen to 21. By 2008 only ten ice plants were referred to in BIM's annual report, producing 14,700 t, and two years later a mere 11,100 t of ice was reported to have been manufactured, 25% of the annual production in 1999. In 2000 one of the declared objectives of the marine services division was:

...to raise the quality of fish supplies through increased use of ice...

The use of ice increased annually while landings rose. According to BIM, the amounts used were never adequate and conditions such as bad weather, which curtailed fishing time, also depressed ice sales, as did colder weather when ice was considered less necessary by fishermen. Ice production appears to have operated most successfully from the mid-1980s to the end of the century. In one year the use of ice approached half of the level recommended in 1964 but improvements in on-board refrigeration in the meantime may have made those earlier recommendations less necessary; larger fishing vessels manufacture their own. In the late 1990s the production of ice might have generated a profit, although it is difficult to be certain from the way the figures were reported. Grants were available to increase the use of ice in fish handling in the later 1990s. After 2004 ice sales declined and losses increased to 40% of turnover in 2005, coinciding with a price increase of 5%. In 2006 the price of ice increased by 10% and sales fell by 17%. In 2010 the ice network was reduced to nine plants under BIM and four on lease. In 2009 it was recommended that the network of ice plants be closed "in favour of a more market-based approach".[19] Ice plants in some ports were shut down.

The reaction of the industry to the withdrawal of subsidised ice production depended, as did many aspects of its operations, on the size of vessel involved. Larger boats now manufacture their own. Boats working close inshore are not, according to recent accounts, icing at sea. Instead they sell their catch on landing, after a brief absence, to buyers who bring ice with them and apply it in the container lorries to which the landings are transferred. Any necessity for subsidised ice production appears to have melted away.

A study of the economics of fishing reported that between 1985 and 1993, the use of ice was an established part of the harvesting routine. Wooden hulled vessels used a greater quantity than steel boats, probably a consequence of more modern vessels being better insulated and equipped with refrigerating facilities. Otherwise, in both groups, the quantity of ice used correlated with both fuel and wage costs, suggesting that ice was an established item in the operational budget.[20]

The availability of ice, like so many aspects of this industry, is another instance of disproportionate advantage conferred by fishery policy on larger vessels. Modernisation grants facilitated the installation of icing and refrigeration facilities on board, this having the duel advantage of allowing those boats to operate further from shore for longer periods, and to do so independently of onshore support facilities such as ice manufacturing plant. Small boats on the other hand, may be totally deprived of sources of ice while the grounds on which they work are accessible to the largest vessels.

How fleet decommissioning and renewal accentuates unfairness and imposes additional fishing pressure on the resource

Decommissioning, as we understand the process, is more than the final act in the life of a fishing vessel. It must be removed from fishing activity and the tonnage (GT) and power (kW) associated with it must be taken out of the fleet at the same time, not to be replaced. By decommissioning, the size of the fleet is permanently reduced. It is recognition of the fact that catching power is too great to be supported by the available fish resource. Or, rather, that is what is supposed to happen.

When decommissioning was mooted as a possibility in the trade press in 1990 the fleet was still expanding: its GRT stood at 62% of the maximum size it would achieve in 2006 while its power was 79% of the level it would reach in 2004. Decommissioning as a concept was, therefore, out of sync with the national appetite for greater catching effort. In real terms, the industry visualised decommissioning as a way of disposing of old, uncompetitive craft without necessarily permanently reducing the size of the fleet.

A European Community scheme was put in place in the early 1990s to shrink the size of the European fleet by compensating owners who removed vessels below 100 GRT with a cessation premium of IR£21,000 plus IR£1,200 per GRT. Thus, a vessel of 50 GRT would qualify for a payment of IR£106,000, half of which would be contributed by the state, the remainder by Brussels.[21] When the scheme was formally announced in 1995, the price per GRT was increased up to IR£1,500.[22]

A national fishing plan prepared by the Department confirmed that decommissioning to the extent of 4,000 GRT would be available between 1994 and 1996. Registered vessels in certain categories, or *segments*, which were regarded as too large would be removed and not replaced. Otherwise, the fleet would be allowed to expand. An additional 3,000 GRT would be added.[23] Allowing for technology creep, the decommissioned 4,000 GRT would almost certainly be over-compensated by 75% GRT in the form of more modern vessels. Of course, it would be argued that the decommissioned boats had been targeting different species but, as demonstrated elsewhere, an expansion in one type of fishery tends to have knock-on effects, intensifying effort in others.

The 1994 - 1996 scheme was the first occasion in the history of Irish fisheries on which fishermen had been paid to destroy their vessels.[24] IR£6 million was to be invested in the destruction of surplus boats. In total, 57 applications were received for decommissioning grants. The boats in question ranged between ten and 60+ years old. However, the scheme did not attract the volume of interest that was expected.[25]

The second decommissioning scheme appeared abruptly in 2005 and, understandably in the circumstances, without very much fanfare. The whitefish renewal scheme, valued at €116 million, had been introduced to populate that fleet with more modern vessels. It had enlarged fishing capacity and in 2005 had not fully run its intended six year course when it was realised that the industry was being conducted on a fraudulent basis;[26] if matters were to be stabilised – and some signs that it would be had to be made apparent, otherwise Ireland would be heavily fined by the EU – then the industry was bankrupt.

The emergency second decommissioning scheme was to be restricted to older and larger vessels, of more than fifteen years and greater than 18 m oal. One of the most prominent groups of candidates to benefit was a fleet of scallop dredgers, rusting and unfit to go to sea. They had taken an effectively persuasive initiative: blockading Rosslare and Waterford ports because their quotas of scallop were being reduced.[27] Buying off actions of this kind has played a major role in the formulation of Irish fisheries policy. The case of the scallop dredgers was even more bizarre: some of the Wexford fleet had received fishing licences just five years before when a cap on the fishing fleet had already been recommended.[28]

Two years later the list of vessels to be scrapped was assembled. There were 120 candidates and the bill would be €66 million which Brussels had to approve. Of this some €58 million would go to compensate for tonnage which would claim between €5,200 and €6,500 per GRT,[29] a considerable increase in value over those in the first decommissioning scheme in 1994-1996. Then there were delays and, when it finally came on stream, cutbacks had intervened and reduced available finance. Eventually, in 2008 and 2009, a total of €36.6 million was spent removing from the fleet 45 boats with a combined capacity of 6,818 GRT and 19,041 kW.[30]

During her brief tenure as Minister for Agriculture Fisheries and Food Mary Coughlan described the combination of whitefish fleet renewal (providing more modern vessels in exchange for older ones) and decommissioning (scrapping the older, less competitive ones) as delivering a modern, internationally competitive whitefish fleet.[31]

At any time, whether decommissioning is available or not, boats cease to operate because they have reached the end of their functioning lives. They may be sold on to work as part of another nation's fishing capacity, or they may be broken up. And when, in the absence of decommissioning, the hull has been destroyed or the vessel moves elsewhere, its tonnage and kiloWattage – then described as dormant or latent – remain within the national fleet segment to which the vessel belonged. Tonnage and kiloWattage are handled by brokers who sell it on to the next purchaser of a fishing boat, so that he can comply with the licensing process. Paradoxically, dormant tonnage may also be bought by anyone who has decommissioned a vessel and wishes to re-enter the industry, an exercise which has been euphemistically described by BIM as "displacing back into the fleet".

An evaluation of the later decommissioning scheme was undertaken in accordance with the Department of Finance "value for money and policy review" initiative for which a steering group, comprising personnel from the Department and various state agencies, was set up.[32] Because a number of factors are involved (fish prices, fuel prices, quota allocations), it was difficult to conclusively demonstrate any benefit to the remaining vessels from the removal of 10,237 GRT (11% of the fleet) and 28,506 kW (13%) although it should be assumed there was some.[33] Those vessels that continued fishing are reported to have improved their performance: they caught 14% more per kW in 2010. However, decommissioning between 2007 and 2010 does not appear to have increased average fishing income to the vessels that remained. The removal of certain vessels, notably beam trawlers, probably conferred an environmental benefit because they drag the heaviest towed gears which inflict great damage on the sea bed, and they have high fuel consumption. Less convincing claims were also made that those vessels which continued fishing were more compliant with regulations but there is no way of establishing whether this is true. The great problem with law enforcement is the defensive atmosphere in which the protection agency, the SFPA, has operated since it came into existence in 2007 together with the unexplained nature of the lawlessness that prevailed in 2004. Something of its extent is widely known but because its *modus operandi* has never been elucidated, there is no way of knowing whether the same kind of behaviour is still going on.[34]

While decommissioning removes boats from a fleet, it does not remove fishermen. An operator who owns a decrepit hulk which is barely operational (technically, it has to be functional to qualify for funding) could obtain more for it than he would on the open market where the vessel might have no value. He is then free to purchase a boat in another uncapped category (fleet segment), or in the same as he has just left, should capacity become available, and continue fishing.

Re-entering fishing with a more modern vessel invariably means being at a competitive advantage over one's fellows who are stuck with the boats which might have been slightly younger than those that were decommissioned but less technologically endowed than the new arrivals. The competitive advantage has been purchased with taxpayers' assistance which thereby favours some citizens over others. A fisherman equipped with more modern technology is further blessed when older, less efficient vessels are removed from the fleet and their potential landings fall to him. Fish stocks may have to contend with greater fishing effort than before despite the removal of capacity. Invariably, should a fisherman choose to resume his calling and should he be excluded from exploiting high value quota species, he will probably end up targeting already highly pressurised inshore stocks; shellfisheries have had to absorb additional fishing pressures in recent years because segments targeting pelagic and certain demersal species were capped and re-entry to them was costly and problematical.

The value for money team examined the possibility of re-entry by fishermen whose vessels were decommissioned and concluded it was negligible. But new vessel purchases were made, within the lower fleet capacity ceiling and their number has not been disclosed. They might have been registered in the names of the fishermen who decommissioned or family members or as companies.[35]

Similar criticisms can be levelled at fleet renewal, a process which, necessarily, favours a minority of boat owners who are thereby granted access to a greater share of finite fish stocks while the majority are obliged to continue labouring with their more primitive vessels and technology, surviving on diminishing returns.

Why it should be necessary to grant-aid vessels belonging to private operators with taxpayers' money at any time and particularly when fishing capacity is excessive and 75% of European fish stocks are over-exploited, must be questioned; more so when the exercise puts fellow fishers out of business. Why such funding should be so discriminatory towards larger boat owners, as was the case during the second decommissioning scheme, is a matter of public concern. If the fleet is too large, it is certainly in the public interest to reduce it, to bring it into balance with the resource on which it depends, but that objective could be secured more cheaply and just as effectively by purchasing tonnage directly from brokers as vessels leave the fleet at the end of their working lives, thus blocking the entry of others. Whitefish tonnage was believed to be available for approximately €4,500 per GRT on the market while its purchase by government through the second decommissioning scheme could have cost the taxpayer up to €7,500 per GRT.

The largest fishing boat in the world

Kevin McHugh was born into a fishing family - his father and two brothers also worked in the industry - in Achill, Co Mayo. He was very bright, industrious and ambitious. McHugh began fishing at the age of fourteen on half deckers using static gears. Two years later he attended the BIM fishing school in Greencastle.[36] His formative years, in the course of which he got his skipper's certificate, included working on Icelandic vessels to learn deep water fishing technique and from there he returned to Ireland. In 1966 he purchased his first boat, second hand, the 21 m wooden general purpose *Wave Crest*, with which he pair-trawled herring with mid-water nets. Throughout his career references to him in the trade press alluded to his fascination with and expertise in fishing technology.[37]

Eight years later he purchased the 27.2 m *Albacore*, a multipurpose steel vessel, for stg£1.2 million. Built in the Netherlands, *Albacore* was the first in the Irish fleet to break the stg£1 million price barrier and one of the first to avail of EEC grant aid. The vessel was capable of carrying 160 t of fish with freezing capacity on board. Crewed by twelve men and equipped with the foremost electronic and hydraulic fishing equipment of its time, its engine power was 750 kW.[38] *Albacore* was capable of purse seining and trawling. Its addition to the fleet was an occasion of much rejoicing by gear providers, who clamoured to advertise their association with Ireland's most modern vessel, and of fascination to other fishermen who swarmed over it, wondering at its technology. It was the kind of reception that would be repeated whenever McHugh made purchases in the future, the amount of public adulation rising in proportion to their price and size. Later acquisitions would be visited by government ministers paying tribute to McHugh's commercial acumen.

Three years after its purchase, the *Albacore* was advertised for sale and McHugh obtained a larger boat, the refrigerated sea water (RSW) *Antarctic*, for IR£2.3 million. It was 40.5 m in length with a 1,500 kW engine. On its first day's pair-fishing, the

Antarctic and a similar sister ship landed 500 t of mackerel.[39] A single haul of their pair trawl could amount to 350 t.[40]

Larger vessels had advantages fishing deeper. When mackerel sought greater depth smaller craft simply could not reach them. In the circumstances, the boats might divert their effort to demersal species like cod. The owners of RSW vessels however enlarged them to even greater dimensions to increase their salt water tank capacity. Eleven of the fleet of 23 RSW mackerel boats had been lengthened at the end of 1985 and their RSW tank capacity was thus enlarged by 90%. The *Antarctic* was sent to a shipyard in Norway to be extended by a further 12 m and its electronics brought up to date. Its new length allowed it to reclaim its place as one of the two largest vessels in the Irish fleet.[41]

Several months later the trade press announced that McHugh had ordered an even bigger vessel, 91.4 m and 4,000 GT, from the Norwegian Hellesoy yard at a cost of IR£12 million. *Veronica* – named after McHugh's wife – Ireland's first factory freezer trawler, its first super trawler processing at sea - would have engine power of 6,300 kW, desalination equipment and ice making capacity of 20 t per day. The vessel would be capable of making 15 knots per hour steaming to market and could accommodate 2,000 t of fish and a crew of 70.[42] Delivered in 1987, it was the most modern fishing vessel anywhere. Paired with another slightly smaller boat belonging to a colleague, *Genesis*, the two dragged a midwater net whose footrope of 270 m (more than a quarter of a kilometre) stretched between them, with a trawl 400 m long.[43] The two were believed to constitute the most powerful pair trawling team in the world.[44] McHugh went on to purchase *Genesis* in 1989.[45] *Veronica* fished mackerel, scad and blue whiting but McHugh also intended to exploit demersal species. The *Veronica* was capable of fishing to a depth of one km and he had his eye on the hitherto untouched grounds west of the Porcupine bank.[46]

When he purchased *Veronica*, Kevin Mc Hugh was 41 years old. He was then regarded as the foremost fisherman in the country but his great aptitude was not confined to this obsession. He was obviously very personable and capable in business matters. In 1976 McHugh was elected vice-president of the IFPO, the first PO, and he also served on the committee and then the board of the IFO.[47] *Who's who* in Ireland listed Kevin McHugh among the top one thousand most influential people in the country in 1985.[48] With others in the KFO, McHugh purchased fish processing factories in Killybegs.[49] His directorship of sea food factories, which had previously failed because of lack of supplies, promised much to the town in the future.[50] He also assembled a property portfolio which included a number of hotels and entertainment venues.[51] The company he formed to manage his financial affairs in the late 1980s was, portentously, named Atlantic Dawn Ltd.

In 1992 *Veronica* caught fire while undergoing repairs in Harland and Wolff in Belfast.[52] The insurers refused to write her off and paid an *ex gratia* sum to the owner instead. McHugh decided to sell the damaged vessel and ordered a replacement from Norway.[53]

The new vessel, *Veronica II*, was slightly larger than her predecessor, 106.3 m, but the same GT. Her total power was 4,500 kW. The vessel would be the largest in Ireland and among the top five in Europe. A crew of 40 would man it. *Veronica II* was capable of freezing 250 t of fish per day, with a carrying capacity of 2,200 t. This, the trade press proudly announced, would provide 10 ten million people with a fish dinner! Built with two cranes to expedite the rapid unloading of its landings, the price of the vessel was approximately IR£25 million and finance was provided by Norwegian banks together with the Bank of Ireland and private sources; no EU money was involved.[54]

McHugh's loyalty to Killybegs, whose fishing community, and particularly the KFO, had supported and would continue to support his endeavours in the future, was an established fact and a vital element of his strength. He would use the port to the greatest possible extent but there was some concern that accommodating *Veronica II* along with the rest of the RSW fleet might cause congestion.

At this point it is apposite to inquire how the rest of the industry anticipated the arrival of yet another, larger McHugh vessel. Some statistical generalisations of industrial trawlers, of which super trawlers were the ultimate development, were in the public consciousness. Making up one percent of world fleet numbers, employing two percent of those working at sea and catching 50% of world landings,[55] they did not offer a great prospect of livelihood to competitors. The trade press, both of whose publications were based in south west Donegal, supported one of their own: "Do as McHugh does. Adapt or perish!" The main editorial thrust of the *Irish Skipper* observed that bigger vessels made an unanswerable case for higher quotas and entitlement to participate in fishing agreements between the EU and third countries. They did not. The fact that the vessel was privately financed was hailed as disdaining the parsimony of BIM.[56] However, concern within the industry, which manifested itself as subdued grumbling, was not initially afforded due recognition on the printed page although it was acknowledged, and dismissed. Instead, the editorials took the line that the amount of quota required to feed the new vessel was nothing more than a reflection on the fact that Ireland had a small quota allocation![57] Irrespective of the size of any quota, there surely had to be some requirement to share it, or at least not concentrate an unduly large proportion of it in a single enterprise, particularly in an industry which was heavily grant aided in the interests of job creation?

In 1998, McHugh, still less than fifty years old, embarked on his greatest gamble. The vessel, due for delivery in 2000, was the *Atlantic Dawn*, the largest fishing boat in the world, four m longer than its nearest Dutch equivalent, functioning as a trawler and purse seiner. Its nets were described as twice the size of London's millennium dome. Its total engine power was 21,100 kW; physical dimensions: 142.8 m oal, 24.3 m, beam, 13,500 GT with capacity for 7,000 t of frozen fish in the hold and the ability to freeze 300 t per day, operated by a crew of 100. Fishing cruises would be of thirty days' duration, each trip consuming US$0.5 million of fuel. The vessel would be built in Norway with the participation of the Vik and Sandvik and the UMOE Sterkoder yards.

The project was to be funded by McHugh himself at a cost of €15 million. The remainder of the €70 million required was to be supplied by a consortium of banks.[58] *Atlantic Dawn* would work off north west Africa out of a base in Puerto del Sol in Gran Canaria where a cold store to accommodate 20,000 t of landings would be constructed; at least that was the plan.[59]

The news of the *Atlantic Dawn*'s construction caused some consternation in the Department of the Marine in Dublin. The project had got underway without any consideration of how a fishing licence would be obtained for the boat. At the time, Ireland's fleet was capped. McHugh was confident there would not be any difficulty. He intended to operate outside EU waters, an activity in which a number of Dutch boats were already engaged. Ireland would be entitled to similar treatment by the Commission in Brussels.

The government duly applied to Brussels for "an international segment" to its fleet in 1999; this would permit the *Atlantic Dawn* to operate as an EU vessel in international waters. The Commission did not respond and when, in August 2000, the giant vessel was about to put in an appearance in Killybegs, among other Irish ports, the minister, Frank Fahey (Fianna Fail, Galway west) and taoiseach Bertie Ahern, both intervened, writing to fisheries commissioner, Franz Fischler, and EU Commission president Romano Prodi who were on holiday at the time. Ireland's commissioner, David Byrne, was enlisted to lobby Fischler who proved elusive. In September 2000, Departmental officials began shuttle diplomacy between Dublin and Brussels to persuade the EU to take the vessel into the European fold.

The consortium of banks and the solicitors involved pressurised marine minister Fahey to grant a temporary fishing licence to the vessel; they probably envisaged their expensive investment rusting away otherwise, unoccupied, unable to engage in the activity for which it was built.

Ahern asked Prodi to allow his minister, Fahey, to grant the vessel a temporary licence. The matter was, he said

...one of very serious national and regional importance...if the Minister is unable to grant the temporary fishing licence sought there is no doubt that the largest fishing company in Ireland will collapse with very serious economic consequences in the region bordering northern Ireland...[60]

The impression created by *Atlantic Dawn* on the fishing community in Killybegs when it made its first visit can only be imagined. Some sixty years before mere fifty footers (15 m) had been described as "seemingly enormous boats".[61]

The ship was duly placed on the mercantile marine, rather than the fishing register, and supplied with the first of several temporary fishing licences by minister Fahey. Thus equipped, and unable to participate in any EU fishing agreement, McHugh clinched a bilateral one with the Mauritanian government instead. Whereas the EU sponsored fleet operating in Mauritanian waters was to some extent regulated, however inadequately, vessels which gained access under private arrangements were allowed greater freedom of action.[62]

The waters off the Mauritanian coast were known to be some of the most productive in Africa and the *Atlantic Dawn* was one of a fleet of super trawlers from various European countries converging on its riches, many under the auspices of EU third country agreements. Mauritania itself was primitive and poorly equipped to deal with the invasion. Slavery was still known to exist although public slave auctions had been discontinued in 1994.[63] Government was by feeble democracy punctuated by coup and its impoverished security forces were easily bribed so that fisheries enforcement was rudimentary. The local population, heavily dependent on fishing for livelihood and food, was obliged to venture ever further offshore in its wooden pirogues to catch sufficient fish as predatory super trawlers ate into local stocks. In the final months of 2006 alone, 200 people were crushed and drowned by giant fishing boats working under cover of darkness in "protected" waters off Noudhibou in the north of the country. A new activity replaced fishing for the Mauritanians: people trafficking to the Canaries, after which the unfortunate Africans took their chances entering mainland Europe.[64]

The advent of the *Atlantic Dawn* received a luke-warm welcome from the Irish fishing industry but scientists and conservationists were dismayed and general commentary was hostile. Frank Doyle, Secretary of the IFO, observed to the press:

It is pretty clear that this vessel has attracted more adverse attention than any other single vessel, perhaps in the world.[65]

The trade press, a little self-consciously, justified the development with an editorial, the second sentence of which was blatantly untrue:

Third countries such as Mauritania depend on these super trawlers (which) have the capacity to exploit these fisheries and therefore provide much needed employment... In recent years Irish input into recovery programmes and conservation measures to redress declining stock issues have been intensive. There should be no doubt that Kevin McHugh's experience and knowledge of the fragile nature of fish species will ensure his presence in west African waters will not upset the balance...[66]

The trust of the trade press in the goodness of fishermen was endearingly innocent. In Mauritania the *Atlantic Dawn* was known as the "sea monster" and "the ship from hell".[67]

Explaining the benefits to Mauritania of the boat's presence there, McHugh's son Karl, marketing manager of Atlantic Dawn Ltd, expanded: at least 10% of the crew would be Mauritanian (i.e. ten jobs?) and there would be royalties also. Atlantic Dawn

Ltd was happy to develop Mauritania's fishing grounds.[68] McHugh himself alluded to wealth created by joint ventures with the people of Mauritania and Morocco.[69]

McHugh's Atlantic Dawn Ltd, had a turnover of €18 million in 2000 and retained profits of €17 million in that year;[70] in 2001 the company registered a turnover of €41 million, €30.5 million in 2002 (operating profits of €13 million and fixed assets worth more than €80 million).[71] In 2003, vessels belonging to Atlantic Dawn Ltd landed fish worth €36 million.

Support for McHugh from the KFO and BIM was understandably fulsome.[72] Many others wondered at the time, and since, what stimulated Ireland's diplomatic gymnastics to "save a business that employed 100 Irish people and enriched one man's fortune?" Frank Doyle of the IFO, who had worked in fisheries for thirty years, told the *Sunday Business Post* he had never known of any single vessel that had been of such interest or claimed such attention from the state.[73] The enigma was even more imponderable when the number of Irish people employed by the venture turned out to be smaller. According to *Phoenix* magazine, during its first year of operations off the coast of west Africa, around 60% of its crew were replaced with lower cost Russian labour. Since the late 1990s, McHugh had hired through a Hong Kong company, Marine Manning Services Ltd., which paid its recruits through an off-shore company.[74]

Out of sight was not out of mind and the EU Commission was not prepared to let the matter rest. At a time when the pelagic fleet was being cut, the addition of the *Atlantic Dawn* was not acceptable. Infringement proceedings, the first stage of litigation, were put in motion against Ireland in 2001.

At the same time, to conform with the terms of its bank loans (the banks included Anglo Irish, IIB bank, Ulster Bank Markets and Bank of Ireland), the lenders were insistent that *Atlantic Dawn* officially belong to the Irish fleet. The compromise reached admitted *Atlantic Dawn* onto the Irish/Community fishing vessel register but insisted that *Veronica II* had to leave.

The latter's fishing rights were transferred to the larger ship. *Veronica II* had been entitled to fish for 240 days a year in EU waters but, because it was larger, *Atlantic Dawn* would have that entitlement reduced to 95 days.[75]

One detail which remained to be clarified was the fate of the tonnage of *Veronica II*. Because the *Atlantic Dawn* was intended to operate outside EU jurisdiction, it was not required to have any, a fact which saved McHugh some €150 million: only the fleet within EU waters was regulated by this mechanism. Within the Department there was recorded debate on the issue, later uncovered by a freedom of information query. In accordance with the regulations and practice of the times, introducing a vessel to the national fleet required that equivalent notional tonnage be removed from it.[76] This was straightforward; McHugh should surrender *Veronica II*'s tonnage. A more senior Departmental opinion favoured his retaining it. Being well in with the political fisheries lobby was a gift that went on giving. The chairman of the Public Accounts Committee, John Perry, later accused the owners of *Atlantic Dawn* of having benefited from an indirect state subsidy of around €100 million[77] and the matter did not end there. Because the tonnage of the total fleet was capped and none was available for purchase, anyone who now wanted to introduce a pelagic vessel was obliged to buy notional tonnage from McHugh. The owners of twelve vessels did and the price, which moved up or down in keeping with supply and demand, appeared to substantially increase. So great was the tonnage, 5,200 t (some 6.3% of the entire fleet), that its release onto the market, although it was confined to pelagic vessels, was said to have distorted the price of demersal tonnage too. McHugh's sale to other owners netted him between €50 and 60 million.[78, 79, 80, 81]

In 2005 Kevin McHugh was reckoned to be worth €101 million and ranked 75th in the *Sunday Times* rich list, partly on foot of the tonnage transactions.

The IFO claimed that the ship's enormous capacity was sufficient to land 60% of the mackerel and herring quota allocated to the country.[82] The *Atlantic Dawn* was supposed to operate only outside EU waters. Within weeks of receiving its EU fishing licence in 2002, it was working off the west of Scotland and south west Ireland (within 60 nm of Loop Head, Co Clare) and landing into Agadir, Morocco. According to *Phoenix* magazine, *Atlantic Dawn* harvested 7,000 t of mackerel, scad and blue whiting conservatively estimated to be worth over €7 million.[83]

The mood of the Irish industry, which had been fairly sanguine about the largest fishing vessel in the world participating in a fishery which impoverished the people of Mauritania, soured. The Department, on being informed that the boat was working close to the Irish coast, predictably shrugged its shoulders and responded it was unaware of the development but, so long as the boat fished within quota, it was entitled to do so. But in the absence of tonnage and kiloWattage qualifications, was this true? Among the complainants was a PO, the IS&WFPO, and the Green Party MEP, Patricia McKenna. Frank Doyle of the IFO put it in a nutshell:

The vessel has arrived on the register in a most untransparent manner. It appears this had a substantial inner track...[84]

McHugh's bilateral agreement with the Mauritanian government allowed the *Atlantic Dawn* to fish there for nine months of the year. Following a coup in 2005, the vessel was obliged to leave. After McHugh's death in 2006, it was sold to a Dutch fishing company for €35 million.[85] In 2006 it was reportedly being equipped in Norway with the latest seismic technology to search for oil and gas.[86] Gone from EU waters at that stage, the vessel's Irish quota was passed on to others acquired by the Atlantic Dawn company. Industry sources complained that the quota should have reverted to the government for redistribution.[87] John McManus, an astute observer of the fishing industry, reviewed yet another anomaly in the way the Atlantic Dawn company was regarded by government: questioned in the Dail by Sinn Fein, the minister John Browne (Fianna Fail, Wexford), assured the house that the quota allocated to the *Atlantic Dawn* was not the property of the owners but reverted to the nation once the boat was sold. Quota for pelagic species was highly valued and very scarce but, in this case, on the contrary, it apparently passed back to the company which purchased other vessels to fish for it. McManus described it as in keeping with precedents set by the government's other dealings on behalf of *Atlantic Dawn*

which generally involved high ranking politicians and civil servants – including the Taoiseach – bending over backwards to get rules changed or ignored to further the commercial endeavours of Atlantic Dawn Ltd...[88]

Banished from its Irish base, *Veronica II* moved to the Canaries Islands and joined the Panamanian fleet register,[89] a flag of convenience whose vessels have a track record of ignoring fisheries regulations. The boat fished Mauritanian waters but, according to one report, lost its licence in September 2004 after local communities lobbied against fishing deals with foreign companies; this report was denied.[90] Australia appeared to be its next option. McHugh's intention was to capture low value pelagic species to make fishmeal for tuna farms. Pelagic species were not abundant in Australian waters and only 5,000 t, a couple of frozen cargoes for the *Veronica II*, had been caught there in 2002. In anticipation of the move, McHugh's operation recruited crews in Australia, paying far above the national rate and thereby alienating local skippers. McHugh also paid A$1.6 million for a licence and set up a local company to manage his business.[91] Australian Greens got wind of the vessel's arrival and protested. *Veronica II* was banned from the waters of New South Wales but that served to keep it only 3 nm off the coast. A spokesman for the vessel, in a typical industry response along the lines "We love fish; it is not in our interest to cause ecological problems" stated lamely:

They (the Greens) are going on about us hoovering up the ocean and this is simply not the case...

New South Wales primary industries minister Ian Macdonald confirmed local fishermen had made representations saying they feared for their livelihoods. He described the vessel, so admired in Killybegs, as "a monstrosity". *Veronica II* apparently took a similar path to *Atlantic Dawn*; it was converted to engage in seismic research.[92]

Two non-fisheries related incidents marred the operation of Atlantic Dawn Ltd. In 2002 a member of the crew died in a deck accident on board *Atlantic Dawn*. The subsequent enquiry by the Marine Casualty Investigation Board reported the ship had not been maintained in accordance with minimum safe manning requirements. Then, in 2004, Price Waterhouse Coopers resigned as auditors because "proper books of account had not been kept by Atlantic Dawn Ltd".[93]

We will never know how much damage the vessels in the Atlantic Dawn company inflicted on the marine environment and fish stocks – no more than the consequences of any fishing enterprise can be precisely quantified - but we can be absolutely certain of the indispensable support they received from the political fisheries establishment in Ireland. Not all fishermen of the nation were cherished equally.

Chapter 19

WHEN DID YOU MEET A HAPPY FISHERMAN?

In the wake of the realisation that the fleet was too large and its quarry too few, the government behaved as it generally does in such circumstances: it commissioned a report to sooth the rawness of the Sea Fisheries and Maritime Jurisdiction Act, 2006. This one, *Steering a new course: Strategy for a restructured, sustainable and profitable Irish seafood industry 2007-2013* had all the essential buzz words. It promised a changed industry, in balance with the resource and generating profit. It would also be known after the name of its chairman, Dr Noel Cawley, who told Lorna Siggins he was determined to be forward-looking and that "when I came up with daft notions I had Joey Murrin (formerly of the KFO and an earlier chairman of BIM) to tell me they were off the wall". BIM which provided secretariat "were also invaluable".[1] The Cawley report is a professionally produced document but, like many on the industry that emenate from the establishment it depended largely on the input of further subsidy to sort out a mess. Many people – I would be one of them – would prefer another approach. This one reads as though the views of the industry, which had got itself into a pickle, had found a new spokesman. The Cawley report stated the need to get rid of a substantial part of the fleet, reducing the over 18 m vessels to 55% of their former capacity. That would be accomplished by decommissioning. Cawley recognised that the fleet would require 45% greater fish resources to maintain itself. There would be lots of work for BIM, organising the industry as it had habitually done, and creating a new "seafood island" image. The bill for the exercise was reckoned at €600 million.

Had all the monies invested through BIM in fleet and industry development, throughout the history of the agency been converted to €, a crude exercise which does not take inflation into account, they might have approximated this sum.[2] Now, contrasting with the pre-industrial scenario, the nation had a massive fleet dependent on insufficient and still declining fish stocks which subsidy could only deplete further.

If, as argued in the Cawley Report, it was imperative that larger vessels reduced capacity, what options had small scale operators?

Squeezing more out of less

One day in the late nineteenth century some companions were rowing to mass from their homes on the Great Blasket Island, off the Co Kerry coast when their oars snagged a cork-bearing rope. "Draw it in" advised one "to see what is the meaning of it all". To the end of the rope was attached a wicker creel containing "a big red crayfish[3] though none of us knew what a crayfish was at the time".

The incident off the Great Blasket island was recounted by Maurice O'Sullivan[4] whose father was one of the crew who made the discovery. Enquiries revealed that some Englishmen were trapping lobsters and crawfish in the vicinity and exporting them to the London market at Billingsgate and, thenceforth, the people of the Great Blasket Island did so too. The export of shellfish to the British market was not unusual at that time. Most mussels gathered in Ireland also found buyers in the UK, as did much of the oyster landings. And yet, it was possible for isolated communities exploiting fisheries close to shore to be unaware of the largesse below their tide line. As communications improved, and global markets established themselves, demand translated more rapidly into export trade. Thus, when French and Spanish supplies of the inshore velvet crab stocks collapsed, as a result of over-exploitation and disease, during the twenty years after 1971, alternatives were quickly sourced in Scotland and Ireland. Velvet crab had not hitherto been harvested commercially although it may have been consumed locally.[5]

In the 1920s crawfish were separately identified in the statistics of capture but their number was a small proportion of lobster, which has an external resemblance to the species. In 1947, a new design of trap (the French cray pot) was launched on the Irish market and thereafter crawfish attracted greater notice. For the twenty years after 1950, the numbers of crawfish landed averaged one third those of lobster and earned half its value. It would appear to have been a sustainable rate of harvest; landings of both species displayed contemporaneous troughs of dearth and peaks of abundance, possibly reflecting only the amount of fishing effort expended on their capture rather than environmental or biological influences.

By the late 1960s, in common with what was happening in inshore waters generally, competition for crawfish intensified in their Co Kerry stronghold. Much of the damage in the vicinity of Puffin Island, Valentia and the Skellig islands was attributed to harvesting by skin diving whose prohibition in 1966 did not immediately stop the activity.[6] Crawfish are believed to be more vulnerable than lobster to collection by this method and there were accounts of large fortunes having been earned for little effort. In 1967, the Department's research vessel, *Cú na Mara*, was dispatched to locate hitherto undiscovered stocks of crawfish at greater depth, using echo sounding to identify suitable shellfish-supporting terrain.

Increasing fishing intensity in the 1970s displaced effort in all directions, facilitating the creation of new fisheries from the by-products of others. The older wooden fishing craft constructed by BIM during its early years, marginalised by more powerful modern vessels, retired to fishing passive gears, gill and tangle nets and pots. By-caught "trash" fish found a market as bait for crustaceans, particularly edible crab which had, to that time, been a poor cousin of lobster and crawfish. Gathered in bulk, brown crab became the basis of a processing industry. Even the green (shore) crab found commercial outlets.

Against this background, the crawfish stock displayed symptoms of increasing stress; scarcity enhanced its value and, in turn, heightened the efforts to capture it. In the early 1970s crawfish accounted for 3% of shellfish landings but 20% of their value. In keeping with the madness of the times, rather than conserve declining stocks, the emphasis focused on capturing ever more of them:

...A study of their habits indicates that specialised cray fishing techniques are essential to the expansion of landings...fast hydraulic power hauling so that traps could be fished as often as possible each day...[7]

There appeared to be an unholy haste to remove the last crawfish from the sea. Ironically, that advice was given by a member of the Department's inspectorate, a qualified biologist; paradoxically, the thrust of the article containing it stated that conservation measures for shellfish should be rigorously respected to avoid hardship for those whose livelihoods depended on them.

Signs that crawfish were coping with exploitation less robustly than lobsters were making themselves apparent: there were biological differences between the species which might have contributed to the explanation. One was the longer larval planktonic phase of crawfish, between six and ten times greater than that of lobster, exposing crawfish to heavier mortality in the course of its development.

While crawfish declined during the 1970s, the circumstances were ideal for the most destructive development in the history of this fishery, the introduction of tangle nets. The meshing of tangle nets is loosely rigged on head and foot ropes and they are set almost as a bundle of meshing scattered over, rather than drawn taut, across the sea bed. Few organisms that wander over a tangle net get to the other side. The meshes are large, 20.5 – 23 cm measured measured from one knot to the next. They are capable of holding a range of crawfish sizes, and other creatures too – up to the largest rays. In the year following their introduction to the waters off Kilmore Quay, the catch of crawfish had fallen by one quarter but nobody was perturbed because the value of landings had risen. By the mid-1970s tangle nets were the method of fishing for the species off the south coast. One gear seller in Glandore regularly advertised:

...The number of crayfish killed in our nets during the last years can only be described as fantastic...

By the end of the decade even the industry recognised how scarce the species had become. The National Salmon and Inshore Fishermen's Association demanded that crustacean traps be fitted with escape hatches to liberate smaller crawfish and recommended the development of "selective" tangle nets, a contradiction if ever there was one. Greater selectivity and more discriminating fishing methods have been a holy grail promised by the industry, and the scientific community, for many years but fishermen insist on practising desctructive fishing methods pending their discovery. The demand to ameliorate over-exploitation did not suggest tangle nets be abandoned. In any case, there was some technical reassurance to hand from BIM:

...The earlier fears that crawfish would be wiped out...receded...as the return from these nets after three or four consecutive years fishing become small enough to make them uneconomic...[8]

In other words, once crawfish have been wiped out, they no longer posed a dilemma. The writer of this particular piece went on to concede that, because tangle nets took very small crawfish, there should be a minimum size but that did not confront the problem either. An animal becomes so bound up in these nets that it can be almost impossible to extricate it in one piece and a large proportion of entangled crawfish would have been dead or dying after disentanglement.

Other commentators in the industry continued to voice disquiet about the use of these gears. In 1996 the Irish Lobster Association expressed the view that tangle netting had brought short-term gain but was an unsustainable method of fishing. In 2003, a contractor working for BIM conceded the nets had decimated the landings and he quoted research indicating that the average size of crawfish in Irish waters had also declined.[9] Interest in catching the species had not receded but the rate of return had fallen at a steady exponential rate of 10% per annum since 1989.[10]

Inshore fishermen were divided on the question of tangle netting but the landings data spoke for themselves. In 2002 the national harvest reached only 25% of its levels thirteen years previously. It was a problem confined to inshore waters and the Department had bigger fish to satisfy so they devised the kind of ideal solution we have come to associate with the administration of fisheries in Ireland. The Crawfish (Fisheries Management and Conservation Order) 2002 came into force on 7 May 2002.[11] Two tiny areas on the west coast, known to host heavy concentrations of the crustaceans, were designated "sanctuary centres" for them. Within their boundaries, tangle nets could still be used but only to capture species other than crawfish. Other fishing methods, such as traps, could also be used. Should a crawfish become entangled in a net it must be released.

In 2002, the first sale price of crawfish averaged €24 per kg – it has been known to exceed €50 per kg on occasion. Had I been a fishermen in 2002 and, had I tangled a crawfish in my nets, I would have put it in a bucket and reported it captured in a trap. Unlike finfish, which carry mesh marks on their outer surface when they are captured by enmeshing or tangle nets, the hard exoskeleton of a crawfish bears no tell-tale signs. However, giving any explanation for the method by which a crawfish was landed would have been unnecessary because the enforcement agencies did not trouble to inquire. The regulation was another

Irish solution to an Irish problem and there has never been a prosecution for infringing it. In the four years following the introduction of this regulation, the heaviest national landings of crawfish were recorded adjacent to the Kerry sanctuary area and local anecdote attributed them to cray nets.

Meanwhile, the use of this type of fishing gear has become a considerable problem for other fish and fishermen. Tralee Bay is a shallow, west-facing water body supporting an unusual fauna. Crawfish is abundant in its vicinity; spider crab is heavily concentrated within it; so are several species of rays, notably angel shark and undulate ray, which are very rare elsewhere. The hinterland is rural, industry sparse, and agriculture, along with fishing and tourism, its economic mainstay. A tourist angling trade is established along the coast and, prudent managagement would ensure it shares resources with the commercial fishing sector. For years attempts have been made by recreational angling and tourism interests to have the tangle nets removed, without success. According to the terms in which the regulations are expressed, it is not feasible to prove any offence is being committed by allowing them to operate. In theory a conservation regulation exists, in reality the commercial fishing community is free to make money uninhibited. The infrequent species of large rays, caught on rod and line and then released, which are the basis of the angling fishery, meanwhile, have gradually been eliminated from the Bay.

Crawfish is, in terms of price per kg, our most valuable marine species. In 2007, it earned €590,000 for the fishing community. In 1990 it earned, in €-equivalent, 3.6 million.

In 2011, the industry requested the size limit of crawfish be reduced to facilitate greater landings; a size limit for the species had been established in 2006.[12] Size limits of this kind are not infrequently imposed on inshore species although, other than for lobster, I have never known them to be enforced. Whether it was or not, a reduction in acceptable, takeable, size would be expected to make a few more available to the market – and this was the reason for the proposal.

As a general rule, fish populations decline in numbers as individuals grow in size with the passing of time. The procedure to change the law in this case was initiated by local deputy Michael Healy-Rae (Independent, south Kerry) who requested the minister to do the necessary in order to "help fishermen…in line with the rest of Europe". The amount of the requested reduction would, actually, have brought the permitted size at capture to below the regulation currently in force elsewhere and might, had it been accommodated, have made crawfish unacceptable on the European market. The inevitable consequence, if the request were granted, would be to reduce the number of breeding crawfish to even lower levels and accelerate the further decline in this valuable species.

On receiving the request from the industry, the minister, Simon Coveney (Fine Gael, Cork south central), referred the matter for technical advice; not however, to the Marine Institute which is charged with providing that service, but to BIM, the agency which had promoted the headlong rush to over-exploitation, precipitating the collapse of crawfish stocks in the first place.

The case of crawfish illustrates the options of small craft operators within the territorial sea. Economic pressures demanding urgent solutions proffer grim alternatives which boil down to how much more of the seed corn is consumed; there is no longer any aspiration to restored, productive, traditionally managed fisheries.

How the other fleet lives

In circumstances where many of the largest and most expensive vessels in the fleet have insufficient fish to support them and the smallest are reduced to coping with immediate survival, is there a future for anyone in sea fisheries today? Your prospects are probably best if you happen to be among the wealthiest with the best appointed vessels and gear, cosily cocooned by the fisheries-political establishment.

Small-scale inshore operators are confined to a limited and very overcrowded area of declining sustenance. The operators of larger vessels face a wider marine prospect. No doubt they have their problems too. Competition with other nations who have an interest in mackerel, for example, is worrisome. But, rather than squeezing the short term maximum out of shrunk inshore resources, the gaze of the operators in the largest vessels is focused on wider and more ambitious concerns.

The common fisheries policy (CFP) evolved from the Common Agriculture Policy in the years following Ireland's accession to the EEC. It was revised on two previous occasions. In April 2009, the European Commission published a green paper on further reform due to take place in 2012 .

The minister for fisheries, Tony Killeen (Fianna Fail, Clare), invited submissions. Sixteen were received from a variety of interested parties and individuals, ranging from the People's Movement to the Federation of Irish Fishermen. A steering group was established under the chairmanship of Dr Noel Cawley[13] to prepare a Response to the green paper. The steering group, I am told, was a Departmental and agencies one (Department of Agriculture Fisheries and Food (DAFF, currently Department of Agriculture Food and the Marine, DAFM), BIM and the Marine Institute.

The Response[14] was launched at a "consultation" in the Glasshouse Hotel, Sligo, on 28 January 2011; the meeting consisted of members of the European Parliament from the north west constituency, including Pat the Cope Gallagher, and a room full of participants in the industry and a large number of school children. It was jointly organized by the European Parliament Office in Ireland and, appropriately, the KFO, one of the constituent members of the FIF. The document was about immediate short-term advantage for powerful commercial interests. No more than lip service was paid to safeguarding the resource on which the industry depends, a strong indictment of the state bodies, the Department and the Marine Institute, which are entrusted with taking a longer and wider perspective. The Response is too long for a detailed analysis here but it is an important official document which sets out Ireland's fisheries aspirations and it would be remiss to ignore it altogether.

Having failed to protect and successfully manage individual commercial finfish species, we were told in the Response document that

Ireland strongly supports the establishment of area based management plans that embrace the ecosystem approach...

The ecosystem approach is a fudge. It presents woolly and imprecisely defined objectives which replace specific, unrealised and probably unattainable, targets of biomass restoration which had aimed at returning species to earlier abundance. However, implementing the ecosystem approach permits scientists to continue working with the industry which, in exchange, is not required to do anything to protect the environment and the resource.

The Response document recommended that "socio-economic" objectives might be instrumental in the way ecosystem plans operated. To reprise these briefly: should fishermen want a greater income than a fishery currently generates (and they always do), socio-economics would justify harvesting greater amounts of fish than recommended on a scientific basis, thus inflicting damage in the longer term.

The commercial sector in Ireland, dominated by the FIF, considers itself the only significant stakeholder concerned with fish stocks, although it represents no more than half of commercial fishers and none of the environmental, wildlife, tourism or recreational angling interests, all of which have a stake in the health of the marine environment and finfish populations, nor the population at large to whom the resource belongs, and those facts were abundantly clear in the Response document. We were told that Ireland supports the full participation of stakeholders in decisions and debates on policy implementation but there is no evidence of any consensus among the disparate interests involved.

The EU green paper on the reform of the CFP suggested that artisanal (inshore) and industrial fleets (those of the FIF) should be administered in different ways. The Response disputed the proposal, rejecting the more enlightened view of the Commission and reinforcing the self obsessed opinions of the more powerful operators at home. Ireland still allows the largest fishing vessels to roam all over its territorial sea, sweeping up whatever species it wishes; there will be no policy to protect smaller commercial fishers or anyone else's interests for that matter.

The Response assures its readers that Ireland supports the extension of the six and twelve nm limits to ten and twenty nm

...(to) facilitate the introduction of management measures including restriction on the type and intensity of fishing activity....

But as things stand, no such management measures have been imposed nationally within the territorial sea.

The inclusion of this proposal to enlarge fishery limits was bizarre. The dimensions of the territorial sea were fixed by international agreement under the 1958 Geneva Convention and there is no mechanism for unilaterally altering them.

As far as the industry is concerned – and the Response is the most recent position paper on its attitude – there should be no constraint on its activities and reasonable behaviour (responsible fishing) would have to be purchased, even though it is in the interest of commercial fishing to manage the resource sustainably. If the right thing is done – which would have to be voluntary – it would be rewarded with additional quota. This is the way the most prosperous fisherman view their unassailable position.

If the content of the Response were not so tragic, the document would be hilarious. Here are a few excerpts:

...The current CFP is based on a top down, detailed control model where it is assumed that fishermen will act irresponsibly unless they are scrutinised and supervised on a constant basis...

The findings of the fraud enquiries of 2004 were sufficiently distant to escape mention.

...this approach has given little incentive to the fishing industry to play a constructive role in the CFP...

Is there a solution to this?

...introduce responsible fishing schemes...including raised quotas...options for delivering defined targets (should be) left to industry...(This) would promote a culture of compliance...

It might but it is more likely to further hasten the disappearance of any remaining fish.

In this new scenario, who would administer fisheries?

...not... Community or member states but...Producer Organisations...

There you are: no further need for the Department or the Marine Institute, not that it would make much difference. And, just in case the message was not crystal clear:

...Ireland considers that producers must have a strong role in all aspects of fisheries management....POs must be given a strong role in delivering improved, environmentally friendly fishing methods and the delivery of recovery and long-term management plans...

The Response loftily tells us

...Policy decisions must be based on robust and sound scientific advice...

But what is the reaction of the industry to such advice as does not suit it? When the FIF troops out to Brussels in December to urge the minister to seek higher TACs in defiance of scientific advice, can its actions be regarded as responsible? Ireland recognises there are problems defining "good health" (my term) in "data-poor" fish stocks and the Response recommends these should be resolved

...in an open and transparent way...

The precautionary approach, a protocol of the CFP, simply states that if you do not know the status of a stock you cannot assume it is being sustainably exploited.

It is an ambition of the CFP that fish stocks should be returned to maximum sustainable yield by 2015 - a feeble target, considerably less ambitious than, for example, cod recovery plans - which is looking increasingly improbable, and the attitudes underlying Ireland's response are not going to help it much either. The ambition was first formulated in the UN conference in Rio de Janeiro in 1992 and restated in 2002 as the Johannesburg Convention. The Response is in agreement with the target which, however, according to the Response, should be introduced on a

...phased basis...beginning in 2010...

There is nothing in the document to laugh about but one detail coaxed a smile. Industrial fishermen are so confident of their position that the Response could make ludicrous statements, oblivious to their irony or contradictions. The document complained:

...Ireland believes that existing CMO (Common Organisation of the Markets in Fishery and Aquaculture products) arrangements

have the direct impact of forcing down prices for certain products...

Indeed they have. A purpose of setting up POs was to achieve lower prices for consumers.

...(This has the effect of)...forcing fishermen to increase levels of landings...

Would increasing the volume of landings not have had the effect of lowering prices?

And there is much, much more. If this Response did not exist it would be impossible to make it up. If nothing else, it is an excellent illustration of the way public policy can be hijacked by powerful commercial interests to the detriment of the resource and society in general. It is about time the industry was reminded that the majority, if not all, of the first sale price of landings obtained by the industry is contributed in subsidy by the Irish and European taxpayer, in spite of which there is no advocate for the public interest among state agencies and the civil service whose briefs confer that responsibility.

While the POs currently enjoy an unassailable dominance of the industry, there is a realisation that it is inappropriate. Minister of state Tony Killeen said of the review of the CFP intended for 2012:

I will not support a policy which promotes the concentration of activity and benefits in the hands of a small number of large companies...[15]

His reservations apparently went out of office with the man.

The price of fishing

A communication from the European Commission reviewed the prospects for the fishing industry throughout the Union on the basis of data up to 2005.[16] The communication acknowledged that depleted fish stocks had provoked the introduction of restrictive management measures and these had increased operational costs and reduced income levels. Operational costs increase when more fuel is expended dragging larger, heavier fishing gears for more prolonged periods. The Commission recognised that the problems had affected all sectors of European fleets but were most acute for vessels towing nets and dredges, which require most fuel to harvest demersal stocks.

Ireland's fisheries were developed through a combination of private effort and taxpayers' support and Ireland's story is not unique. Fisheries are, in theory, public resources so that some degree of government involvement in their administration is inevitable and essential. Worldwide, the industry has been adept at persuading governments to subsidise its activities and governments have never been able to make up their minds whether fisheries, in common with other primary sector activities like agriculture, should be managed as an industry or a social service. Membership of the EEC and its subsequent reincarnations escalated demand for a variety of financial supports and, according to rules of membership, all nations had to be treated equally.

Public investment in fisheries is considerable although few people realise how much money is expended on them. In an article in 2006, the Green party deputy Eamon Ryan observed that in 2004 the entire value of the catch was worth €200 million while the cost of fishery protection services alone was half of that. Add to that the amount invested by government in fleet development over the previous five years, plus the sum allocated to decommissioning, and the value of landings in 2004 was effectively cancelled.[17] But that was only part of the story.

Worldwide, subsidisation of fisheries followed a broadly similar path. In the 1930s and 1940s, financial investment was intended to be seed capital to an "infant sector". But subsidies have since mushroomed into an issue of contention, creating problems of over-capacity which ultimately provoked their discontinuance.

Three categories of subsidies are currently recognised, handily and memorably grouped under "good", "bad" and "ugly" labels, with reference to their impact on the fish and marine environment; the industry is dependent on both for its productivity.[18] Good subsidies enhance careful husbandry. They include research into and rational management of fisheries. Bad subsidies, the category absorbing most expenditure, embraces fleet construction, renewal and modernisation, fishery development and services supporting that objective, fishing port construction and renovation, marketing support, processing and fish storage programmes, tax exemption schemes, including fuel subsidies and foreign access agreements. In Ireland's case, the vast majority of fisheries-associated expenditure was invested in this category. The third group, ugly subsidies, can have either good or bad outcomes, depending on how they are operated. Decommissioning schemes are a typical example. Where they lead to permanent removal of fishing capacity they benefit the resource. In circumstances where they enhance the value of vessels prior to their removal and thus release more capital for reinvestment on other craft, they are very problematical.

In 2003 the amount invested in fisheries subsidies (all, excluding fuel) came to €94 m on landings valued at €213 m (44%). In other words, two fifths of the value of the first sale price of landings was contributed to the industry by the Irish and European taxpayer.[19]

The fuel element of modern fishing operations is highly significant. In common with agriculture, fisheries are entitled to avail of subsidised diesel in their work. Attempts to calculate its contribution to subsidy in 2003[20] were based on the assumption that 80% of all landings were captured by towed gears. Approximately 611 litres of diesel were burned to capture one t of marine produce and the contribution of subsidy to each t worked out at €226. When the fuel subsidy is added to the others, the total value of subsidy climbed to 75% of the first sale value of landings!

Bad subsidies have a number of consequences. Most obviously, subsidised operations allow fisheries to work uneconomically. In the extreme, much or all of the profit generated in an operation of this kind may reside in the subsidy. Prolonged fishing, sieving more water, intensifies discarding and habitat destruction. The typical sequence of events begins with the removal of, for example, whitefish such as cod in the Irish Sea. Populations of the fodder species on which they live, like *Nephrops*, then proliferate for a time until they suffer the same fate. The 2011 TAC negotiations in Brussels were confronted with scientific advice that *Nephrops* required conservation measures: curtailment of landings. In the event, the industry had its way and the advice was, as so frequently happens, brushed aside. There is no shortage of evidence for the decline of predatory marine fish species as a result of over-fishing during the past century. It is relevant to ask: what fisheries remain when the invertebrate fodder species go the same way as their erstwhile predators?

The irony of subsidies is that the contributor, the taxpayer, has no effective voice in the way they are spent nor control on the fishing effort which is destroying the marine environment in which the taxpayer has an interest and should have strong representation as a stakeholder. Few members of the general public would approve of a policy which has had such disastrous environmental consequences but the technicalities of the industry, combined with the facts of fishing being a hard life and the sympathetic regard for Irish fishermen being put upon by competition with other European nationals, combine to befuddle the issue. The Irish fleet takes no responsibility for stock depletion and is in denial about illegal landings, another supposedly foreign practice.

Solutions to the economic difficulties identified by the industry are likely to worsen the fishing environment in the longer term. High fuel prices featured prominently on its list of problems when the French premier, Nicolas Sarkozy, visited Ireland in 2008. Financial assistance for tying up vessels in port was demanded, along with the perennial plea for larger quota allocations and decriminalisation of "less serious" fisheries offences, and an end to the forced discarding of over-quota marketable fish. Some 400 skippers from "every port on the compass" gathered in Athlone to threaten demonstrations and port blockades. Their armory of threats involvded voting against the adoption of the Lisbon Treaty, in support of their demands.[21]

The reckoning

One of the most dramatic demonstrations of the extent of loss as a result of fishing was a practical experiment described by George Englehard.[22]

An antique sailing trawler was commissioned to embark from the port of Lowestoft, crewed by scientists from the Centre for Environment, Fisheries and Aquaculture Science (CEFAS), the government agency which deals with marine fisheries in the UK. The vessel was equipped with a beam trawl, a replica of those used in the 1880s. The crew fished it for a week and caught nothing, but the experiment was deemed an eloquent indication of the decline in fish abundance over the intervening period. In their time, such vessels, thus equipped, were competent to earn a living for their crews and owners.

In the UK, fisheries have been documented in greater detail than in Ireland and it is feasible to quantify some of the changes in productivity by some of the vessel designs which made most of the landings over certain periods. For example, motorised otter trawlers are estimated to have had fifty times the fishing power for cod than a century before while modern twin beam trawlers were 100 times more effective at catching plaice.

In recent years, many scientists have been preoccupied by the decline in abundance of finfish as a result of over-fishing. A sophisticated enquiry, published in 2008, standardised fishing effort over 118 years and suggested that the biomass of demersal fish species in the waters around England and Wales declined in this period by 94%.[23] It might, were sufficient diligence devoted to it, be possible to construct a similar history for whitefish species in the waters around Ireland. Less detailed available information allows us to do so over a shorter time scale.[24]

From the beginning, two elements vied for prominence in the formulation of fishing policy: the traditional, more conservative and initially far larger one was made up of boats using static gears but the more influential and productive one towed mobile nets and dredges. Trawling was separately monitored and quantitatively described annually in government reports. In 1900, for example, the fleet in Ireland (north and south) consisted of 6,341 vessels, crewed by more than 26,000 men and boys. We know nothing of what full-time job equivalents these figures represented but much of the employment is likely to have been seasonal and temporary. Trawling was undertaken by 500 boats whose activities were reported in some detail. This method of fishing was initially carried out by sailing vessels but, by 1900 there were more than 1,000 wooden steam-driven trawlers in Britain. A census in 1907 placed 60% of western Europe's steam-propelled fleet in the UK.[25]

In Ireland, steam trawlers were operated by the Dublin Trawling Ice and Cold Storage Co. Ltd, based at Ringsend, close to the Dublin fish market, the only organised one in the country. In comparison with Great Britain, the Irish fleet was very small, averaging about ten vessels. When the Dublin market was over-supplied they landed into Milford Haven, in Wales, instead, so that what we know of their landings to Ireland is probably an incomplete account of their productivity. For purposes of calculating the amount of effort invested in their landings, I estimated, generously, that each steam trawler averaging at 100 GT and propelled by 80 kW landed some 5 t of plaice, 48 t of cod and 6 t of haddock annually between 1921 and 1939. Put these landings alongside those of the current estimated towing effort of 141,000 kW in the year 2009 and dramatic reductions in abundance during the interim emerge: 97% for plaice, 99%, cod and 58%, haddock.

Over the shorter term, the diminution in catches is hardly discernible; over the longer term it is stark. And the decline is not confined to whitefish species. Landings of mackerel, Ireland's single most financially important species peaked in 1981 and have tended downwards since. Twenty years after their peak, landings had fallen by 50%. In that case, the northern displacement of the mackerel shoals was part of the explanation.[26]

In addition to data generated within the commercial industry, there are other indicators of fish decline. The recollections of anglers have not always been attributed high credibility but when the details of their captures are carefully recorded, they present us with a cogent and corroborative narrative.

The Irish Specimen Fish Committee (ISFC) was founded in 1955 to record the incidence of rod-caught fish above threshold weights. The IFSC database is one of the longer enduring time series of data on marine fishes in the history of the state. It operates on an "island" rather than "republic" basis. Its contents illuminate the history of species from a variety of perspectives. It is an eloquent advertising window for recreational tourist anglers, to which end it is supported by the tourist industry which makes use of its annual reports to attract angling business to the country. The information it contains has, for example, revealed changes associated with warming of the marine environment and the immigration of species, formerly rare, from waters to the south of Ireland. And it tells us much of changes within our fish populations and of the behaviour of those who angle for them.

The ISFC's procedure is to publish a threshold weight (known as a "specimen" weight) for a species; anglers authenticate captures which attain this and their achievements are recognised and published.[27] If the threshold weight is set too low, the administrative committee is inundated with specimen claims; if too high, claims may be few or, in the extreme, non-existent. In order to sharpen interest in catching these fish, threshold weights are reviewed annually and adjusted to ensure that specimens are not too common. All but one marine recreational species are shared with the commercial industry, the exception being sea bass which is reserved for tourist and angling interests. One of the effects of commercial fishing is to reduce the average size of the fish and invertebrates it targets. Adjustment of specimen weights in recent years has, in keeping with this generalisation, been downwards.

Enthusiasm for specimen fish angling is high; some anglers specialise in catching them. News of locations containing large numbers of specimen fish spreads rapidly among the angling community. When, in 2006, hitherto undiscovered wrecks which had somehow escaped BIM's magnetometer, yielded high numbers of large saithe, that species provided one third of all specimens recorded. This phenomenon is also expressed in the concentrations of mixed species of qualifying weights at certain locations which, for whatever reason, have escaped exploitation by the commercial sector. In 2008, for example, Red Bay, in Co Antrim, provided more than one third of all specimens. The existence of isolated "islands" of diversity of this kind exemplifies the vulnerability of Ireland's marine fishes.[28]

Decline in fish abundance over a single lifetime has been dramatic. Inter-generationally, the sharpness is accentuated. I can remember the enormous local quantities of cod in the Irish Sea during my youth but my father would have had memories of even greater abundance in his time. The fisheries biologist Daniel Pauly labelled the phenomenon *shifting baseline syndrome*:

*...successive generations of naturalists, ecologists or ..... nature lovers use the state of the environment at the beginning of their conscious interactions with it as **the** reference point...*

with which they will compare its condition in their later lives.[29]

The baselines against which succeeding generations observe the fall in abundance, shift down, so that change over the longer term is less disturbing, becomes more acceptable. Hence, the alarm which should greet a review of change over the past century is less shrill than it should be.

It is appropriate to leave the closing commentary to those who themselves worked at the business of catching fish. There is no shortage in the trade press of recollections of earlier fishery abundance and subsequent loss. In his book on the first vessels in the modern fleet, the fifty footers, Pat Nolan assembled a number of them.[30] Allowing that good memories might be more persistent survivors of the past, things appeared to be much happier then, with stronger camaraderie throughout the catching sector. We know, of course, because it was recorded at the time, of the frustration and impatience that were the growing pains of an emerging industry. For the working crew, insufficient consumer demand and limited markets for landings was a large element of their discontent.

One contributor, speaking of the mid-1950s, remembered:

...There were plenty of whiting, haddock, plaice, cod and hake to be caught around Donegal Bay and Tory Island (Co Donegal)...I knew the fish (to cover the purchase of a new vessel) were out there...

Other recollections follow:

...From a base where a thriving fishing industry existed thirty years ago, today there is scarcely any fish landed at Burtonport (Co Donegal)...Only crabs are now landed, no white fish. Only for the Arranmore ferry there would be nothing at all happening on the pier...

...The mid-water men took over and put us out of business. They came in too close and cleaned the herring out...There should have been at least a five mile limit on the mid-water boats. Nobody cared. Who was in control?

...the 50-footers and other smaller boats would still be working profitably had realistic control limits been imposed on the mid water operations...

...It's sad to see it all gone. We had the best industry in the world if it had been properly looked after...

White fishing is now unbelievably bad, you could go ten miles off where the water is sixty to seventy fathoms deep and still there is nothing...When I spoke to him he was preparing to go crawfish fishing, using tangle nets, north-west of Inishturk (Co Mayo). When I asked if there were many others fishing out there, he replied "the whole country is there".

...we had a good thing going and it was allowed to slip away through lack of control and general mismanagement...

...Fish were so plentiful that one day, "ringing" with ropes, the Ros Beag *caught seventy boxes of plaice fishing off Kells in Dingle Bay (Co Kerry). There isn't a fish there now...*

...there was plenty of fish to be got but nobody wanted it. Prices and demand were terrible...they got worse as the 1960s progressed...because there was no market for the fish we caught, the likes of haddock, cod and whiting, we filled them into bags and sold them for cat food at seven shillings a bag. Then there were times when fish was just dumped...

...the one major problem that bedevilled the industry – poor returns for fish landed...There were times during the 1960s when the agents would come down to the pier and tell us not to go fishing as there was no market for them; the fish couldn't be sold...

...loads of fine whiting and haddock, for which the fishermen were paid next to nothing, being carted up the road to the fishmeal factory at Killybegs (Co Donegal)...

There were 36 boats, including half deckers and 50-footers fishing out of Roundstone (Co Galway). You'll never see so good again. Now there are no herring, mackerel, salmon, crawfish, whitefish or any other kinds of fish caught, whereas you could fill a boat less than 30 years ago...

...I remember trawling in a half-decker. We used a small trawl and hand hauled the doors. We had no problem landing ten boxes of ray wings per day. The skate were so plentiful we could see them on the bottom below the boat as we fished. A situation was allowed to develop that slowly but surely forced inshore fishing out of business...

And it is not as though premonitions of disaster were not available. When the Bjuke report, which designated certain harbours specifically for fishery development, was published in 1959, the trade press greeted it as:

...based entirely on a misconception...that there exists in Irish waters stocks of white fish that can be exploited economically by middle water vessels...The fishing community has no faith in the kind of development outlined by the minister...(He) is going to build big boats and big harbours. Then he will come back to the Dail with the second verse of the lament. We cannot get any fish.[31]

But the imperative to prosper, supplemented with subsidies and liberally conditioned with greed and self delusion, did the rest. The balance between potential and demand was always out of kilter; either the market was too small to absorb what was caught or the supply of fish was insufficient to satisfy demand. Each end of this spectrum exerted its own pressures on operators. Where, within it, could you locate a happy fisherman?

References

Additional references to legislation and newspapers are given in the chapter notes.

Aldecoa, I. (1958) *Gran Sol* Santillana Ediciones Generales S.L., Madrid Pp 304

Anon (1889-2008) Annual reports of the government department responsible for fisheries

Anon (1820) Fishery *Dublin Journal*, 25 September 1820

Anon (1958) *Programme for economic expansion.* Dublin, Stationery Office. Pp 50.

Anon (1962) *Programme for sea fisheries development.* White paper. Dublin, Stationery Office. Pp 14

Anon (1963) *Second programme for economic expansion.* Dublin, Stationery office, Pp 340

Anon (1964) *Second programme for economic expansion, progress report.* Dublin, Stationery Office, Pp 122

Anon (1965) *Second programme for economic expansion, progress report.* Dublin, Stationery Office, Pp 124

Anon (1969) *Third programme for economic and social development, 1969 – 1972.* Dublin, Stationery office. Pp 275

Anon (1999) *Irish Inshore Fisheries Sector – Review and recommendations.* Dun Laoghaire, Bord Iascaigh Mhara, Pp 74

Anon (2009) *Report of special group on public service numbers and expenditure,* Volume 2, detailed papers Pp. 221

Anon (2011) *Ireland's response to the Commission's green paper on the reform of the common fisheries policy.* Dublin, Department of Agriculture Fisheries and Food. Pp 40

Armstrong, M.J., D. Smyth and W. McGurdy (1991) How much *Nephrops* is eaten by cod in the western Irish Sea? *International Council for the Exploration of the Sea* CM 1991/G:15

Bates, R. (1983) Shellfish have greater earnings potential. *The Irish Skipper* June 1983: 3

Bjuke, C.G. (1959) *Report on the project of improvement of harbour facilities in Ireland.* Dublin, Stationery office. Pp 59.

Blake, J.A. (1868) *The history and position of the sea fisheries of Ireland and how they may be made to afford increased food and employment.* Waterford, J.H. McGrath Pp, 133

Board of Agriculture and Fisheries (1908) *Return of the number of steam trawlers registered at ports in the states of Western Europe in the year 1907.* London, HMSO. Pp 2

Bord Isacaigh Mhara (BIM) (1952 – 2010) *Annual report and statement of accounts.* Dublin.

Bord Iascaigh Mhara (1999) *BIM Seafood Industry Agenda 2000-2006*: Realising the regional potential of the Irish seafood industry. Dublin. Pp: 39.

Boyer, D. C. and H. J. Boyer (2004) Sustainable utilisation of fish stocks – is this achievable? a case study from Namibia: discussion paper 9 in *Overcoming factors of unsustainability and overexploitation in fisheries: selected papers on issues and approaches.* Pp 18, FAO, Rome

Brabazon, W. (1848) *The deep sea and coast fisheries of Ireland with suggestions for the working of a fishing company.* Dublin, James McGlashan.

Bratton, S. and S.M. Hinz (2002) Ethical responses to commercial fisheries decline in the Republic of Ireland. *Ethics and the environment* 7 (1): 54-91

Brunicardi, D. (1966) Helga – the research ship that has a place in history. *The Irish Skipper*, September 1986: 18

Butt, I. (1871) *Resolution: That the continual refusal to comply with the strong and repeated recommendations of committees of the house of commons.....* At a meeting of the home government association of Ireland, held in their rooms, No 212, Great Brunswick Street on 17 October 1871. Pamphlet.

Caird, J., T.H. Huxley and G.S. Lefevre (1866) Report of the Commissioners appointed to enquire into the sea fisheries of the United Kingdom. London, HMSO, 1: Pp 108.

Carroll, M. (1992) *The second Spanish Armada*, Bantry: Private publication, Pp 80.

Cawley, N. (chairman) (2006) *Steering a new course: Strategy for a restructured, sustainable and profitable Irish seafood industry 2007-2013*. Report of the seafood industry strategy review group. Dublin, BIM, Pp196.

Cawley, N. (Chairman) (2010) *Ireland's response to Commission's the green paper on the reform of the common fisheries policy.* Department of Agriculture, Fisheries and Food. Pp 40.

Commissioners of Fisheries, Ireland (1843) First Annual Report, 1 May 1843

Crummey, C. (1997) *The physical interactions between grey seals and fishing gear*. Draft final report. Ref: PEM/93/06. Mimeo, Pp 59 + Figs.

Dalhousie, E. Majoribanks, W. Sproston Caine and T.F. Brady (1884-85) Report of the Commissioners appointed to enquire and report upon the complaints that have been made by line and drift net fishermen of injuries sustained by them in their calling owing to the use of the trawl net and beam trawl, in the territorial waters of the United Kingdom; with minutes of evidence and appendix. Pp 564

David, E. (1994) *Harvest of the cold months: the social history of ice and ices*. London Pp, 440.

de Callao, C. (1997) Organising around the parish pump. *The Irish Skipper*, February 1997: 28

de Courcy Ireland, J. (1981) *Irish sea fisheries: a history*. Dublin, Glendale Press. Pp 183.

de Courcy Ireland, J. (1986) *Ireland and the Irish in maritime history*. Dublin, Glendale Press. Pp 489.

Department of Agriculture Food and the Marine (2011) *Value for money review: Fisheries decommissioning schemes* 2005-2008. Pp. 102

Dickens, C. (1914) *Oliver Twist*. Collins pocket classics No 33. Glasgow, Collins Clear-type Press. Pp 528

Dinneen, P.S. (1904) *An Irish-English dictionary*. Dublin. M.H. Gill & Son. Pp 803

Edwards, E. (1979) *The edible crab and its fishery in British waters*. Farnham, Fishing News Books Pp.142

Ekin, D. (2008) *The Stolen Village*. Sparkford, O'Brien Press, Pp. 477

Englehard, G.H. (2008) One hundred and twenty years of change in fishing power of English North Sea trawlers. In: *Advances in fisheries science 50 years after Beverton and Holt* (eds. A. Payne, J. Cotter and T. Potter) Oxford, Blackwell: 1-25

European Commission (2006) Communication from the Commission to the Council and the European Parliament on improving the economic situation in the fishing industry. COM (2006) 103 final. Pp17

European Commission (2010) Consultation on fishing opportunities for 2011 COM (2010) 241 final, Pp 19.

European Communities (2009) *The Common Fisheries Policy, a user's guide* Belgium, Pp 40.

Fahy, E. (2008)a Happy silver anniversary? *Marine Times*, September 2008: 2

Fahy, E. (2008)b Performance of an inshore fishery in the absence of regulatory enforcement. *Marine Policy* 32 (6): 1037-1042

Fahy, E. (2008)c Gordon Ramsay's sea urchin stunt again exposes a law unfit for purpose. *Marine Times*, July 2008: 12

Fahy, E. (2009)a Monkfish: an icon of our times *The Irish Skipper*, November 2009: 38

Fahy, E. (2009)b Failure to reverse the collapse of the Irish Sea cod stock raises serious question about the way the industry is managed. *Marine Times*, September: 30

Fahy, E. (2009)c What the "Red Bay phenomenon" might tell us about the status of fish stocks. *Irish Angler's Digest*, 11(2): 60-61

Fahy, E. (2010)a Does Palourde hold the record for brief life among Irish boom-to-bust fisheries? *Marine Times* June 2006: 2

Fahy, E. (2010)b A tale of three trawlers. *Marine Times,* September 2010: 2

Fahy, E. (2010)c What a difference a century makes. *Marine Times,* October 2010: 2

Fahy, E. (2010)d The fate of the north west crab fishery indicts both management and research. *The Irish Skipper,* November 2010: 30

Fahy, E. (2011)a Expanding horizons. *Marine Times* July 2011: 14

Fahy, E. (2011)b Eaten fish is soon forgotten. *Marine Times* February 2011: 2

Fahy, E. (2011)c Of seals and men. *Marine Times* June 2011: 2

Fahy, E. (2011)d Blooming jellies! *Marine Times,* April 2011: 2

Fahy, E. (2011)e *Déjà vu* all over again? *Marine Times,* August 2011: 14

Fahy, E. (2011)f A tax mechanism to encourage compliance with marine fisheries management regulations. Report commissioned by *Smart Taxes* – a development network led by *Feasta, the Foundation for the Economics of Stability*. Dublin. Pp 51.

Fahy, E. and J. Carroll (2009) Vulnerability of male spider crab *Maja brachydactyla* (Brachyura: Majidae) to a pot fishery in south-west

Ireland. *Journal of the Marine Biological Association of the United Kingdom* 89 (7): 1353-1366

Fahy, E., J. Carroll and S. Clarke (2008) Prudent management essential to sustain a velvet crab fishery. *Marine Times* May 2008: 16

Fahy, E., Carroll, J. and D. Stokes (2002) The inshore pot fishery for brown crab (*Cancer pagurus*), landing into south east Ireland: estimate of yield and assessment of status. *Irish Fisheries Investigations*. No 11. Pp. 26.

Fahy, E., J. Carroll, M. O'Toole, C. Barry, L. Hother-Parkes. (2005) Fishery associated changes in the whelk *Buccinum undaturm* stock in the south west Irish Sea 1995-2003. *Irish Fisheries Investigations*, 15: Pp 26.

Fahy, E., J. Carroll, J. Rafferty, V. Roantree, C. Reid, M. Norman and S. Clarke (2010) Observations on the local distribution, biology and ecology of palourde *Tapes decussatus* (L) of relevance to its exploitation in the Republic of Ireland. *Biology and Environment: Proceedings of the Royal Irish Academy* 110B (2): 95-108.

Fahy, E., D. Fee, S. O'Connor and T. Smith (2007) Activity patterns of some inshore fishing vessels in 2006-2007. *Irish Fisheries Bulletin*, 29: Pp 8.

Fahy, E., P. Green, L. Borges, A. Power and E. McGuinness (2009) Sequel of the directed fishery for spiny dogfish (spurdog) in the 1980s off the west coast of Ireland. In, *Biology and management of dogfish sharks*: eds, V.F. Gallucci, G.A. McFarlane and G.C. Bargmann, American Fisheries Society: 391-399

Fahy, E., E. Healy, S. Downes, T. Alcorn and E. Nixon (2008) An atlas of fishing and some related activities in Ireland's territorial sea and internal marine waters with observations concerning their spatial planning. *Irish Fisheries Investigations* (19) Pp 33.

Fitzgerald, J., I. Kearney, E. Morgenroth and D. Smyth (1999) *National investment priorities for the period 2000-2006*. Policy research series 33. Dublin. Economic and Social Research Institute. Pp 318

Fisheries Consolidation Act (1959) no 14 of 1959

Foley, N., T. van Rensburg and C.W. Armstrong (2010) *The rise and fall of the Irish orange roughy fishery: an economic analysis.* Working paper no. 156. Department of Economics, National University of Galway. Pp 24.

Frangoudes, K. (2001) France. Chapter 8 in *Inshore Fisheries Management* eds. D. Symes and J. Phillipson. Dordrecht, Kluwer Academic Publishers Pp: 318

George III (1773-1774) 13&14 Geo 3 c. 41. An Act for renewing and continuing several temporary statutes and to prevent the destructive practice of trawling for fish in the Bay of Dublin.

George III (1819) 59 G 4 c. 109. An Act for the further encouragement and improvement of the Irish fisheries.

George IV (1821) 1 Geo IV C.82. An Act for the further encouragement and improvement of the Irish fisheries.

George IV (1825) 5 Geo IV C. 64 & C.67. An Act for the further encouragement and improvement of the Irish fisheries.

Glude, J.B., R.J. Lavell, J.W. Slavin and K.A. Smith (1964) *Recommendations for the improvement of the sea fisheries of Ireland by American survey team.* Dublin, Stationery office Pp 47

Grescoe, T. (2008) *Bottomfeeder: How the fish on our plates is killing our planet* London, MacMillan Pp, 389

Hall, S., and C. Hall (1841-1843) Ireland, its scenery, character etc. Vol 3. London, Hall, Virtue and Co.

Hardy, A. (1959) *The open sea: II Fish and Fisheries.* London, Collins. Pp 322

Healey, E. (1978) *Lady unknown, the life of Angela Burdett-Coutts.* New York, Coward, McCann and Geoghegan, Pp 253.

Hérubel, M.A. (1912) *Sea fisheries: their treasures and toilers.* London, T. Fisher Unwin. Pp 380

Higgins, M.D. (2006) *Causes for concern:Irish politics, culture and society.* Dublin, Liberties. Pp 340

Higgins, W. (1957) Baltimore had a fishery school seventy years ago. *Cork Examiner* (details unknown) reprinted in the *Irish Fishing and Fish Trades Gazette* 5 (18): 31 August 1957.

Hillis, J.P. (1971) The Whiting Fishery Off Counties Dublin and Louth On the East Coast of Ireland: Research Vessel Investigations, *Irish Fisheries Investigations* Series B. Pp 54

Holden, M. and D. Garrod (1994) *The common fisheries policy.* Oxford, Fishing News Books. Pp 288.

Holt, E.W.L. (1909) Report of a survey of trawling grounds on the coasts of counties Down, Louth, Meath and Dublin. *Department of Agriculture and Technical Instruction, Fisheries Branch, Scientific Investigations* (1): Pp 538, 2 plates. Dublin. H.M.S.O.

Holt, S.J. (1956) How exploration and research help the fisheries. *Irish Fishing and Fish Trades Gazette,* 4 (18) September 1956.

Huxley, T.H. (1884) *Inaugural address, International Fisheries Exhibition, 1883, Literature* 4: 1-19

Irish Fishing and Fish Trades Gazette (1954 – 1962) Discontinuous series. After issue VII (9) in 1959, the journal changed its title to *Irish Fishing and Fishing Trades Gazette.*

Irish Sea Fisheries Association – Comlachas Iascaigh Mhara nah-Eireann (1930-1952) Reports of its ordinary annual meetings.

Irish Specimen Fish Committee (ISFC) annual reports, 1955-2011

Johnson, T., M. Moore, P. Moylett, J. Murphy, A.J. Nicolls, J.P. O'Shea, J. Sweetman and R.N. Tweedy (1921) *Report on Sea Fisheries. Commission of Inquiry into the Resources and Industries of Ireland.* Dublin, American Chambers, Pp 104.

Khan, A.S., U.R. Sumaila, R. Watson, G. Munro and D. Pauly (2006) The nature and magnitude of global, non-fuel fisheries subsidies. Chapter 1 in *Catching more bait: a bottom-up re-estimation of global fisheries subsidies.* U.R. Sumaila and D. Pauly, eds. Fisheries Centre Research Reports 14 (6): 5-37

Kurlansky, M. (1998) *Cod, a biography of the the fish that changed the world*. London, Jonathan Cape, Pp 294.

Lawlor, A.T. (1945) *Irish Maritime Survey*. Dublin. Parkside Press, Pp 371.

Lee, T. (1978) Trawling survey in south-west, BIM issues report. *The Irish Skipper*, November 1978: 7

Lei, B. V. (2006) The impact of the increase of the oil price in European Fisheries. Directorate General Internal Policies of the Union. Policy Department Structural and Cohesion Policies, Fisheries. IP/B/PECH/2005 – 142

Lewis, M. (1989) *Liar's Poker*. London, Hodder and Stoughton, Pp 298

Lynam, C.P., M.K.S. Lilley, T. Bastian, T.K. Doyle, S.E. Baggs and G.C. Hays (2011) Have jellyfish in the Irish Sea benefited from climate change and overfishing? *Global change biology* 17: 767-782

Mac Laughlin, J. (2010) *Troubled Waters, a social and cultural history of Ireland's sea fisheries*. Four Courts Press, Dublin. Pp 397.

Marine Institute (1998-2011) *Stock book*(s) Galway, Marine Institute, Fisheries Science Services Division.

Maritime Jurisdiction Act 1959, No 22 of 1959

McArthur, J.S. (1959) Report to the government of Ireland on the development of the sea fishing industries. Rome, FAO, mimeo, Pp 63

McGuire, J. and J. Quinn (2009) *Dictionery of Irish biography to the year 2002*. Cambridge University Press.

Meany, R.A. (1980) BIM to continue fisheries development programme, main aim remains to profitably diversify efforts of the fishing fleet. *The Irish Skipper*, March 1980: 11

Meredith, D. (1998) *Geographical impacts of the common fisheries policy: a review of Ireland's fishing industry 1983-1997*. Thesis, M. Litt. (Research), NUI Dublin (University College Dublin), Pp 161.

Molloy, J. (2004) *The Irish fishery and the making of an industry.* Killarney, Killybegs Fishermen's Organisation and the Marine Institute, Pp, 245.

Molloy, J. (2006) *The herring fisheries of Ireland (1900-2005): Biology, research, development and assessment.* Galway, Marine Institute, Pp 235

Muus, B.J. and P. Dahlström (1994) *Guía de los peces de mar del Atlántico y del Mediterráne.* Barcelona, Omega, Pp, 259

Nolan, C. P. (2004) A technical and scientific record of experimental fishing for deepwater species in the north east Atlantic, by Irish fishing vessels in 2001. *Fisheries Resource Series.* Dun Laoghaire, 535 Mhara, Pp 189.

Nolan, P. (2008) *Sea change – the rise of the BIM 50 footer and its impact on coastal Ireland.* Dublin, Nonsuch, Pp 192

Orwell, G. (1937) *The Road to Wigan Pier.* Penguin Books, Pp 191

O'Connor, R., J.A. Crutchfield, B.J. Whelan and K.E. Mellon (1980) *Development of the Irish sea fishing industry and its regional implications.* Dublin, Economic and Social Research Institute (ESRI) Paper No. 100, Pp 386

O'Donnell, P. (1927) *Islanders.* Cork, Mercier Press. Pp 126

O'Riordan, C.E. (1964) *Nephrops norvegicus* the Dublin Bay prawn, in Irish waters. *Scientific Proceedings of the Royal Dublin Society* series B, 3: 131-157

O'Sullivan, M. (1983) *Twenty years a growing.* Oxford, OUP, Pp 298.

Partridge, J.K. (1977) Studies on *Tapes decussatus* (L) in Ireland – natural populations, artificial propagation and mariculture potential, with an investigation of the major European fisheries and a full annotated bibliography. Ph.D. thesis, National University of Ireland. 2 vols.

Pauly, D. (1995) Anecdotes and the shifting base-line of fisheries. *Trends in Ecology and Evolution* 10(10): 430

Pérez-Rubín, J. (2006) Los primeros acuerdos pesqueros con Marruecos (siglos XVI-XVIII). *Mar*: 452: 20-27.

Purcell, J.E. and M. N. Arai. (2001) Interactions of pelagic cnidarians and ctenophores with fish: a review. *Hydrobiologia* 451: 27-44

Roberts, C.M. (2007) *The unnatural history of the sea. The past and future of humanity and fishing*. London, Gaia. Pp. 448.

Rogan, E. and S.D. Berrow (1995) The management of Irish waters as a whale and dolphin sanctuary in *Whales, seals, fish and man* eds A.S. Blix. L. Walløc and Ø. Ulltang (eds): 671-181. Amsterdam, Elsevier.

Rozwadoski, H.M. (2002) *The sea knows no boundaries*. A century of marine science under ICES. USA, ICES, Pp 419

Salmon Review Group (1987) *Report of the salmon review group for the Minister of the Marine.* Government Publications, Dublin, Pp 103

Sea Fisheries Act (1952), No 7 of 1952.

Sea Fisheries (Amendment) Act (1963), No 21 of 1963.

Sea Fisheries and Maritime Jurisdiction Act (2006), No 8 of 2006

SFPA (Sea Fisheries Protection Authority) 2006-2009. Annual reports

Shewan, J.M. (1941) Preservation of fish by the use of nitrite ice. *Nature*, 166: 613-614.

Smylie, M (1999) *Traditional fishing boats of Britain and Ireland, design, history and evolution*. Shrewsbury, Waterline, Pp 256

Sparks, C., E. Buecher, A. S. Brierley, B. E. Axelsen, H. Boyer and M. J. Gibbons (2001) Observations on the distribution and relative abundance of the scyphomedusan *Chrysaora hysocella* and the Hydrosoan *Aequorea aequorea* in the northern Benguelan ecosystem. *Hydrobiologia* 451 (Dev Hydrobiol. 155): 275 – 286.

Sumaila, U.R. and D. Pauly (eds) (2006) *Catching more bait: a bottom-up re-examination of global fisheries subsidies* Fisheries Centre Research Reports 14 (6): Appendices: 78-118.

Symes, D. and J. Phillipson (2001) *Inshore Fisheries Management*, London, Kluwer Academic Publishers, Pp 318.

Times of London (1880) Parliamentary intelligence: Sea fisheries (Ireland) Bill debate: 8, 8 July 1880.

Thurstan, R.H., S. Brockington and C.M. Roberts (2010) The effects of 118 years of industrial fishing on UK bottom trawl fisheries. *Nature Communications* 1 (15) Pp.13

Tully, O. (2003) What future has the Irish craw fishery? *The Irish Skipper*, June 2003: 20

Tully, O. (2010) Long term trends in the NW crab fishery. *The Irish Skipper*, September 2010: 21

Victoria (1842) 5th and 6th c.106 An act to regulate the Irish fisheries.

Villasante, S. and U.R. Sumaila (2010) Estimating the effects of technological efficiency on the European fishing fleet. *Marine Policy* 34 (3): 720-722.

von Brandt, A. (1964) *Fish catching methods of the world*. London, Fishing News (Books) Pp. 191

Waind, K.W. (1974) BIM survey reveals extensive fish stocks off south west coast. *The Irish Skipper*, February 1974: 8

Went, A.E.J. (1976) Review of fisheries bye-laws. Dublin, Department of Agriculture and Fisheries, Fisheries Division. Mimeo, Pp 119

Whelan, K., (1991) *The angler in Ireland* Dublin, Country House Pp, 161

White, P. (2005) *Decommissioning requirements for Ireland's demersal and shellfish fleets*: a report to marine minister Pat the Cope Gallagher T.D. Dun Laoghaire, Bord Iascaigh Mhara, July 2005, Pp. 40

Wildlife Act (1976), no 39 of 1976

Appendix 1
ABBREVIATIONS AND ACRONYMS

BIM	Bord Iascaigh Mhara
CEFAS	Centre for Environment, Fisheries and Aquaculture Science, an agency of the UK ministry DEFRA (Department of Environment, Fisheries and Rural Affairs).
CFP	Common Fisheries Policy
CLAMS	Co-ordinated Local Aquacultue Management Systems
EEC	European Economic Community
EEZ	Exclusive Economic Zone
ESRI	Economic and Social Research Institute
FAO	Food and Agriculture Organisation of the United Nations
FEOGA	European Agricultural Guidance and Guarantee Fund
FIF	Federation of Irish Fishermen
FIFC	Federation of Irish Co-operatives
FIFG	Financial Instrument for Fisheries Guidance
IAOS	Irish Agricultural Organisation Society
ICES	International Council for the Exploration of the Sea
IFO	Irish Fishermen's Organisation
IFRG	Inshore Fishereis Review Group
IFSC	Irish Specimen Fish Committee
INFC	Irish National Fisheries Council
IS&EFO	Irish South & East Fishermen's Organisation
IS&WFPO	Irish South & West Fish Proucers' Organisation
ISFA	Irish Sea Fisheries Association
KFO	Killybegs Fishermen's Organisation
LAC	Local Aqdvisory Committee
LE	Long Eireannach = Irish ship (precedes the name of naval vessels).
LEADER	EU Rural Development Programme
m	metre
MAGP	Multi Annual Guidance Programme
NFDA	National Fishermen's Defence Association
NFO	National Fishermen's Organisation
nm	nautical mile(s)
NSIFA	National Salmon and Inshore Fishermen's Association
NSRG	National Strategy Review Group
oal	overall length
PESCA	European fisheries fund, originating in the Commission
PO	Producer Organisation
RSW	Refrigerated salt water vessels
SAG	Species Advisory Group
SFPA	Sea Fisheries Protection Authority
STCCI	Salmon and Trout Conservaton Council of Ireland
STECF	Scientific Technical and Economic Committee on Fisheries
t	tonne or ton
TAC	Total Allowable Catch
UN	United Nations
WORD	Wexford Organisation for Rural Development

Appendix 2

LIST OF COMMERCIAL "VARIETIES" LANDED BY THE IRISH FISHING FLEET: COMPILED FROM RECENT FISHING YEARS.

English name	Scientific (Latin) name	Fishery assemblage (not necessarily targeted)	Capture regulated, as by TAC / quota species in 2010?
Albacore tuna	Thunnus alalunga	Pelagic	Yes
Alfonsino	Beryx spp	Deep water	Yes
Angler fish	Lophius spp	Demersal	Yes
Argentines	Argentina silus	Deep water	Yes
Atlantic wolffish	Anarhichas lupus	Demersal	
Black dogfish	Centroscyllium fabricii	Deep water	
Black scabbardfish	Aphanopus carbo	Deep water	Yes
Black sole	Solea solea	Demersal	Yes
Blackmouth dogfish	Galeus melastomus	Deep water	
Blonde ray	Raja brackyura	Demersal	Yes
Blue ling	Molva dypterygia	Deep water	Yes
Blue mussel	Mytilus edulis	Shellfish	
Blue shark	Prionace glauca	Pelagic	
Blue skate	Dipturus batis	Demersal	
Blue whiting	Micromesistius poutassou	Deep water	Yes
Bluefin tuna	Thunnus thynnus	Pelagic	Yes
Bluemouth	Helicolenus dactylopterus	Deep water	
Boarfish	Capros aper	Deep water	Yes
Bonito tuna	Sarda sarda	Pelagic	
Brill	Scophthalmus rhombus	Demersal	
Cardinalfish	Epigonus spp	Deep water	
Chub mackerel	Scomber japonicus	Pelagic	
Cod	Gadus morhua	Demersal	Yes
Common cockle	Cerastoderma edule	Shellfish	
Common mora	Mora moro	Deep water	
Common squid	Loligo forbesi	Demersal	
Conger eel	Conger conger	Demersal	
Crawfish	Palinurus spp	Shellfish	
Cuckoo ray	Raja naevus	Demersal	Yes
Cuttlefish	Sepia officinalis	Demersal	
Dab	Limanda limanda	Demersal	
Edible crab	Cancer pagurus	Shellfish	
European squid	Loligo vulgaris	Demersal	
Flounder	Platichthys flesus	Demersal	
Greater forkbeard	Phycis blennoides	Deep water	
Green crab	Carcinus maenas	Shellfish	
Greenland halibut	Reinhardtius hippoglossoides	Demersal	Yes
Gurnard	Trigla spp	Demersal	
Haddock	Melanogrammus aeglefinus	Demersal	Yes
Hake	Merluccius merluccius	Demersal	Yes
Herring	Clupea harengus	Pelagic	Yes
Horse mackerel	Trachurus trachurus	Pelagic	Yes
John dory	Zeus faber	Demersal	
Leafscale gulper shark	Centrophorus squamosus	Deep water	
Lemon sole	Microstomus kitt	Demersal	
Ling	Molva molva	Deep water	Yes
Lobster	Homarus gammarus	Shellfish	
Longnosed skate	Raja oxyrinchus	Deep water	
Mackerel	Scomber scombrus	Pelagic	Yes
Manila clam	Tapes semidecussata	Shellfish	

English name	Scientific (Latin) name	Fishery assemblage (not necessarily targeted)	Capture regulated, as by TAC / quota species in 2010?
Megrim	*Lepidorhombus* spp	Demersal	yes
Mullet	*Mugilidae*	Pelagic	
Norway lobster	*Nephrops norvegicus*	Demersal/shellfish	yes
Norwegian skate	*Dipturus nidarosiensis*	Deep water	
Nursehound	*Scyliorhinus stellaris*	Demersal	
Octopus	*Eledone cirrhosa*	Demersal	
Orange roughy	*Hoplostethus atlanticus*	Deep water	Yes
Oyster	*Ostrea edulis*	Shellfish	
Palaemon shrimp	*Palaemon serratus*	Shellfish	
Palourde	*Ruditapes decussatus*	Shellfish	
Periwinkle	*Littorina littorea*	Shellfish	
Pilchard	*Sardina pilchardus*	Pelagic	
Plaice	*Pleuronectes platessa*	Demersal	Yes
Pollack	*Pollachius pollachius*	Demersal	Yes
Poor cod	*Trisopterus minutus*	Demersal	
Porbeagle shark	*Lamna nasus*	Pelagic	Yes
Pouting	*Trisopterus luscus*	Demersal	
Queen scallop	*Aequipecten opercularis*	Demersal/shellfish	
Rabbit fish	*Chimaera* spp	Deep water	
Rays Bream	*Brama brama*	Demersal	
Razor clams	*Ensis* spp	Shellfish	
Red crab (deep sea)	*Chacon (Geyron) affinis*	Shellfish	
Red mullet	*Mullus surmuletus*	Demersal	
Redfish	*Sebastes* spp	Deep water	Yes
Roughhead grenadier	*Macrourus berglax*	Deep water	
Round sardinella	*Sardinella aurita*	Pelagic	
Roundnose grenadier	*Coryphaenoides rupestris*	Deep water	Yes
Saithe	*Pollachius virens*	Demersal	Yes
Sand eel	*Ammodytes* spp	Demersal	
Sand gaper	*Mya arenaria*	Shellfish	
Sand sole	*Pegusa lascaris*	Demersal	
Scad	*Trachurus trachurus*	Pelagic	Yes
Scallop	*Pecten maximus*	Shellfish	
Scorpionfishes	*Trachyscorpia cristulata*	Deep water	
Sea bass	***Dicentrarchus labrax***	**Pelagic**	**Non-commercial species**
Shagreen ray	*Raja fullonica*	Deep water	
Shortfin mako shark	*Isurus oxyrinchus*	Pelagic	
Siki shark	*Centroscymnus coelolepis*	Deep water	
Silver scabbardfish	*Lepidopus caudatus*	Deep water	
Smooth hound	*Mustelus mustelus*	Demersal	
Spider crab	*Maia brachydactyla*	Shellfish	
Spiny lobster	*Palinurus elephas*	Shellfish	
Spotted ratfish	*Chimaera monstrosa*	Deep water	
Sprat	*Sprattus sprattus*	Pelagic	
Spurdog	*Squalus acanthias*	Demersal	Yes
Stingrays	*Dasyatis* spp	Demersal	
Surf clam	*Spisula* spp	Shellfish	
Swordfish	*Xiphias gladius*	Pelagic	
Thornback ray	*Raja clavata*	Demersal	Yes
Thresher shark	*Alopias vulpinus*	Pelagic	
Tope	*Galeorhinus galeus*	Demersal	
Turbot	*Scopthalmus maximus*	Demersal	
Torsk	*Brosme brosme*	Demersal	Yes
Undulate ray	*Raja undulata*	Demersal	
Velvet crab	*Necora puber*	Shellfish	
Whelk	*Buccinum undatum*	Shellfish	
Whiting	*Merlangius merlangus*	Demersal	Yes
Witch	*Glyptocephalus cynoglossus*	Demersal	

Appendix 3

DIRECT SUBSIDIES TO THE FISHING INDUSTRY BETWEEN 1952 AND 2009.

The contribution of taxpayers' money directly to the industry is only one aspect of a more generous consideration. Funds invested in infrastructure – the construction of fisheries harbours, for instance - are not part of the monies considered here. Nor is finance allocated for administration, protection and law enforcement or research. The allocations detailed here were sourced from the exchequer and the European Community and channelled through BIM, the state's development agency.

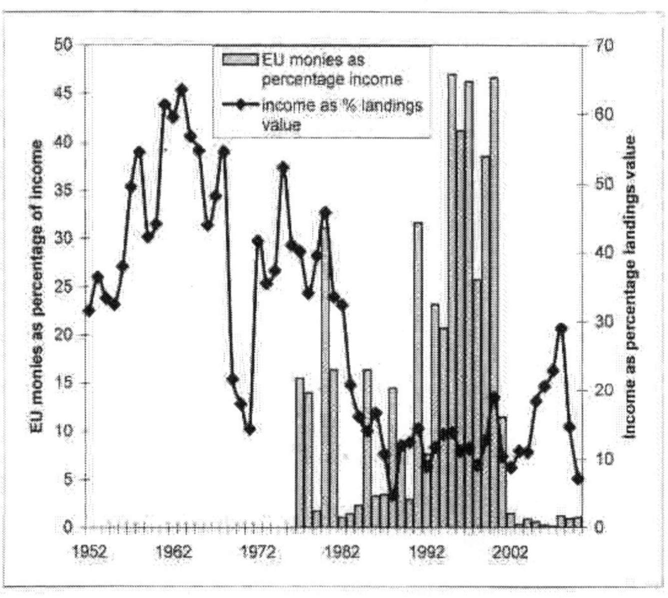

Fig. 3. Investment in Ireland's marine fisheries through its development agency (BIM), expressed as a percentage of landed value, and the percentage of the total contributed by the European Community.

Over the period under consideration there were three media of exchange: the pound sterling (stg£) was the Irish currency until 1979 when the link was broken and the Irish punt (IR£) replaced it. Its value was pegged against the Euro (€) in 1999. The most straightforward way to represent subsidy is to express the amount as a percentage of the value of landings, the principal product of the fleet.

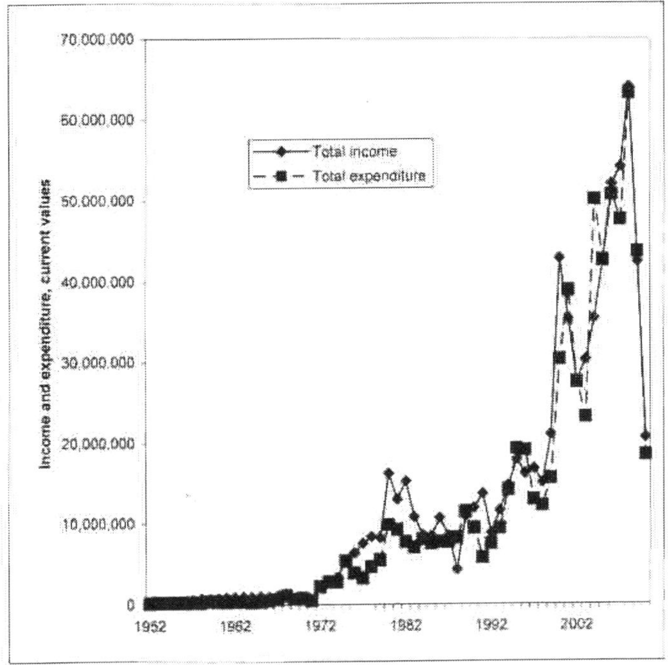

Fig 4. Annual income and expenditure of BIM, between 1952 and 2009, much of it spent on the development of the catching sector.

For a brief period in the 1960s, subsidy to the industry amounted to more than 60% of the monies obtained from the first sale of produce (Fig 3). The contribution from Europe commenced shortly after joining the EEC and it rose to almost 50% of total input during the decade preceding the new millennium. While this interpretation of the graph covers the early years of BIM, the 1990s were progressively more devoted to aquaculture which became an alternative focus of development as fish stocks were depleted. The capture sector of the industry was, therefore, relatively more generously supplied with taxpayers' funds during the earlier years of industrialisation.

Income and expenditure through the development agency are expressed most straightforwardly, as current values, in Fig 4.

Expenditure increased dramatically just before Ireland joined the EEC and declined in the new millennium. Converting these amounts to rough €-equivalent, indicates a total investment of some €812 million between 1952 and 2009. However, factoring in inflation, raises the total to approximately €2.9 billion.

The relative distribution of the main categories of spending in the two periods 1952-1977 and 1978-2010, demonstrates the growth of interest in aquaculture as capture fisheries declined. Precisely categorising some of the funds was problematical, hence the totals do not round to 100%.

Categories	1952-1977	1978-2010
Administration	19%	19%
Aquaculture	0%	16%
Fleet and fisheries	63%	56%
Others (including training, marketing and processing).	14%	10%

Appendix 4

DEVELOPMENT OF THE IRISH FISHING FLEET, 1929-2009

Estimates of the approximate number and gross tonnage (GT) of the Irish fishing fleet were made from published data reported between 1929 and 2009 inclusive. Government departments issued the information until 1962 when BIM assumed that task. No estimates of fleet size were available between 1986 and 1988 inclusive after which a more detailed and selective account of vessel numbers, GT and engine power (kW) were reported annually, compiled from registered individual boat details, in accordance with European Commission requirements to cap the fleet exploiting quota species.

Until 1989, the customary way of expressing fleet size was to group vessel numbers in size categories. In order to estimate total fleet tonnage the mid-point value of each size category was multiplied by the number of vessels within it. A marginal category (such as above or below a particular tonnage) was attributed a weight 25% above or below although a more precise figure might be substituted, depending on information available.

Estimates of fleet size and tonnage prior to 1990 are approximations and they refer to all vessels included in the annual census. BIM issued more detailed information in 1972 where 1,020 vessels were attributed a combined GT of 19,256. My figures for that year were 2,265 vessels totalling 23,194 GT. Had I based my calculations only on vessels exceeding 18 feet keel length (5.5 m) and making up the "motorised fleet", my totals would have been 1,065 boats with a combined burden of 21,567 GT.

A similar comparison was feasible in 1973. BIM's published breakdown of fleet numbers was 2,347 with a combined estimated total of 24,302 GT. Excluding vessels of less than 5.5 m keel length, the total estimated GT would have been 22,642 distributed among 1,125 boats. BIM's fleet size was 1,095 boats with a combined displacement of 20,670 GT.

Accepting that my estimates are on the high side, I think it legitimate to include all boats which engage in fishing in any calculation of fleet size. Boats powered by outboard motor or simply propelled by oars or wind, are capable to exerting fishing effort through the use of passive gears although these are considerably less productive than mobile ones.

One further comparison: in 1978 BIM provided a fleet estimated number of 1,443 vessels with a GT of 30,000. Working from the fleet breakdown my estimate was 33,424 GT among 2,919 boats. Had I referred only to the "motorised fleet" I would have had 1,460 vessels with a combined 31,374 GT.

The absence of data for three years, 1986-1988 covers a period when efforts were being made to quantify vessel numbers, GT and kW more accurately.

As it became more obvious that the expanding fleet was inflicting damage to fish stocks, a number of Multi-Annual Guidance Programmes (MAGP) were introduced by the Commission to bring the fleet into balance with the resource. The methodology of fisheries management in Europe required only that vessels larger than 12 m be considered part of the regulated fishing fleet although all vessels engaged in fishing and aquaculture were eventually required to hold licences. Initially, details of fleet size were proffered by the Department to DG XIV (the relevant division within the Commission) which were, later, admitted to be guesstimates. In 1983 there were probably 1,614 boats with engines, having a capacity exceeding 51,000 GT. Following criticism by the industry, the Department obtained permission from the Commission to revise its figures. The conclusion reached was that the Irish fleet in 1983 was made up of 45,300 GT and 181,200 kW. This is 24% above my own estimate of GT in that year.

The data included in Fig. 5 include sectors of the fleet which are not regarded by the Commission as part of Ireland's fishing effort for fleet regulation purposes. One of these is the aquaculture sector whose large steel vessels dredge mussels inshore and transplant them in sheltered embayments such as Wexford Harbour and Belfast Lough for on-growing. The underlying logic suggests that this is essentially an agriculture-like process rather than a wild fishery because the procedure mimics sowing, transplanting and harvesting. However, wild mussel reefs are centres of biological diversity, supporting juvenile finfish species on which capture fisheries depend. They should, in my view, be regarded as part of the natural productive cycle. Boats exploiting them, together with all sizes and types of fishing vessels, should be included in one register of their numbers, individual GT and kW.

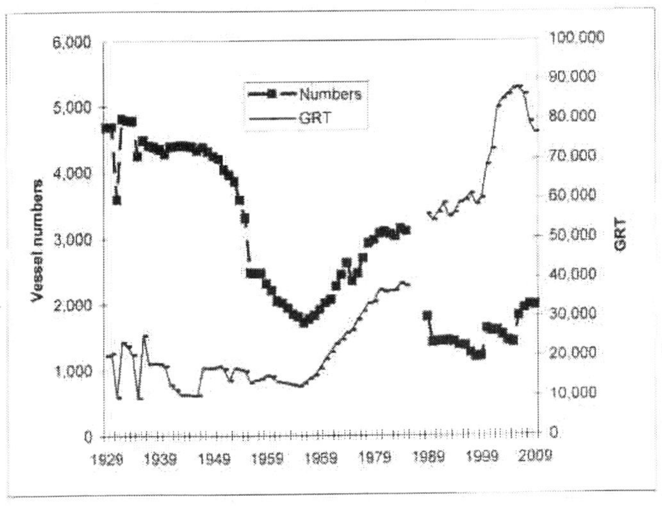

Fig. 5. The evolution of the Irish fishing fleet in numbers and displacement, gross registered tonnes (GRT), between 1929 and 2009.

Official fleet numbers post-1989, are likely to be an understatement because they omit vessels of less than 12 m oal.

The Irish fishing fleet expanded considerably and rapidly during the phase of industrialisation. Fishing effort also became concentrated – centralised – in fewer, larger vessels.

COMPOSITION AND VALUE OF LANDINGS BY THE IRISH FLEET.

The recent species composition of fish landings to Ireland is provided in Appendix 2. The terminology in the book refers to them as "varieties" because many are not distinguished at species level. The number of varieties recorded over the past century is set out in the Fig.6. They are grouped within the three main categories of produce, demersal, pelagic and shellfish.

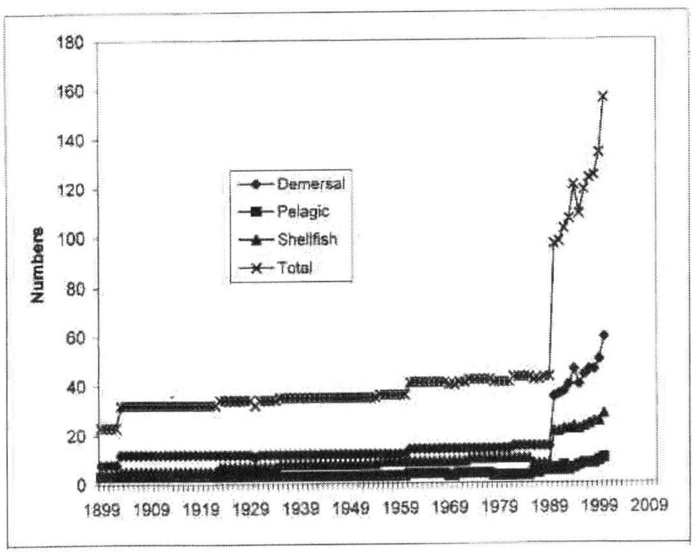

Fig. 6. Varieties of sea fish recorded in the Irish landings, 1899 - 2010 inclusive.

The tonnage of landings over the same period is shown in Fig. 7, again according to the three main categories. Their combined tpmmage peaked in 1995 and trended downwards thereafter.

The current value of the four categories of landings is set out in Fig 8.

Although the earnings from sea fish landings appear to be increasing, this is not the case when the figures are standardised at € value and adjusted for inflation (Fig 9). Peak value was reached in 2001 and has tended downwards since.

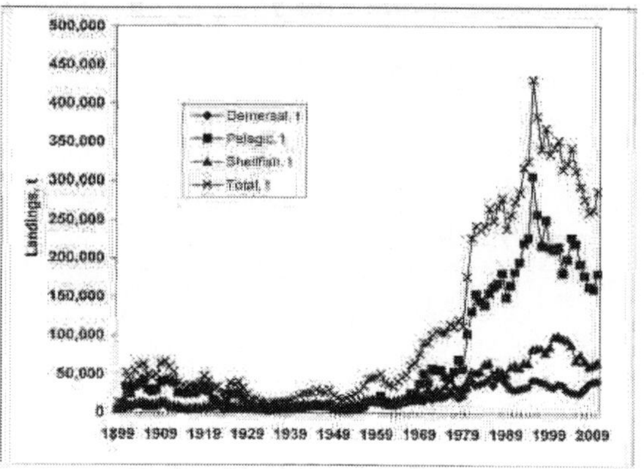

Fig. 7. Landings of sea fish by Irish vessels, 1899-2010 inclusive.

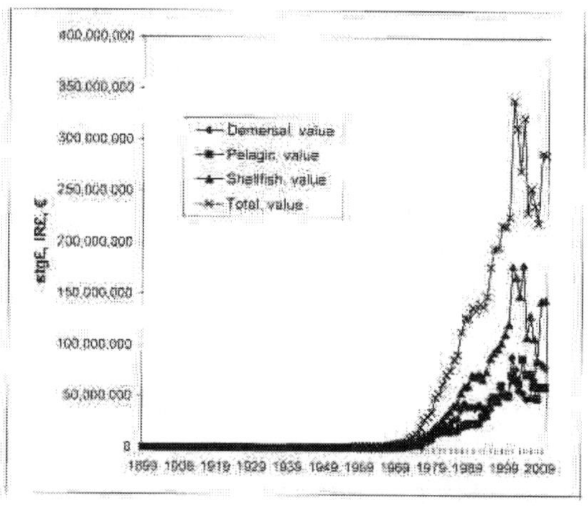

Fig 8. The current value of landings to the Irish fleet, 1899 – 2009.

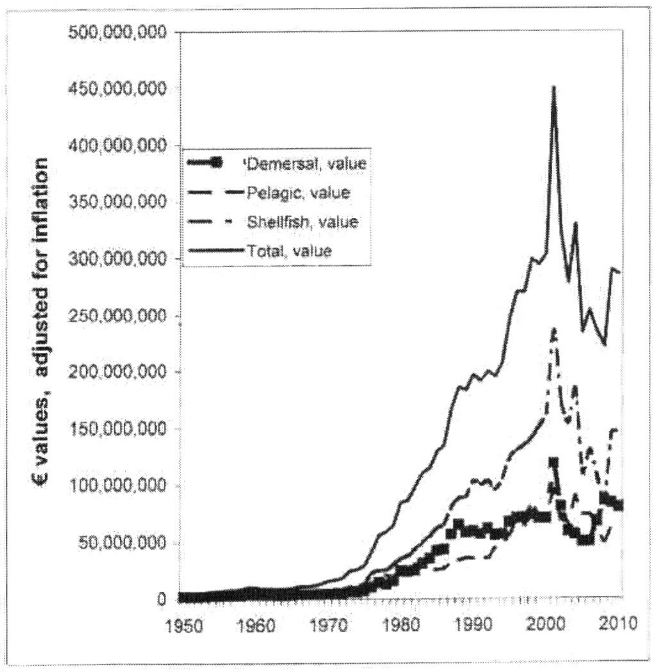

Fig. 9.   Value of landings, standardised €, adjusted for inflation, 1950 – 2010.

THE OFFICIAL FATE OF FOUR WHITEFISH SPECIES.

White fish species are selected to illustrate the pattern of exploitation which accompanied the industrialization of Ireland's fisheries. Two of them, cod and whiting show similar patterns of increasing landings as the fleet and fishing effort built, followed by decline.

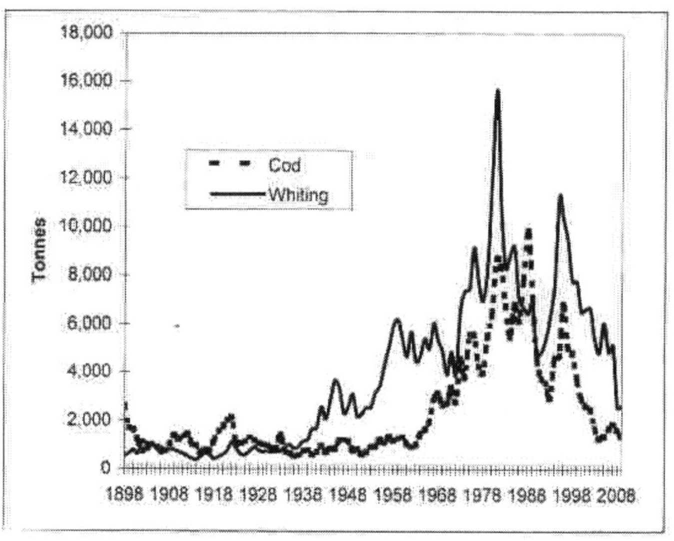

Fig. 10. Landings of cod and whiting to Ireland, 1898 – 2008.

Haddock and cod are believed to interact; when cod declines, haddock populations tend to expand (Fig. 11).

The final graph (Fig. 12) is of plaice whose landings are presented alongside the displacement of the fleet (Appendix 4).

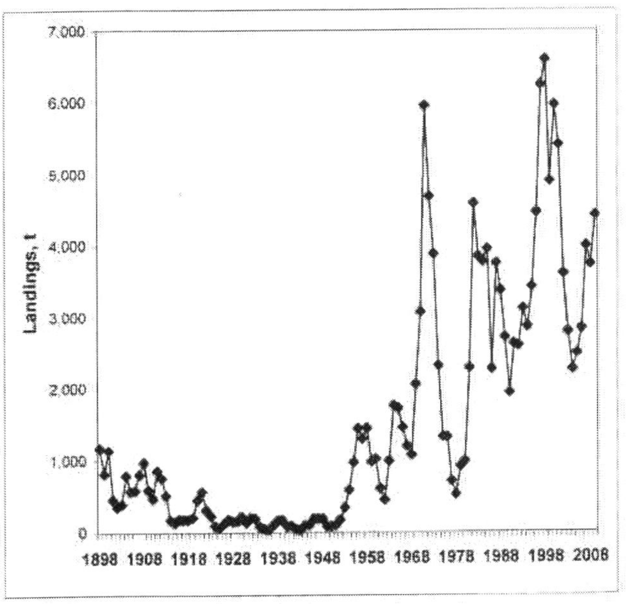

Fig 11. Landings of haddock to Ireland, 1898 – 2008.

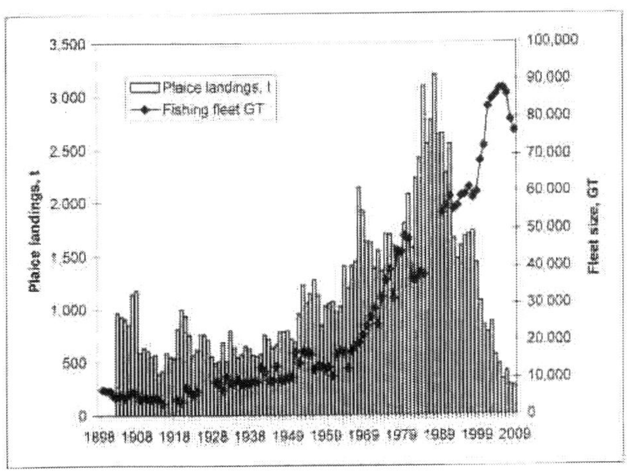

Fig. 12. Landings of plaice to Ireland, compared with the growth of the Irish fleet (GT), 1903 – 2009.

Index

Vessel names, document titles and non English items are in italics

Abalone.. 158

Abyssal plain.................................... 271

Africa...2, 48, 108, 109, 135, 171, 260,
    377, 378, 379, 380

    Algiers.. 28

    Mauritania 109, 378, 379, 380, 382,
        383

    Morocco............................380, 382

    Namibia....................273, 350, 410

Ahern, Bertie. 325, 326, 328, 329, 377,
    378

Aisling L.E....................................... 133

Albacore 373

Albacore tuna..25, 156, 204, 210, 262,
    263, 264

America.. 48

American parlour pot...................... 252

American survey team ...162, 163, 164,
    165, 179, 231, 252, 415

Angel shark 391

Angler species134, 135, 136, 137, 138,
    204, 221, 261, 262, 263, 276

Angling ..151, 157, 188, 268, 299, 301,
    391, 394, 404, 405

Angling Council of Ireland 194

Anglo Irish Bank.............................. 380

Anglo-Irish Treaty, 1922................. 88

Antarctic.............................. 373, 374

Antibiotics as preservative 361, 362

Aquaculture14, 50, 151, 153, 155, 158,
    160, 164, 168, 169, 174, 190, 201,
    211, 239, 249, 256, 268, 288, 290,
    291, 292, 296, 297, 299, 358, 402,
    418, 425, 426, 428, 429

Ard Colm 112, 117

Ard Mhuire......................................112

Argentine..........................262, 264, 272

Artificial baits 157

Atlantic Challenge 264

Atlantic Dawn. 20, 108, 375, 376, 377,
    378, 379, 380, 381, 382, 383, 384

Atlantic Fish Industries 114

Atlantic Fisheries Ltd......................181

Australia......48, 92, 116, 143, 273, 383

    New South Wales......................383

    Tasmania 109

Azores..271

Baltic Sea 3, 128

Bank of Ireland 375, 380

Barrett, Sean...................................289

Bartley, Gerald........................ 103, 105

baselines .92, 93, 94, 96, 193, 303, 405

Basking shark....................................64

Baylor University study 315

Beam trawling...................30, 371, 402

Belgium 46, 94, 115, 126, 130, 144,
   279, 360, 412

Billingsgate market, London... 44, 282,
   386

Bjuke, C.G. 408, 410

Black scabbard fish.........264, 265, 273

Black Sea..350

Blake, J.A. 72, 73, 410

Blue ling...........................265, 272, 273

Blue whiting.. 156, 258, 259, 260, 262,
   264, 382

Bluefin tuna.....................156, 187, 267

Board of Works........................... 69, 77

Boatyards, BIM. 14, 84, 139, 140, 145,
   150, 194, 232, 233, 234, 235, 236,
   246

Brabazon, Wallop 40, 59, 60, 61, 62,
   63, 64, 65, 76, 79, 410

Brandon and Cloghane Farmers and
   Fishermen's Improvement
   Association....................................90

Brehon law..1

Brennan, Joseph.............................178

Brill....................................30, 142, 276

Brotherhood of Mutual Assistance...58

Browne, John 382

Brussels19, 94, 96, 118, 137, 144, 153,
   156, 185, 202, 207, 210, 213, 214,
   215, 218, 229, 237, 238, 241, 242,
   244, 264, 288, 296, 302, 317, 323,
   326, 337, 346, 347, 368, 369, 377,
   393, 397, 398, 401, 412, 427

BT Group plc.....................................294

Bulgaria..133

Buller, Sir Reginald Manningham ...92

Burdett Coutts, Angela . 74, 75, 76, 77,
   415

Butt, Isaac ..5, 69, 70, 71, 73, 356, 357,
   358, 411, 457

Byrne, David 377

Byron, Lord George...........................74

Cafolla, Mario 280

Canada 92, 120, 343, 360

   Newfoundland 3, 4, 336, 337

   Ontario 109

Cannon, Pat 318, 319, 320

Caribbean..21

Cawley, Noel.152, 296, 330, 359, 360,
   385, 386, 393, 411, 457

Celtic Sea.21, 135, 190, 197, 198, 217,
   219, 224, 261, 262, 284, 286, 309,
   341

Central Fisheries Board...................299

Central Statistics Office.................161

Ceylon...92

Charr 158

Childers, Erskine............................ 222

Chile..............................91, 92, 274

China.....................................280, 350

    Hong Kong 380

Clam boom..................................... 174

CLAMS 299

Clear, J.K................ 108, 109, 223, 412

Cliona, L.E. 88

Co Antrim

    Red Bay 405, 413

Co Clare.......... 198, 237, 254, 314, 393

    Burren 312

    Cappa 198

    Carrigaholt................................. 283

    Loop Head 382

Co Cork74, 75, 78, 79, 84, 89, 90, 110,
134, 138, 173, 188, 204, 220, 258,
282, 313, 320, 324, 333, 360, 363,
392, 416, 418

    Ballycotton 182, 362, 363

    Baltimore...2, 28, 75, 76, 104, 110,
       232, 416

    Bantry...........................74, 89, 411

    Bull Rock..................................... 90

    Cape Clear 74, 75

Castletownbere...90, 123, 133, 134,
    138, 155, 220, 230, 258, 263,
    320, 322, 362, 363, 364

Crookhaven 74

Kinsale.....................................5, 80

Schull.91, 113, 114, 117, 313, 362,
    363, 364

Sherkin Island 75

Skibbereen........................... 75, 204

Youghal 87, 315

Co Donegal .2, 27, 78, 84, 85, 89, 110,
170, 172, 178, 191, 200, 211, 220,
264, 267, 283, 285, 303, 304, 305,
318, 319, 320, 328, 329, 376, 406,
407

    Arranmore 406

    Bunbeg...328

    Bundoran 89

    Burtonport 283, 285, 315, 406

    Donegal Bay................................406

    Gortnasate.....................................85

    Greencastle.......143, 155, 264, 283,
       320, 364, 373

    Inishowen 182

    Inver Bay 208, 267

    Killybegs61, 78, 84, 100, 110, 113,
       114, 117, 123, 175, 178, 181,
       182, 183, 193, 195, 203, 204,
       208, 209, 211, 220, 230, 232,
       235, 258, 259, 262, 264, 299,
       302, 319, 320, 322, 328, 329,

330, 331, 361, 362, 363, 364,
375, 377, 378, 384, 407, 418

Malin Head...................... 262, 299

Mevagh 78, 84, 110

Moville 84

the Rosses..............................170

Tory Island 406

Co Dublin

Dublin Bay 38, 41, 47, 79, 335, 418

Howth 104, 117, 123, 217, 220,
226, 230, 341, 364

Lambay Deep 335

Loughshinny................................84

Ringsend......................38, 279, 280

Sandymount...............................217

Co Galway..... 2, 21, 27, 41, 42, 43, 47,
65, 66, 78, 84, 85, 86, 103, 108,
113, 114, 117, 123, 173, 220, 230,
252, 254, 262, 267, 282, 300, 302,
314, 320, 324, 361, 362, 363, 364,
377, 415

Barna..41

Carna.............................. 117, 300

Cladagh.....................................66

Clarinbridge..................... 175, 252

Cleggan.............................. 362, 363

Connemara 297

Glenninagh 41

Inishmore 117

Kilcolgan....................................174

Rossaveal ..220, 259, 267, 320, 322

Roundstone 408

Slyne Head..............................117

Co Kerry 27, 55, 78, 79, 84, 85, 89, 90,
115, 173, 188, 252, 283, 297, 299,
300, 320, 387, 391, 392, 407

Annascaul.....................................42

Blasket Islands...........................346

Brandon............................42, 55, 90

Caherciveen 117, 362

Connemara............. 85, 86, 265, 301

Cromane................ 85, 86, 115, 252

Dingle.....2, 42, 43, 47, 84, 89, 110,
117, 177, 283, 320, 322, 362,
363, 407

Fenit..283

Great Blasket Island 386

Magharees...................................299

Portmagee 117

Skellig Islands 387

Tralee Bay...................................391

Valentia 178, 387

Waterville..................................188

Co Limerick...........................290, 362

Co Louth

    Carlingford 252, 258

    Clogherhead...............315, 362, 364

Co Mayo 2, 59, 62, 78, 84, 85, 174, 200, 208, 209, 254, 257, 344, 345, 346, 373

    Achill 84, 364, 373

    Clew Bay 60, 65, 85

    Erris...........................344, 345, 348

    Inishkea Islands 62, 63, 343, 344, 345, 346

    Inishturk Island.......................... 407

    Killala Bay.................................. 208

    Killary...................................... 85

    Murrisk 84, 174, 361, 362, 363

    Newport 60, 61

Co Meath................ 182, 224, 252, 319

    Boyne estuary............................ 252

    Mornington....... 182, 183, 224, 253

Co Sligo

    Sligo 208, 324, 393

Co Waterford....... 27, 72, 84, 220, 267, 324, 369, 410

    Dunmore East.... 80, 117, 162, 204, 220, 267, 315, 341, 362, 363, 364

    Helvic................. 84, 200, 315, 361

    Passage East...................... 123, 230

Co Wexford. 27, 29, 54, 252, 258, 297, 300, 369, 382, 429

    Courtown 226

    Kilmore Quay...............29, 364, 389

    Rosslare 369

    Saltee Islands.............................133

Co Wicklow....................................54

    Arklow............................... 226, 364

    Wicklow 226

Cockle...........22, 29, 58, 164, 173, 207

Cod 3, 18, 22, 23, 25, 28, 33, 34, 45, 48, 51, 61, 63, 92, 128, 130, 147, 165, 172, 187, 221, 222, 223, 224, 225, 258, 259, 272, 276, 277, 279, 280, 308, 317, 335, 336, 337, 338, 339, 340, 341, 343, 350, 374, 401, 402, 403, 405, 406, 407, 410, 413, 417, 434

Cod recovery programme....... 338, 339

Collins, Stephen...............322, 323, 325

Combat Poverty Agency 190, 191

Common Agriculture Policy... 11, 125, 393

Common Fisheries Policy. 11, 14, 118, 125, 126, 128, 129, 130, 131, 132, 133, 141, 147, 151, 185, 189, 198, 202, 213, 215, 225, 234, 237, 242, 244, 248, 255, 256, 277, 291, 301, 393, 395, 396, 397, 398, 412

Common Organisation of the Markets in Fishery and Aquaculture products 397

Community pond 127

Confradías de pescadores.................58

Conger eel.............63, 64, 68, 347, 348

Congested districts................. 5, 77, 78

Connolly, James...........................80

Consumer Price Index 174

Consumption of fish9, 12, 99, 143, 167

    abstinence 4

    per capita consumption... 6, 89, 143

Continental shelf..9, 89, 146, 259, 264,
271, 272

Continental slope 9, 14, 21, 35, 271,
273

Control of Landings Regulations....137

Convention on International Trade in
Endangered Species....................287

Coriolis effect 21

Cot...53

Coughlan, Mary 329, 330, 370

Coughlan, Steve 94

Coveney, Simon............................392

Cowen, Brian 329

Crab ...22, 26, 36, 41, 48, 64, 142, 157,
190, 209, 244, 252, 255, 299, 303,
304, 305, 387, 413, 414, 420

Crawfish 22, 25, 64, 85, 117, 190, 313,
386, 387, 388, 389, 390, 391, 392,
407, 408

Crawfish (Fisheries Management and
Conservation Order), 2002 390

Creating employment.....................353

Criminal Assets Bureau..................322

Crotty, J.F. (Judge)....................90, 91

Crummey, Brian............................201

Cú Feasa...........88, 116, 117, 124, 167

Cú na Mara 88, 167, 387

Curragh 5, 83, 119, 242

Dail, the......81, 98, 199, 211, 313, 325,
326, 328, 382, 408

Dalhousie, Earl of...............iv, 46, 411

Daly, Brendan.........................237, 345

Davitt, Michael...............................75

de Courcy Ireland, John...97, 187, 355,
356, 357, 358, 359, 412, 457

Decca Navigator Co Ltd...................27

December council meeting of
ministers............214, 242, 317, 331

Decommissioning vessels16, 153, 245,
296, 370

DeepNet programme......................137

Deirdre, L.E. 310

Dempsey, Noel......137, 319, 324, 325,
326, 328, 329

Denmark 11, 46, 79, 115, 127, 129,

130, 207, 216, 257, 360, 364

Faroe Islands................................ 259

Greenland....................34, 265, 343

Department for Industry and
    Commerce..................................... 84

Department of Agriculture Food and
    the Marine.................................. 212

Department of Agriculture and
    Technical Instruction.....33, 53, 87,
    416

Department of Agriculture Fisheries
    and Food............................393, 409

Department of Agriculture Food and
    the Marine...................177, 393, 412

Department of Communications
    Marine and Natural Resources 210,
    319

Department of Defence................... 144

Department of Education 117

Department of Environment..314, 402,
    414

Department of Fisheries 281, 282

Department of Fisheries and Forestry
    345

Department of the Marine..... 169, 209,
    317, 346, 377

Department of the Marine and Natural
    Resources................................... 347

Department of Tourism Fisheries and
    Forestry...................................... 345

Development Commission of the UK

....116

Dickens, Charles...............75, 279, 412

Discards...51

Disraeli, Benjamin............................74

Dogfish..47

Donegan, Paddy 195

Doyle, Frank . 195, 198, 202, 203, 379,
    380, 382

Dredging............................36, 256, 369

Drift nets......25, 32, 47, 48, 50, 54, 63,
    117, 119, 148, 180, 184, 187, 188,
    189, 190, 191, 196, 200, 204, 207,
    210, 215, 256, 264, 282, 283, 310,
    329, 344, 411

Dublin Bay prawn............................335

Dublin Retail Fish Merchants'
    Association.................................193

Dublin Trawling Ice and Cold Storage
    Co Ltd...279

Dublin Trawling Ice and Cold Storage
    Co. Ltd.......................................403

DuPont Corporation 25

Easter Rising.....................................87

Echinoderm 22, 158, 312, 313, 314,
    412

Economic and Social Research
    Institute..............293, 294, 295, 418

Ecuador 91, 92

Edible crab 24, 25, 142, 226, 227, 242,
    252, 254, 256, 263, 285, 299, 303,

304, 305, 306, 388, 406, 412

Edward III, King 37

Eel1, 7, 38, 158, 254, 347, 348

Eiranova 90, 134, 138

England1, 3, 28, 30, 31, 37, 41, 45, 46,
    47, 50, 52, 53, 67, 68, 71, 77, 173,
    278, 279, 364, 402, 412

    Brixham 30, 38

    Cornwall 45, 90

    Devon..38

    Grimsby......................................48

    Hull..92

    Lowestoft..................................402

    Norfolk 26, 75

    North Shields..............................31

Englehard, George...........................402

Esat Telecom Group plc.................294

Essex, Lord .... iv, 272

European Agricultural Guidance and
    Guarantee Fund 144, 234, 235, 236,
    238, 240

European Communities' logbook...131

European Court of Justice 130, 320

European Economic Community.9, 11,
    12, 59, 89, 94, 96, 124, 125, 126,
    127, 128, 130, 131, 132, 133, 135,
    138, 143, 144, 146, 149, 150, 152,
    187, 191, 192, 195, 202, 206, 215,
    224, 261, 272, 357, 359, 373, 393,

399, 425, 426

European ombudsman 347, 348

European Parliament..... 216, 324, 393,
    412

European Parliament Office 393

European Union 17, 18, 19, 59, 89,
    141, 155, 157, 158, 160, 168, 169,
    176, 177, 188, 214, 220, 228, 245,
    246, 247, 248, 249, 258, 259, 275,
    277, 279, 286, 290, 293, 296, 298,
    304, 310, 317, 318, 326, 339, 342,
    365, 369, 375, 376, 377, 378, 380,
    381, 382, 395

Exclusive Economic Zone...11, 89, 95,
    96, 128, 130, 132, 133, 191, 192,
    194, 197, 241, 257

Exports........6, 9, 12, 68, 142, 143, 148,
    156, 161, 162, 166, 167, 173, 237,
    291, 292, 311

Fahey, Frank 300, 301, 377, 378

Federation of Irish Fishermen205, 212,
    325, 327, 355, 393, 394, 395, 397

Federation of Irish Fishing Co-
    operatives 191, 197

Federation of Marine Industries.... 192,
    193, 201

Ferrari, Peter...................................280

Fianna Fail.15, 94, 103, 130, 178, 185,
    195, 211, 222, 237, 286, 289, 300,
    313, 319, 323, 325, 328, 329, 345,
    377, 382, 393

Fifie 53

Fifty mile exclusive Irish fishery limit
    .... 192, 194, 195, 196

Fifty mile exclusive UK fishery limit 280

Financial Instrument for Fisheries Guidance 249

Fine Gael .15, 103, 179, 195, 200, 203, 212, 289, 290, 324, 328, 333, 345, 392

Finucane, Michael 290

First World War ... 76, 77, 87, 280, 358

Fischler, Franz 339, 377

Fish and chip trade 33, 280, 281

Fisheries Consolidation Act, 1959 313, 414

Fisheries Prices Support Board 120

Fishery districts 39, 47, 189

Fishery harbours 12, 84, 123, 158, 167, 220, 230, 267

Fishing company 40, 60, 134, 410

Fishing Industry Development Committee 118

Fishing Industry Development Fund 123, 363

Fishing traps 24, 26, 30, 36, 40, 58, 65, 119, 157, 226, 302, 388, 389, 390

Fishmeal 13, 80, 99, 123, 124, 181, 182, 206, 224, 253, 254, 272, 275, 383, 407

Fitzgerald, Garret 131, 414

Fitzpatrick, Tom 345

Fjords 171, 172, 173, 174

Flanagan, Oliver J 103, 106, 179

Fleet development measure ... 247, 248, 249

Fleet renewal ... 15, 244, 246, 247, 248, 266, 270, 272, 289, 295, 331, 353, 367, 370, 372, 400

Fleet segments 186, 256, 370, 371, 377

Fleetwood Fishing Vessel Owners' Association 116

Flinn, Hugo 80

Flounder 64

Food and Agriculture Organisation of the United Nations ... 119, 141, 287, 410

Forkbeard 273

Fort Rannoch 88

France .. 2, 4, 31, 38, 46, 47, 57, 58, 67, 71, 73, 89, 90, 92, 94, 119, 126, 130, 136, 143, 144, 172, 173, 174, 207, 216, 223, 232, 233, 262, 263, 272, 274, 279, 282, 289, 320, 326, 350, 357, 387, 401, 415

Fraud squad 320

French cray pot 387

Gael Linn 117

Gaelic overlords 2, 3

Gaeltacht 111, 118, 121, 229, 299

Gallagher, Eamon 132

Gallagher, Pat the Cope .211, 302, 319, 323, 324, 393, 420

Garda 319, 329

Genesis 374

Geneva Convention, 1958 8, 10, 11, 94, 95, 96, 125, 128

Gentian violet....................................216

Germany....6, 31, 34, 46, 94, 104, 114, 126, 130, 143, 144, 181, 272

Ghost fishing....................................26

Gill nets...24, 25, 36, 63, 64, 119, 137, 208, 210, 226, 257, 261, 283, 284, 286, 340, 346, 347, 350, 387

Gladstone, William...........................69

Global positioning system (GPS)....26, 90

Gonzales, Felipe 138

Granton Steam Company.................46

Grattan's Parliament.........................67

Great Grimsby Steam Trawling Company 31

Great Sole Bank....................... 89, 263

Greece....................................92

Green crab.............................. 263, 388

Green Party ... 323, 324, 325, 328, 329, 382, 383, 399

Grenadier..................34, 264, 265, 273

Grescoe, Taras 135, 415

Griffin, Dan....................................134

Griffith, Arthur 357, 358, 359

Grimsby..92

Gronland..181

Gross National Product....18, 166, 167, 168, 311

Gulf of Mexico.................21, 271, 350

Gulf Stream 21

Gundelach, Finn 149, 196

Gurnard 61, 64

Habitats Directive 215

Haddock 23, 33, 45, 47, 48, 61, 63, 64, 115, 142, 165, 223, 224, 225, 276, 277, 279, 280, 282, 308, 336, 338, 350, 403, 406, 407, 434, 435

Hake 2, 24, 61, 68, 116, 132, 134, 165, 198, 204, 223, 225, 262, 276, 282, 308, 347, 406

Halibut 34, 48, 65, 156, 265

Hall, Mr and Mrs..............................59

Hand lines.......................................64

Harfon..181

Hatton Bank....................................265, 271

Haughey, C.J. 129, 209

Healy-Rae, Michael 392

Helga....................................87, 88, 410

Herring......5, 13, 23, 25, 28, 29, 32, 41, 48, 51, 52, 54, 60, 61, 63, 64, 67, 68, 71, 77, 79, 82, 89, 99, 100, 113, 114, 115, 117, 124, 128, 142, 143, 146, 162, 165, 170, 180, 181, 182, 183, 190, 197, 198, 204, 217, 219,

221, 223, 225, 235, 252, 253, 254, 255, 257, 282, 299, 309, 357, 363, 373, 381, 406, 408

Hickling, E.G. 116

Higgins, Michael D........................ 300

Hilliard, Michael............................ 94

Hire-purchase 82, 84, 100, 101, 104, 110, 111, 112, 120, 122, 123, 140, 163, 164, 229, 230, 231, 233, 245

Holt, S.J. 116, 251, 412, 416

Holy See 93

Home Government Association of Ireland 5

Hook and line 50, 60, 65, 156, 252, 264, 272, 301, 338

Hooker.. 53, 74

Horgan, John 310, 311

Hotchkiss quick firing gun 87

Houses of Parliament, England. 67, 71, 74, 356

Howley, Martin............... 289, 329, 331

Hull 252, 257

Huxley, T.H..... 23, 41, 46, 50, 51, 109, 411, 457

Ice 3, 61, 79, 99, 124, 151, 164, 352, 353, 360, 361, 362, 363, 364, 365, 366, 367, 374, 411, 419

Iceland 92, 93, 117, 130, 158, 181, 259, 271, 336, 337, 373

Iceland-Faroe Ridge 271

IIB bank..380

Illicium 134

India 92, 108, 280

Indian Ocean 274

Indonesia 350

Industrial Schools Act........................75

INMARSAT.....................................319

Inshore Fisheries Advisory Committee 298

Inshore Fisheries Review Group 303

International Conference on the Law of the Sea............................. 91, 191

International Council for the Exploration of the Sea 213, 304, 337, 410, 419

Intervention market support.... 13, 122, 216, 220

Irish Agricultural Organisation Society Ltd...................................... 197, 202

Irish Country Women's Association 118, 144

Irish Fish Processors and Exporters' Association........................ 193, 201

Irish Fish Producers' Organisation 193, 202, 203, 204, 205, 217, 218, 339, 375

Irish Fishermen's Association 186

Irish Fishermen's Organisation......129, 130, 136, 149, 191, 192, 193, 194, 195, 196, 197, 198, 199, 200, 201, 202, 203, 204, 207, 208, 218, 235,

262, 291, 375, 379, 380, 381, 382

Irish Fishing and Fish Trades Gazette
.... 102, 103, 105, 280, 416

Irish Lobster Association 390

Irish Marine and Coastwatching
Service 88

Irish Marine Service 88

Irish National Fisheries Council 190,
191

Irish Sea 5, 6, 21, 40, 62, 64, 82, 86,
165, 220, 221, 224, 226, 227, 228,
253, 255, 258, 261, 263, 267, 272,
278, 279, 283, 319, 335, 336, 337,
338, 339, 340, 341, 343, 401, 405,
410, 413, 414

Irish Sea Fisheries Association 6, 82,
83, 84, 85, 86, 97, 98, 100, 101,
108, 110, 113, 115, 152, 226, 361,
362

Irish Skipper.. 182, 183, 194, 202, 332,
359, 376, 410, 411, 413, 420

Irish South and East Fish Producers'
Organisation 205

Irish South and East Fishermen's
Organisation 205

Irish South and West Fish Producers'
Organisation 204, 205, 382

Irish South and West Fishermen's
Organisation 136

Irish South and West Producers'
Organisation 138

Irish Specimen Fish Committee 286,
404

Irish Times 294, 310, 325, 329, 331,
332, 343, 351

Irish Transport and General Workers'
Union 194

Irish Wildlife Federation 345

Irminger Sea 271

Islanders 170, 418

Isle of Man 41, 117

Italy 126, 172, 280, 281, 282, 350

Japan 260, 267, 350, 362

Jellies18, 348, 349, 350, 351, 413, 417

Jigging machine 255, 298, 301

Johannesburg Convention 397

John dory 61, 64

Johnson, T.R. 80

Kennedy, John F. 162

Kenny, Enda 345

Killala 208

Killeen, Tony 393, 398

Killybegs Fisheries Association 182

Killybegs Fishermen's Association
.... 195, 208

Killybegs Fishermen's Organisation
.203, 205, 208, 212, 214, 218, 236,
289, 304, 317, 318, 319, 321, 328,
329, 339, 375, 380, 385, 393

kiloWattage 241, 289, 301, 370, 382

Kirk, Captain Kent............................ 129

Korea 227, 350

K-strategy 108, 151, 172, 175, 223, 226, 271, 274, 283, 415, 417, 418, 420

Labour Party........ 80, 94, 188, 310, 313

Labrador Sea 271

LEADER 297, 300

Lemass, Sean.............................95, 162

Lemon sole 156

Lenihan, Brian.................130, 185, 195

Lent.. 4, 99

Lewis, Michael................................. 291

Liar's Poker 291

Licensing of vessels........120, 165, 179, 193, 197, 210, 228, 370

Ling..................... 61, 63, 262, 272, 273

Liverpool Crown Court 327

Lobster........22, 24, 64, 85, 86, 90, 101, 117, 123, 142, 158, 165, 190, 226, 230, 232, 252, 258, 260, 282, 297, 298, 299, 313, 386, 387, 388, 391

Local Advisory Committee 300, 303

Loch Laoi.. 104

Loch Lein.. 104

Loch Lorgan................................... 104

Long-lining 24, 85, 137, 156, 208, 254, 263, 282

Luxembourg.....................................126

Lynch, Jack 138

Ma Bretonne 90

Mac Laughlin, Jim.............. 4, 170, 457

Mac Laughlin, Jimi..........................457

Macari, Dominic..............................280

Macdonald, Ian 383

Macha, L.E.......................................88

Mackerel 22, 23, 29, 51, 61, 64, 77, 79, 82, 89, 119, 128, 148, 165, 181, 204, 212, 216, 217, 218, 219, 225, 237, 255, 260, 280, 298, 299, 317, 322, 324, 374, 381, 382, 393, 404, 408

Maddock, Joey.................................201

Madeiras, the.....................................68

Maeve, L.E. 88, 90

Magnetometer......................... 261, 405

Mahon Tribunal 329

Mallin, Michael................................100

Manila clam 158, 165, 174

Mare clausum 3

Mare liberum 2, 3

Marine Casualty Investigation Board 384

Marine Institute.......204, 210, 211, 291, 293, 309, 321, 341, 392, 393, 396, 457

Marine Times 320, 412, 413, 414

Maritime Institute of Ireland... 98, 358, 359

Matassa, Antoni 280

McArthur, J.S.119, 120, 121, 122, 162, 163

McDowell, Michael........................328

McGinley, Dinny 328

McHugh, Kevin 373, 374, 375, 376, 377, 378, 379, 380, 381, 382, 383

McIntosh, W.C............................ 45, 52

McKenna, Patricia...........................382

McManus, John...............294, 382, 383

Mediterranean Sea 57, 58, 128, 131, 171, 172, 312, 335

Megrim....................134, 262, 263, 276

Mercantile Marine Act, 1955..........138

Mercator..181

Mid-Atlantic Ridge.........................271

Mill, John Stuart 70, 356

Molloy, John 309, 418

Monofilament nets..........................200

Moramora.......................................265

Morpeth, Lord...................................70

Muintir na Tire 178

Muirchu 88

Mulcahy, John A. 188

Mullet..25, 65

Multi-Annual Guidance Programmes .185, 186, 240, 241, 243, 246, 249, 428

Munro, Hugh 195

Munster plantation 4

Murphy, Michael Pat 188, 193, 203, 313

Murrin, Joey ..193, 195, 196, 203, 207, 208, 209, 210, 211, 235, 236, 289, 385

Mussel. 22, 29, 37, 41, 63, 85, 86, 102, 115, 139, 142, 155, 156, 164, 182, 207, 252, 254, 256, 258, 297, 357, 386, 429

Naomh Simon 85, 117

Nardone family...............................280

National Bureau of Criminal Investigation 320, 322

National development plans...........290

National development programmes160

National Federation of Fish Friers.280

National Fisheries Advisory Council 201

National Fishermen's Organisation182

National Fishermen's Defence Association 188, 189, 192

National Parks and Wildlife Service 346

National Salmon and Inshore Fishermen's Association..189, 190, 191, 192, 193, 196, 197, 200, 201, 203

National Strategy Review Group..290, 292, 293, 294, 295

Nephrops .34, 134, 142, 148, 165, 198, 212, 218, 220, 224, 225, 262, 263, 267, 335, 336, 338, 340, 341, 401, 410, 418

Netherlands2, 3, 38, 46, 53, 73, 79, 85, 90, 92, 94, 126, 130, 143, 144, 181, 207, 232, 233, 244, 272, 279, 317, 320, 373, 376, 377, 382

New Zealand 48, 273, 274

Nicky 53

Nobby 53

Nolan, Pat... 406

North Atlantic gyre........................... 21

North Sea.30, 31, 33, 48, 89, 106, 171, 221, 283, 286, 350, 412

Northeast Atlantic Fisheries Commission............................... 307

Northern Ireland...... 95, 116, 132, 143, 152, 180, 195, 338

Belfast............. 62, 80, 95, 360, 429

Harland and Wolff.................... 375

Kilkeel.................................. 327

Norway 3, 6, 10, 46, 73, 77, 92, 93, 127, 128, 171, 232, 258, 259, 272, 282, 317, 335, 374, 375, 377, 382

Hellesoy boatyard 374

UMOE Sterkoder boatyard 377

Vik and Sandvik boatyards 377

Norway lobster..................................335

O'Brien, Denis....................................294

O'Brien, William 80

O'Connor, Jim202, 203, 204, 414, 418

O'Connor, T.P. 73

O'Donnell, Peadar ..170, 171, 172, 418

O'Donoghue, Sean 319, 339

O'Driscoll, Donal 104

O'Keeffe, Jim....................................333

O'Kelly, Brendan.................... 139, 150

Ó'Mealláin, Seamus 100

O'Neachtain, Sean 286

O'Neill, Terence 95

O'Sullivan, Maurice........................386

O'Toole, Paddy..............200, 345, 414

Oliver Twist............................. 280, 412

Operational Programme 168, 169, 245, 247, 290, 291

Orange roughy264, 273, 274, 275, 415

Orwell, George 280

Ovoca 33

Oyster ..5, 37, 48, 51, 64, 85, 142, 156, 164, 252, 254, 300, 357, 386

Pacific Ocean 34, 227, 274

Palourde.164, 170, 171, 172, 173, 174, 175, 413, 414

Panamanian fleet register 383

Pandora..75

Partridge, Karl................174, 175, 418

Pauly, Daniel....................................405

Payment by share.............................42

Peace and Reconciliation fund........155

People's Movement..........................393

Perch..158

Periwinkle 85, 170

Permanent Commission of the
   International Fisheries Convention
  307

Perry, John 324, 381

Peru......................................91, 92, 181

PESCA programme 157, 158, 297

Pescanova..134

Phoenix magazine...................380, 382

Pilchard 2, 28, 41, 51, 73, 182

Plaice.18, 23, 25, 46, 61, 64, 223, 225,
   254, 276, 277, 279, 308, 341, 402,
   403, 406, 407, 434, 435

Poland.................................34, 94, 232

Pollock..............................65, 165, 301

Polyvalent Pot licence....................302

Porbeagle shark 282

Porcupine Bank 259, 262, 263, 264,
   265, 374

Portugal.........2, 63, 130, 131, 172, 279

Pot24, 26, 27, 32, 53, 123, 208, 226,
   230, 255, 258, 263, 302, 388

Pot hauler..........26, 123, 157, 230, 236

Potato Famine....5, 6, 61, 70, 170, 355,
   356, 358

Power block.......................................34

Precautionary principle....40, 213, 242,
   397

Price Waterhouse 288

Price Waterhouse Coopers 384

Prime fish...........................23, 134, 276

Prodi, Romano................................377

Producers' organisation13, 15, 59, 189,
   202, 203, 205, 206, 207, 208, 209,
   210, 211, 212, 213, 214, 216, 218,
   220, 236, 267, 291, 304, 317, 319,
   320, 325, 326, 339, 375, 382, 396,
   398

Programme for economic and social
   development.......................167, 409

Programmes for economic expansion
   .... 101, 140, 151, 162, 165, 409

Progressive Democrats 325, 328

Protestant 67

Prud'homies 57, 58

Public Accounts Committee...........381

Purse seining.....29, 58, 181, 197, 252, 373, 376

Quadragesimo Anno encyclical....... 34

Quota .13, 15, 128, 130, 131, 135, 136, 137, 138, 146, 147, 149, 152, 168, 196, 198, 202, 204, 207, 212, 213, 215, 216, 219, 220, 224, 237, 242, 243, 246, 256, 260, 266, 267, 275, 277, 279, 289, 293, 295, 304, 317, 318, 321, 327, 329, 331, 333, 338, 369, 370, 371, 376, 381, 382, 395, 396, 401, 402, 427

Ramsay, Gordon 314, 412

Ray .. 22, 25, 48, 61, 64, 142, 257, 281, 389, 391, 408

Razor clam.....................22, 24, 29, 256

Red Sea.................................. 108, 171

Redfish.. 265

Refrigerated sea water vessels 35, 208, 209, 260, 264, 299, 317, 373, 374, 376

Reproductive Loan Fund............. 72, 76

Rerum Novarum encyclical.............. 34

Response to the green paper on the reform of the Common Fisheries Policy .99, 393, 394, 395, 396, 397, 398

Restriction on Fishing orders......... 136

Revenue, Department of Finance... 318

Reynolds, Arthur............183, 202, 359

Rio de Janeiro convention.............. 397

Roberts, Callum .. 37, 48, 49, 419, 420, 457

Rock hopper gear...............................33

Rock salmon 281

Rockall 262, 263, 265, 271, 273

Roman Café 280

Roman Catholic....................................3

Ros Beag...407

Royal Dublin Society 79, 117, 418

r-strategy 172

Russia 34, 94, 133, 327, 380

Ryan, Eamon....................323, 325, 399

Saithe 61, 225, 261, 405

Salmon...1, 6, 7, 12, 23, 25, 38, 50, 51, 64, 65, 89, 109, 119, 143, 148, 151, 168, 182, 184, 187, 188, 189, 190, 191, 196, 197, 200, 201, 207, 215, 256, 258, 260, 283, 285, 286, 299, 310, 311, 329, 344, 345, 347, 348, 408, 419

tinned 6, 89

Salmon and Trout Conservation Council of Ireland 188

Salvesens...183

Sandeel 224, 254, 347

Sardine...350

Sarkozy, Nicolas.............................401

Sars, G.O. 45

Saudi Arabia 92

Scad (horse mackerel)...260, 262, 317, 374, 382

Scallop.....29, 63, 64, 85, 86, 114, 117, 120, 136, 155, 248, 249, 252, 254, 258, 295, 297, 299, 302, 369

Schooner..62

Scientific Technical and Economic Committee for Fisheries...304, 305

Scotland3, 5, 31, 41, 46, 52, 53, 67, 68, 69, 70, 71, 73, 77, 267, 272, 286, 317, 322, 339, 344, 357, 387

Aberdeen 49, 69, 260, 361

Clyde, the 267

Hebrides.....................................344

Moray Firth 48

Orkney Islands 282

Peterhead 322

Shetland Islands ..63, 264, 265, 282

St. Kilda.....................................264

Scuba diving 155, 313, 314

Sea Fisheries (clam and bait beds) Act 52

Sea Fisheries (Protection of Immature Fish) Act , 1937...........................86

Sea Fisheries and Maritime Jurisdiction Act, 2006325, 326, 385

Sea Fisheries Protection Authority211, 326, 327, 328, 332, 371, 419

Sea Shepherd Conservation Society 345

Seafood Industry Agenda 290, 410

Seal.... 1, 266, 343, 344, 345, 346, 347, 348, 411, 413, 419

Second World War.... 8, 54, 83, 86, 88, 91, 97, 98, 170, 221, 282

Seining.. 2, 28, 29, 42, 45, 65, 89, 119, 197, 221, 222, 224, 252, 254, 257, 263

Sephardic Jews................................279

Shellfish framework297, 302, 303, 306

Shifting baseline syndrome 405

Shiny clam 265, 266

Shlûter, Poul................................129

Shrimp22, 24, 27, 38, 46, 52, 252, 254, 261, 262, 357

Siggins, Lorna 385

Siki 265

Silver smelt.....................................262

Sinn Fein...382

Six mile fishery limit..92, 93, 94, 180, 186, 200, 228, 298, 303

Smack Owners Association of Hull.46

Small scale coastal fisheries....58, 176, 185, 197, 201, 297, 298

Society of Friends 76

Sole 48, 64, 105, 134, 142, 150, 276

Sonar 26, 236

Sonia 132

South and East Coast Fishermen's Co-operative Society Ltd................ 204

Soviet Union...............33, 34, 129, 133

Spain 2, 4, 12, 34, 53, 57, 58, 63, 67,
89, 90, 130, 131, 132, 133, 134,
135, 137, 138, 143, 156, 170, 172,
174, 198, 204, 208, 262, 263, 264,
272, 279, 289, 320, 325, 326, 327,
387, 411

Canaries Islands 377, 379, 383

Ondorroa................................... 132

Vigo.................................... 134, 137

Species Advisory Group................. 303

Spider crab.............. 261, 262, 299, 391

Sprat........150, 182, 183, 224, 253, 255

Spurdog23, 25, 65, 255, 279, 281, 282,
283, 284, 285, 286, 287, 349, 414

Squat lobster............................156, 267

Squid....................................... 156, 262

Stanton Bank 262

Steam propulsion.... 26, 29, 30, 31, 32,
47, 48, 50, 53, 54, 77, 78, 87, 101,
109, 180, 190, 278, 279, 358, 403,
410

Stockholm syndrome...................... 342

Structural funds...... 131, 202, 234, 240

Subsidy 5, 7, 10, 12, 17, 61, 67, 68, 69,
70, 71, 73, 77, 126, 139, 162, 163,
199, 215, 218, 219, 229, 237, 245,
247, 248, 269, 296, 334, 341, 357,
363, 364, 381, 398, 399, 400, 401,
408, 419, 424, 425

Sunday Business Post......................380

Sunday Independent 174, 195

Sunday Times 381

Surf clam 22, 265

Sutherland, Duke of............................70

Sweden 46, 117, 233

Taiwan...350

Tangle nets 25, 36, 136, 137, 254, 388,
389, 390, 391, 407

Technology creep....27, 159, 199, 236,
237, 265, 358, 368

Temple, Sir William.................. iv, 272

Territorial sea.... 11, 13, 19, 90, 91, 92,
93, 94, 95, 96, 129, 176, 182, 183,
185, 198, 209, 262, 277, 304, 392,
395, 414

Thatcher, Mrs Margaret...................129

The Bell 170

The Sea Fisheries Act, 1952 103

The Sea Fisheries and Maritime
Jurisdiction Act, 2006....... 322, 323

Three mile fishery limit...8, 87, 90, 91,
92

Tonnage...14, 19, 35, 36, 77, 128, 135,
149, 177, 213, 225, 241, 246, 289,
294, 301, 337, 342, 367, 369, 370,
372, 381, 382, 427, 431

Tope...282

Torry Research Centre 260

Torsk...273

Total Allowable Catch ..128, 135, 136, 198, 207, 212, 213, 214, 215, 224, 242, 259, 275, 277, 278, 279, 286, 293, 321, 337, 340, 341, 397, 401

Trammel nets 25, 39, 65, 208

Trawling ..6, 10, 20, 30, 34, 35, 36, 37, 38, 40, 41, 42, 43, 44, 45, 46, 47, 48, 49, 50, 51, 52, 53, 54, 55, 62, 63, 64, 79, 85, 87, 89, 116, 117, 136, 156, 176, 177, 179, 180, 190, 201, 207, 224, 252, 254, 255, 259, 260, 261, 262, 263, 267, 278, 279, 316, 322, 340, 373, 374, 408, 415, 416

Treaty of Rome 13, 125, 127, 130, 206

Trout 1, 7, 27, 38, 190

Turbot.....................25, 30, 61, 64, 156

Twelve mile fishery limit....89, 91, 92, 93, 94, 145, 183, 198, 303, 395

Twin rigging 34

Ulster Bank Markets........................380

Ulster plantation 4, 203, 380

Undulate ray.....................................391

Union with England..............5, 68, 356

United Kingdom ..2, 6, 7, 8, 10, 31, 33, 40, 41, 47, 48, 51, 69, 72, 73, 86, 87, 88, 89, 90, 92, 93, 94, 95, 98, 100, 116, 120, 127, 129, 130, 143, 144, 176, 177, 180, 192, 193, 194, 195, 198, 207, 209, 223, 227, 233, 241, 267, 271, 276, 278, 279, 280, 281, 282, 327, 335, 336, 345, 346, 348, 351, 357, 360, 386, 402, 403, 412, 419, 420

United States of America.....92, 93, 94, 115, 162, 274, 280, 362, 365, 419

University of British Colombia......130

van Geldern, Gerrit 343, 348

Vega 319

Velvet crab...................... 263, 387, 414

Veronica 374, 375, 376, 380, 381, 383, 384

Vessel decommissioning .15, 158, 245, 248, 249, 295, 322, 353, 367, 368, 369, 370, 371, 372, 385, 399, 412

Vessel monitoring system, VMS ..319, 324, 353

Vivier boats 263, 304

V-notching scheme 297

Vocational Education Committee .117, 118

Voisinage agreement.........................96

Wales ...31, 41, 45, 278, 383, 402, 403

Milford Haven 279, 403

Warty venus 263

Wave Crest.......................................373

Wellington, Duke of 75

West Indies 68

West of Scotland fishing grounds...21, 135, 259, 262, 304, 382

Western Endeavour..........................264

Whale 1, 52, 183, 209, 419

Whale and Dolphin Conservation
    Society 209

Whelan, Ken..................................... 269

Whelk...10, 24, 41, 225, 226, 227, 228,
    258, 261, 299, 414

Wherry... 67

Whitaker, T.K. 151

White Fish Authority98, 100, 193, 252

White paper on sea fisheries, 1962 122

White, Padraic...98, 99, 100, 151, 193,
    248, 249, 291, 295, 409, 420

Whitefish renewal scheme 154, 169,
    248, 249, 265, 273, 294, 295, 296,
    369

Whiting18, 64, 65, 119, 120, 147, 220,
    221, 222, 223, 224, 225, 258, 259,
    264, 336, 374, 406, 407, 416, 434

Whooley, Jason........................ 138, 296

Wildlife Act, 1976 314

Wind, propulsion by 29, 31, 36, 428

Wondyrechaun 37

Woods, Michael..............289, 290, 293

WORD.. 297

Wyville Thomson Ridge 271

Yawl... 53

Zodiac.. 31

Chapter notes

Chapter 1: From inauspicious beginnings

[1] Mac Laughlin, 2010
[2] De Courcy Ireland, 1981
[3] Ekin, 2008
[4] Mac Laughlin, 2010
[5] de Courcy Ireland, 1986
[6] Mac Laughlin, 2010
[7] Pérez-Rubín, 2006.
[8] Butt, 1871
[9] Mac Laughlin, 2010
[10] Fisheries grant assistance from BIM in 2010 benefits over 700 vessel owners. BIM press release 18 April 2011
[11] Cawley, 2006

Chapter 2: Harvesting the fauna of Ireland's shelf waters

[1] Marine Institute, stock books
[2] Huxley,1884
[3] Edwards,1979
[4] BIM, annual report 1971-1972
[5] Fahy, Carroll and Stokes, 2002
[6] Villasante and Sumaila, 2010
[7] In this book overall length measurements (oal) of vessels are in metres (m); older measurements were in feet; some boat designs such as the wooden "fifty footers" were customarily referred to by this name, a convention also used here.
[8] Vessel weight is measured in gross registered tonnes (GRT) or gross tonnes (GT)
weight is measured in gross registered tonnes: GRT or GT
[9] Engine power is measured in units of horse power (HP) or, more recently, kilowatts (kW): 1 HP=0.735 kW
[10] Roberts, 2007
[11] Ekin, 2008
[12] von Brandt, 1964,
[13] Roberts, 2007
[14] Roberts, 2007
[15] Hérubel, 1912
[16] Smylie, 1999

[17] Board of Agriculture and Fisheries, 1908
[18] Roberts, 2007
[19] de Courcy Ireland, 1981: 71
[20] Roberts, 2007
[21] Roberts, 2007
[22] *Irish Fishing and Fish Trades Gazette*, 3 (6), 12 March 1955
[23] Roberts, 2007
[24] Data abstracted from official reports, see Appendix 3
[25] Roberts, 2007
[26] *Ibid*
[27] Holt, 1909
[28] George III, c. 41, 1773-74
[29] Victoria, 1842
[30] *The Irish Skipper*, March 2003: 1
[31] Commissioners of Fisheries, Ireland, 1843
[32] Roberts, 2007: chapter 10
[33] Brabazon, 1848
[34] Caird *et al*, 1866
[35] Huxley, 1884
[36] Reprised in Dalhousie *et al*, 1884-1885
[37] Dalhousie *et al*, 1884-1885
[38] Roberts, 2007
[39] Quoted by Roberts, 2007
[40] Huxley, 1884
[41] Rozwadowski, 2002
[42] Rozwadowski, 2002
[43] Victoria, 1842
[44] Fahy, 2010c
[45] Smylie, 1999
[46] Went, 1976: 6
[47] The "Department", without further qualification, refers to the government ministry responsible for sea fisheries matters. It has gone through a long list of title changes in the history of the state.
[48] Went, 1976: 7
[49] Went, 1976: 8-11
[50] Went, 1976: 10
[51] Fahy, Healy, Downes, Alcorn and Nixon, 2008

Chapter 3: A matter of size

[1] Aldecoa, 1958
[2] Frangoudes, 2001

[3] de Callao, 1997
[4] Brabazon, 1848
[5] Anon, 1820
[6] Caird *et al*, 1866

Chapter 4: For want of capital

[1] Johnson *et al*, 1921
[2] 59 Geo 111 C.109, 1 Geo IV C. 82, 5 Geo IV C.64, 67
[3] Dinneen's Irish-English dictionary translates this term as whiting or saithe of more than two years old.
[4] *The Times* of London, 8 July 1880
[5] de Courcy Ireland, 1981
[6] Butt, 1871
[7] The legislation which followed was the Irish Reproductive Loan Fund Amendment Act, 1882, Vic. 46th and 47th c.22
[8] *The Times* of London, 8 July 1880
[9] Healey, 1978
[10] Higgins, 1957
[11] Johnson *et al*, 1921
[12] Brabazon, 1848
[13] Johnson *et al*, 1921
[14] Johnson *et al*, 1921
[15] de Courcy Ireland, 1981
[16] Johnson *et al*, 1921
[17] McGuire and Quinn, 2009
[18] de Courcy Ireland, 1981
[19] Lawlor, 1945

Chapter 5: Widening sea, narrowing ocean

[1] Brunicardi, 1986
[2] de Courcy Ireland, 1981
[3] *Irish Fishing and Fish Trades Gazette* 3 (2), 15 January 1955
[4] Carroll, 1992
[5] *Irish Fishing and Fish Trades Gazette* 5 (24), 23 November 1957
[6] *Irish Fishing and Fish Trades Gazette* 6 (6), 15 March 1958
[7] *Irish Fishing and Fish Trades Gazette* 6 (7), 29 March 1958
[8] *Irish Fishing and Fishing Trades Gazette* 8 (4), April 1960
[9] Fahy, Healy, Downes, Alcorn, and Nixon, 2008
[10] Fahy, Healy, Downes, Alcorn and Nixon, 2008
[11] *The Irish Skipper*, March 1978

[12] S.I. No 4/1960 – Licensing of sea-fishing vessels regulations, 1960
[13] *Irish Fishing and Fishing Trades Gazette* 8 (1), January 1960

Chapter 6: A brief flirtation with reality

[1] de Courcy Ireland, 1981: 100-102
[2] *Irish Fishing and Fish Trades Gazette* 2(4), 6 Nov 1954
[3] *The Times* of London, 8 July 1880
[4] Anon, 1958
[5] *Irish Fishing and Fish Trades Gazette* 4 (6), 24 March, 1956
[6] *Irish Fishing and Fish Trades Gazette* 2 (5), 20 Nov, 1954
[7] *Irish Fishing and Fish Trades Gazette* 3 (14), 3 July, 1955
[8] *Irish Fishing and Fish Trades Gazette* 3 (15), 16 July 1955
[9] Fahy, 2010b
[10] *Irish Fishing and Fish Trades Gazette* 1 (26), 11 September 1954
[11] *Irish Fishing and Fish Trades Gazette* 4 (13), 30 June 1956.
[12] *Irish Fishing and Fish Trades Gazette* 5 (4), 16 Feb, 1957.
[13] *Irish Fishing and Fish Trades Gazette* 5 (28), 18 Jan, 1958
[14] *Irish Fishing and Fish Trades Gazette* 6(11), 24 May, 1958
[15] *Irish Fishing and Fish Trades Gazette* 6 (10), 10 May, 1958.
[16] Huxley, 1884
[17] *Irish Fishing and Fish Trades Gazette* 4 (21), 20 Oct, 1956
[18] *Irish Fishing and Fish Trades Gazette* 6 (9), 26 April, 1958.
[19] Sea Fisheries Act, 7, 1952, section 18 (1)
[20] Sea Fisheries (Amendment) Act, No 30, 1956
[21] Sea Fisheries (Amendment) Act, No 28, 1959
[22] Until 1975, the accounting year straddled two calendar years, running from 1 April to 31 March. For brevity the year in question referred to is the first of the two.
[23] Molloy, 2006
[24] Holt, 1956
[25] Holt, 1956
[26] McArthur, 1959
[27] Anon, 1962

Chapter 7: The elaborate regime that failed

[1] Meredith, 1998, chapter 4
[2] Holden and Garrod, 1994, chapter 2
[3] Meredith, 1998, p 51
[4] Fahy, 2008a

[5] Kurlansky, 1998: chapter 10
[6] http://111.eurocean.org/np4/80.html consulted on 7 April 2012.
[7] Meredith, 1995: 55-57
[8] John Cooney, *Irish Times*, 29 October 1980
[9] European Commission, 2010
[10] Lorna Siggins, *Sunday Tribune*, 11 December 1983.
[11] Arthur Reynolds, *Irish Times*, 22 October 1984
[12] Ella Shanahan, *Irish Times*, 23 October 1984
[13] Richard Wigg, *Irish Times*, 29/30 October 1984
[14] Arthur Reynolds, *Irish Times*, 1 October 1976
[15] *The Irish Skipper*, September, 1984
[16] *The Irish Skipper*, July 1985
[17] Mac Laughlin, 2010
[18] Grescoe, 2008, chapter 1
[19] O'Connor et al, 1980, P42
[20] Fahy, 2009a
[21] Arthur Reynolds, *Irish Times*, 11 March 1983
[22] Mairin de Burca, *Sunday Tribune*, 11 July 1982

Chapter 8: Good news and subsidies

[1] Appendix 5
[2] Appendix 3
[3] Landings statistics are constantly being revised so there may be discrepancies between what is reported here and elsewhere within this account. However this chapter is essentially about how BIM saw itself and the data used here are those in its annual reports.
[4] Victoria, 1842, 5th and 6th c. 106.
[5] *The Irish Skipper*, March 1978
[6] http://www.finfacts.ie/irishfinancenews/article_1014 consulted 20 March 2012
[7] http://investing.businessweek.com/research/stocks consulted 20 March 2012
[8] http://teagasc.ie/aboutus/teagasc_board.asp consulted 20 March 2012
[9] BIM, annual report, 1987
[10] BIM, annual report, 1996
[11] BIM, annual report, 1997
[12] Anon, 2009

Chapter 9: Guidance, guesstimates and the first casualty of many

[1] Anon, 1964; 1965

[2] Glude *et al*, 1964
[3] McArthur, 1959
[4] Anon, 1969
[5] BIM, annual report, 1994.
[6] Chapter 15
[7] Mac Laughlin, 2010.
[8] *Irish Fishing and Fish Trades Gazette* 4 (13), 30 June 1956
[9] Partridge, 1977
[10] Chapter 14
[11] Glude *et al*, 1964
[12] Fahy, 2010a
[13] Partridge, 1977
[14] Fahy *et al*, 2010a

Chapter 10: Meanwhile, back in the territorial sea

[1] Anon, 1999
[2] Cecil Beamish, Assistant Secretary, Department of Agriculture, Food and the Marine September 2011, Mespil Hotel, Dublin: International fisheries journalists' conference.
[3] *Irish Fishing and Fish Trades Gazette*, 1 (21), 3 July 1954
[4] *Irish Fishing and Fish Trades Gazette*, 1 (22) 17 July 1954
[5] *Irish Fishing and Fish Trades Gazette*, 2 (5) 20 November 1954
[6] *Irish Fishing and Fish Trades Gazette* 1 (24) 14 August 1954
[7] *Irish Fishing and Fish Trades Gazette* 1 (24) 14 August 1954
[8] *Irish Fishing and Fish Trades Gazette* 3 (8) 27 August 1955
[9] *Irish Fishing and Fish Trades Gazette* 7 (7) July 1959
[10] Sea Fisheries Act, No 7, 1952, section 9 sub-section 5
[11] Went, 1976 stated: "there is already a condition that certain vessels shall not fish within the three mile limit"
[12] Went, 1976: 6-11
[13] Molloy, 2006: 100
[14] *Irish Fishing and Fishing Trades Gazette* 7 (12) December 1959
[15] *The Irish Skipper*, June 1968: 8
[16] *The Irish Skipper*, January 1969: 8
[17] De Courcy Ireland, 1981
[18] *The Irish Skipper*, October 1986
[19] *The Irish Skipper,* December 1979
[20] *The Irish Skipper*, December 1980
[21] See Chapter 12
[22] Meredith, 1998: 82-83
[23] De Courcy Ireland, 1981: 96-97

[24] Control of fishing for salmon order, number 298, 1972

[25] *The Irish Skipper*, June 1981

[26] George Burrows, *Irish Times*, 22 May 1980

[27] *The Irish Skipper,* December 1983

[28] *The Irish Skipper,* June 1974

[29] *The Irish Skipper,* January 1975

[30] *The Irish Skipper,* February 1975

[31] *The Irish Skipper,* April 1975

[32] *The Irish Skipper,* August 1975

[33] *The Irish Skipper,* January 1976

[34] *The Irish Skipper,* February 1976

[35] *The Irish Skipper,* April 1976

[36] *The Irish Skipper,* September 1977

[37] *The Irish Skipper,* December 1977

[38] *The Irish Skipper,* March 1978

[39] *The Irish Skipper,* April 1978

[40] *The Irish Skipper,* August, 1979

[41] *The Irish Skipper,* June 1978

[42] *The Irish Skipper,* October 1978

[43] *The Irish Skipper,* May 1981

[44] *The Irish Skipper,* August 1984

[45] George Burrows, *Irish Times,* 9 March 1984

[46] Salmon Review Group, 1987

[47] *The Irish Skipper,* July 1979

[48] *The Irish Skipper,* March 1986

[49] *The Irish Skipper,* November 1980

[50] *The Irish Skipper,* September 1983

[51] *The Irish Skipper,* December 1984

[52] European Communities, 2009

[53] *The Irish Skipper,* November 1975

[54] *The Irish Skipper,* January 1976

[55] *The Irish Skipper,* April 1976

[56] *The Irish Skipper,* October 1976

[57] *The Irish Skipper,* December 1975

[58] *The Irish Skipper,* July 1979

[59] *The Irish Skipper,* May 1977

[60] *The Irish Skipper,* April 1993: 4

[61] *The Irish Skipper,* August 1993: 5

[62] *The Irish Skipper,* January 1995: 4

[63] Molloy, 2006: 54

[64] *The Irish Skipper*, August 2002: 2

[65] *The Irish Skipper*, April 2003: 5

[66] *The Irish Skipper*, February 2004: 5
[67] *The Irish Skipper*, January 2006: 24
[68] *The Irish Skipper*, December 2006: 5

Chapter 11: Fish producers' organisations take possession

[1] Chapter 10
[2] *The Irish Skipper*, September 1975
[3] Symes and Phillipson, 2001
[4] Symes and Phillipson, 2001: 129
[5] Molloy, 2006: 100
[6] *The Irish Skipper*, December 1990
[7] *The Irish Skipper*, January 1991
[8] Went, 1976: 11
[9] *The Irish Skipper*, November 1991
[10] Rogan and Berrow, 1995
[11] Colette Sheridan, *Sunday Tribune*, 9 June 1991
[12] *The Irish Skipper*, July 1991
[13] *The Irish Skipper*, August 2002: 1
[14] *The Irish Skipper*, October 1992
[15] *Sunday Journal, Donegal on Sunday*, 17 October 2004: 13
[16] *The Irish Skipper*, June 2004
[17] European Commission, 2010
[18] EC Habitats Directive: Council Directive 92/43/EEC
[19] *The Irish Skipper*, December 1976
[20] *The Irish Skipper*, June 1979
[21] *The Irish Skipper*, August 1979
[22] *The Irish Skipper*, November 1981
[23] *The Irish Skipper*, December 1981
[24] John McAleese, *Sunday Tribune*, 14 March 1982
[25] Paul Murray, *Irish Times*, 14 October 1982
[26] *The Irish Skipper*, July 1982
[27] *The Irish Skipper*, December 1982
[28] *The Irish Skipper*, December 1981
[29] *The Irish Skipper,* February 2002: 35
[30] *The Irish Skipper*, February 1982
[31] *The Irish Skipper*, January 1982
[32] *The Irish Skipper*, August 1984
[33] *The Irish Skipper,* July 1981
[34] Molloy, 2006: 18
[35] *The Irish Skipper*, February 1992: 9
[36] *The Irish Skipper*, November 1992: 15

[37] *The Irish Skipper*, June 1998: 21
[38] *The Irish Skipper*, January 2002: 11
[39] *The Irish Skipper*, May 2002: 6
[40] Hardy, 1959: 228-229
[41] Hillis, 1971
[42] Fahy, 2011b
[43] *The Irish Skipper*, February 1972
[44] Johnson et al, 1921: 34-35
[45] Chapter 9
[46] Fahy, Carroll, O'Toole, Barry and Hother-Parkes, 2005
[47] Fahy, 2008b

Chapter 12: A fleet for all seasons

[1] Chapter 6
[2] Anon, 1962
[3] Glude *et al*, 1964
[4] Appendix 4
[5] BIM, annual report, 1965
[6] The other three are market, external relations and conservation. See Holden and Garrod, 1996
[7] *The Irish Skipper*, August 1978
[8] *Irish Times*, 24 December 1983
[9] Lorna Siggins, *Irish Times*, 2 July 1982
[10] Leonard Doyle, *Irish Times*, 1 July 1982
[11] Council regulation (EC) 2980/83
[12] BIM, annual report, 1984
[13] *The Irish Skipper*, June 1985
[14] BIM, annual report, 1985
[15] Meredith, 1998
[16] Regulation 4028/86 EC
[17] A reason for the enthusiasm for entering the fishery was the success of whitefish sales in newly joined member states, particularly Spain, which had considerable interest in purchasing species like angler (monk) fish for which there was little appetite in Ireland at the time. The case of monk is described in chapter 7.
[18] Meredith, 1998, p 73
[19] Meredith, 1998, p 73
[20] The Precautionay Principle has wide application in biological resource management. It was first expressed in 1982 in relation to "straddling stocks" (fish stocks which occur on both sides of geographical boundaries) as part of the Agreement for the

implementation of the provisions of the United Nations convention on the Law of the Sea. Its integration into the common fisheries policy took place after 2000.

[21] BIM, annual report, 1990

[22] Chapter 15

[23] Meredith, 1998

[24] Holden and Garrod, 1994

[25] BIM, annual report, 1999

[26] I say "recently" because the consequences of rising competition for mackerel, our single most important species, by the Faeroes and Iceland is as yet unknown.

[27] Chapter 15

[28] Council regulation (EC) No 1263/1999

Chapter 13: The scramble for fish

[1] Glude *et al*, 1964

[2] Molloy, 2006

[3] Chapter 10

[4] BIM, annual report, 1972

[5] Waind, 1974

[6] Chapter 15

[7] BIM, annual report, 1999

[8] Fahy, Fee, O'Connor and Smith, 2007

[9] Appendix 5

[10] E.S.R.I.

[11] Appendix 5

[12] BIM, annual report, 1976.

[13] *The Irish Skipper*, April 1973: 10

[14] *The Irish Skipper*, January 1974: 5

[15] *The Irish Skipper*, June 1974: 1

[16] *The Irish Skipper*, April 1976: 15

[17] Marine Institute, 1998-2011

[18] *The Irish Skipper,* January 1977: 9

[19] *The Irish Skipper*, June 1977: 9

[20] *The Irish Skipper*, September 1977: 17

[21] *The Irish Skipper*, June 1976: 1

[22] *The Irish Skipper*, October 1977: 7

[23] Waind, 1974

[24] *The Irish Skipper*, June 1978: 5

[25] Lee, 1978

[26] *The Irish Skipper*, October 1978: 3

[27] Meaney, 1980

[28] *The Irish Skipper*, March 1986: 1

[29] Chapter 7

[30] Mac Laughlin, 2010

[31] Fahy and Carroll, 2009

[32] *The Irish Skipper*, October 1984

[33] Brabazon, 1848.

[34] BIM, annual report, 1991

[35] Statistics of landings, species weight and value, Department of the Marine, 1990-1994.

[36] http://www.ukmarinesac.org.uk/communities/seapens/sp1_3htm consulted on 4 April 2012

[37] Whelan, 1991: 43.

Chapter 14: The official fate of some fish stocks

[1] BIM, annual report, 2001

[2] Marine Institute, Stock book, 2011

[3] An introduction to *K*-strategy life cycles is contained in Chapter 9

[4] Marine Institute, Stock book, 2001

[5] Muus and Dahlström, 1994

[6] Foley et al, 2010

[7] Marine Institute, Stock book, 2001

[8] Nolan, 2004

[9] SFPA, statistical data on line

[10] Marine Institute, Stock book, 2011

[11] Victoria, 1842

[12] Hardy (1959): 221-227

[13] Cod resurgence in Canadian waters
http://www.queensu.co/news/articles/cod-resurgence-canadian-waters.
Consulted 7 April 2012

[14] de Courcy Ireland, 1981: 98

[15] Fahy, 2010c further explores these data.

[16] http://britishfood.about.com/od/diningdrinkingtradition/a/fishandchips.htm, consulted 26 July 2012

[17] Orwell, 1937

[18] *The Irish Skipper*, December 1977

[19] Johnson et al (1921)

[20] Brabazon, 1848

[21] Salmon Review Group, 1987

[22] Fahy, Green, Borges, Power and McGuinness, 2009

[23] *The Irish Skipper*, March 1991: 3

[24] Chapter 15
[25] *The Irish Skipper*, July 2008: 6

Chapter 15: Two last casts of the dice

[1] Lorna Siggins, *Irish Times*, 30 June 1997
[2] *The Irish Skipper*, November 1990
[3] *The Irish Skipper*, March 1996
[4] *The Irish Skipper*, May 1997
[5] *The Irish Skipper*, July 1998
[6] *The Irish Skipper*, December 1998
[7] *Irish Times*, 4 June 2002
[8] *Dáil Eireann Debate* 504 (1): 7
[9] White, 2005
[10] Fitzgerald *et al*, 1999
[11] John McManus, *Irish Times*, 18 June 2001
[12] White, 2005
[13] *The Irish Skipper*, August 2005: 9
[14] Lorna Siggins, *Irish Times*, 27 June 2008
[15] Appendix 3
[16] Lorna Siggins, *Irish Times*, 24 June 2008
[17] BIM, annual report, 1997
[18] *The Irish Skipper*, November 1999: 12
[19] BIM, annual report, 2000
[20] Higgins, 2006: 73 - 96
[21] *The Irish Skipper*, December 2006: 45
[22] Fahy, 2010d
[23] Tully, 2010

Chapter 16: A rot on the coast

[1] *Irish Fishing and Fish Trades Gazette* 6, 16 December 1958
[2] *Irish Fishing and Fish Trades Gazette* 5, (28) 18 January 1958
[3] *Irish Fishing and Fish Trades Gazette* 7, 7 July 1959
[4] *Irish Fishing and Fishing Trades Gazette* 9, (6) June 1961
[5] *Irish Fishing and Fishing Trades Gazette* 9, (7) August 1961
[6] *Irish Fishing and Fishing Trades Gazette* 9, (10) October 1961
[7] Molloy, 2006: 75
[8] *Irish Times*, 26 August 1992
[9] Skindiving for shell fish bye-law No. 533 of 14 June 1966
[10] Went, 1976
[11] Fahy, 2008[c]

[12] Bratton and Hinz, 2002

[13] Lorna Siggins, *Irish Times*, 1 & 2 January 1997

[14] Stephen O'Brien, Sunday Times, 5 February 2006

[15] John Downes, *Irish Times*, 20 January 2007

[16] Scott Millar, *Sunday Times*, 15 December 2002

[17] *Sunday Business Post,* 10 October 2004

[18] Marie O'Halloran, *Irish Times*, 28 February 2003

[19] *Sunday Journal, Donegal on Sunday,* 17 October 2004: 11-13

[20] *Sunday Journal, Donegal on Sunday*, 17 October 2004: 11-13

[21] Sean Hargrave, *Sunday Times*, 24 December 1995

[22] Lorna Siggins, *Irish Times*, 20 April 1998

[23] *Sunday Journal, Donegal on Sunday*, 17 October 2004: 11-13

[24] *Irish Times,* 13 October 2004

[25] Stephen Collins, *Irish Times*, 9 February 2006

[26] *Irish Times,* 18 November 2004

[27] Enda Leahy, *Sunday Times*, 31 October 2004

[28] Enda Leahy, *Sunday Times*, July 2005

[29] *Sunday Times* 24 July 2005

[30] *Sunday Tribune,* 27 November 2005

[31] *Irish Times,* 30 November 2005

[32] *Sunday Times,* 24 July 2005

[33] Lorna Siggins, *Irish Times,* 19 October 2005

[34] *Sunday Tribune,* 27 November 2005

[35] Stephen O'Brien, *Sunday Times*, 5 February 2006

[36] *Sunday Tribune,* 30 October 2005

[37] *Irish Times,* 10 February 2006

[38] Stephen Collins, *Irish Times*, 9 February 2006

[39] Stephen Collins, *Irish Times*, 9 February 2006

[40] Lorna Siggins, *Irish Times*, 9 February 2006

[41] *Irish Times,* 13 February 2006

[42] Stephen O'Brien, *Sunday Times*, 5 February 2006

[43] Mark Hennessy and Lorna Siggins, *Irish Times*, 23 February 2006

[44] *Irish Times,* 1-2 January 2007

[45] SFPA, annual report, 2007

[46] Lorna Siggins, *Irish Times*, 16 April 2008

[47] *Irish Times*, 13 January 2007

[48] SFPA, annual report 2007

[49] Lorna Siggins, *Irish Times*, 24 April 2008

[50] *Irish Times,* 3 March 2006

[51] *Irish Times,* 6 April 2007

[52] *Donegal Democrat,* 14 June 2007

[53] *Irish Times,* 11 August 2007

[54] Eamon Ryan, *Irish Times*, 17 January 2006
[55] Cawley, 2006
[56] *The Irish Skipper*, February 2008: 1
[57] *Irish Times,* 7 December 2006
[58] *Irish Times,* 15 September 2007
[59] *Irish Times,* 29 December 2007
[60] Paul O'Doherty, *Irish Times*, 15 November 2006
[61] *The Irish Skipper*, February 2008: 1
[62] Lorna Siggins, *Irish Times*, 7 February 2008
[63] Lorna Siggins, *Irish Times*, 5 February 2008
[64] Lorna Siggins, *Irish Times*, 18 June 2009
[65] Michael O'Regan, *Irish Times*, 13 May 2010

Chapter 17: The treacherous reassurance of fisheries science

[1] Appendix 3
[2] O'Riordan, 1964
[3] Hardy, 1959: 144
[4] *Ibid*
[5] Armstrong *et al*, 1991
[6] Departmental statistics
[7] Kurlansky, 1998
[8] Fahy, 2009b – this publication supplies more detail
[9] *The Irish Skipper*, June 2003: 1
[10] *The Irish Skipper*, January 2006: 1
[11] *Irish Times*, 12 November 1979
[12] Wildlife Act, 1976
[13] This is based on Fahy, 2011c
[14] Crummey, 1997
[15] http://www.ombudsman.europa.eu/decision/en/030754.htm
[16] Purcell and Arai, 2001
[17] Boyer and Boyer, 2004
[18] Sparks *et al*, 2001
[19] Fahy, 2011d
[20] Lynam *et al*, 2011

Chapter 18: Anatomy of a take-over

[1] *Irish Fishing and Fish Trades Gazette* 4 (11), 2 June 1956
[2] Sean Hargrave, *Sunday Times*, 24 December 1995
[3] Jan Battles, *Sunday Times*, 12 March 2000
[4] Cawley, 2006

[5] *Marine Times*, May 2012: 1

[6] De Courcy Ireland, 1981

[7] Butt, 1871

[8] Cawley, 2006

[9] David, 1994

[10] Johnson et al, 1921: para 139 et seq

[11] *Irish Fishing and Fish Trades Gazette,* 4 (7), 7 April, 1956

[12] *Irish Fishing and Fish Trades Gazette,* 5 (10), 11 May, 1957

[13] *Irish Fishing and Fish Trades Gazette,* 5 (17), 17 Aug, 1957

[14] *Irish Fishing and Fishing Trades Gazette,* 8(9), Sept, 1960

[15] *Irish Fishing and Fish Trades Gazette,* 3(24), 18 Nov, 1955

[16] *Irish Fishing and Fish Trades Gazette,* 4 (7), 7 April, 1956

[17] Glude et al, 1964

[18] BIM, annual report, 1966-1967.

[19] Anon, 2009

[20] Meredith, 1998

[21] *The Irish Skipper*, February 1990

[22] *The Irish Skipper*, August 1995: 2

[23] *The Irish Skipper*, November 1993

[24] *The Irish Skipper*, December 1995: 5

[25] *The Irish Skipper*, July 1996: 25

[26] Chapter 16

[27] John McManus, *Irish Times*, 18 July 2005

[28] Eamon Ryan, *Irish Times*, 17 January 2006

[29] *The Irish Skipper*, October 2007: 1

[30] BIM, annual report, 2008

[31] *The Irish Skipper*, March 2008

[32] Department of Agriculture Food and the Marine, 2011

[33] The White report introduced one scheme which extended into another as a result of the Cawley report. The first decade of the millennium therefore saw two.

[34] Chapter 16

[35] John Mooney, *Sunday Times*, 8 July 2012

[36] *The Irish Skipper,* September 2000

[37] *The Irish Skipper,* June 1992

[38] *The Irish Skipper,* February 1977

[39] *The Irish Skipper,* November 1980

[40] *The Irish Skipper,* December 1980

[41] *The Irish Skipper,* January 1986

[42] *The Irish Skipper,* September 1986

[43] *The Irish Skipper,* October 1990

[44] *The Irish Skipper,* March 1993

[45] *The Irish Skipper,* November 1989
[46] *The Irish Skipper,* December 1987
[47] *The Irish Skipper,* August 1976
[48] *The Irish Skipper,* January 1985
[49] *The Irish Skipper,* June 1987
[50] *The Irish Skipper,* December 1987
[51] *The Irish Skipper,* July 1991
[52] *The Irish Skipper,* July 1992
[53] *The Irish Skipper,* October 1993
[54] *The Irish Skipper,* November 1993
[55] Brian O'Riordan, *The Irish Skipper*, December 2002:7
[56] *The Irish Skipper,* October 1986
[57] *The Irish Skipper,* February 1995
[58] Simon Carswell, *the Sunday Business Post*, 22 June 2003
[59] *The Irish Skipper,* August 1998
[60] Simon Carswell, *the Sunday Business Post*, 22 June 2003
[61] Nolan, 2008
[62] *The Irish Skipper*, December 2002: 7
[63] *Sunday Times,* 17 November 1996
[64] *Irish Times,* 3 September 2008
[65] Simon Carswell, *the Sunday Business Post*, 22 June 2003
[66] *The Irish Skipper,* September 2000
[67] *Sunday Times,* 5 July 2009
[68] *The Irish Skipper ,* September 2000
[69] Ann Cahill, *Cork Examiner*, 24 February 2003
[70] Aine Coffey, *The Sunday Tribune*, 9 June 2002
[71] *Sunday Business Post,* 3 October 2004
[72] Lorna Siggins, *Irish Times,* November 2001.
[73] Simon Carswell, *the Sunday Business Post*, 22 June 2003
[74] *Phoenix* Magazine, 18 January 2003
[75] Lorna Siggins, *Irish Times*, 31 December 2002
[76] *Irish times,* 2 July 2004
[77] Brian Dowling, *Irish Independent*, May 2004.
[78] *Sunday Times,* 5 July 2009
[79] *Irish times,* 2 July 2004
[80] *Irish Times,* 8 December 2003
[81] Lorna Siggins, *Irish Times*, 11 February 2003
[82] John Lee, *Sunday Times*, 2 June 2002
[83] *Phoenix* magazine, 18 January 2003
[84] Aine Coffey, *The Sunday Tribune*, 9 June 2002.
[85] *Sunday Times,* 5 July 2009
[86] *Sunday Times,* 26 November 2006

[87] Lorna Siggins, *Irish Times*, 26 October 2007

[88] John McManus, *Irish Times*, 5 November 2007

[89] *Irish Times,* 4 June 2002

[90] *Sunday Business Post*, 10 October 2004

[91] Douglas Dalby, *Sunday Times*, 19 September 2004

[92] Lorna Siggins, *Irish Times*, 26 October 2007

[93] *Irish times,* 2 July 2004

Chapter 19: When did you meet a happy fisherman?

[1] Lorna Siggins, *Irish Times*, 2 February 2007

[2] Appendix 3

[3] Crayfish is the term frequently used for this species although it should more correctly be crawfish, abbreviated to "craw".

[4] O'Sullivan, 1983

[5] Fahy, Carroll and Clarke, 2008

[6] Skin diving for Shell Fish Bye-Law, number 533 of 14 June 1966

[7] *The Irish Skipper*, April 1970: 14

[8] Bates, 1983

[9] Tully, 2003

[10] Fahy, 2011e

[11] Crawfish (Fisheries Management and Conservation Order) 2002, S.I. No 179 of 2002.

[12] Crawfish (Conservation of stocks) regulations 2006, S.I. No 232/2006

[13] Cawley, 2006, previous work

[14] Cawley, 2010

[15] Lorna Siggins, *Irish Times*, 6 October 2008

[16] European Commission, 2006

[17] Eamon Ryan, *Irish Times*, 17 January 2006

[18] Khan et al, 2006

[19] Sumaila and Pauly, 2006

[20] Fahy, 2011f

[21] Lorna Siggins, *Irish Times*, 1 July 2008

[22] Englehard, 2008

[23] Thurstan *et al*, 2010

[24] Fahy, 2010c

[25] Fahy, 2010c

[26] Molloy, 2004

[27] Irish Specimen Fish Committee reports, 1955-2011

[28] Fahy, 2009c

[29] Pauly, 1995

---

[30] Nolan, 2008

[31] *Irish Fishing and Fishing Trades Gazette*: IX, 7 July 1961

About the author

Edward Fahy was born and educted in Dublin. He received his primary degree in zoology and an M.Sc in entomology from Trinity College and his Ph.D. in freshwater entomology from the National University of Ireland, Galway. He has worked in the civil service and various state agencies on environmental planning and conservation and, latterly fisheries management. He currently works as a journalist.

Previous books were *Child of the Tides, A sea trout handbook*, The Glendale Press, 1985 and *The Brown Trout in Ireland, A fragile history* Immel Publishing, 1995.

3144505R00257

Printed in Great Britain
by Amazon.co.uk, Ltd.,
Marston Gate.